Fundamentals of
RENEWABLE ENERGY SYSTEMS

FUNDAMENTALS OF RENEWABLE ENERGY SYSTEMS

D Mukherjee
Professor
Electronics and Tele-Communication Engineering Department
and
Active Member of the Centre of Excellence in Green Energy &
Sensor Systems (CEGESS)

S Chakrabarti
Assistant Professor
Department of Mechanical Engineering
Bengal Engineering and Science University
Shibpur, West Bengal

Publishing Globally

NEW AGE INTERNATIONAL (P) LIMITED, PUBLISHERS
LONDON • NEW DELHI • NAIROBI
Bangalore • Chennai • Cochin • Guwahati • Hyderabad • Kolkata • Lucknow • Mumbai
Visit us at **www.newagepublishers.com**

Published by New Age International (P) Ltd., Publishers
First Edition: 2004
Reprint: 2018

GLOBAL OFFICES

- **New Delhi** **NEW AGE INTERNATIONAL (P) LIMITED, PUBLISHERS**
  7/30 A, Daryaganj, New Delhi-110002, (INDIA)
  Tel.: (011) 23253771, 23253472, **Telefax:** 23267437, 43551305
  E-mail: contactus@newagepublishers.com • Visit us at www.newagepublishers.com
- **London** **NEW AGE INTERNATIONAL (UK) LTD.**
  27 Old Gloucester Street, London, WC1N 3AX, UK
  E-mail: info@newacademicscience.co.uk • Visit us at www.newacademicscience.co.uk

BRANCHES

- **Bangalore** 37/10, 8th Cross (Near Hanuman Temple), Azad Nagar, Chamarajpet, Bangalore- 560 018
  Tel.: (080) 26756823, **Telefax:** 26756820, **E-mail: bangalore@newagepublishers.com**
- **Chennai** 26, Damodaran Street, T. Nagar, Chennai-600 017, **Tel.:** (044) 24353401, **Telefax:** 24351463
  E-mail: chennai@newagepublishers.com
- **Cochin** CC-39/1016, Carrier Station Road, Ernakulam South, Cochin-682 016, **Tel.:** (0484) 2377303, **Telefax:** 4051304
  E-mail: cochin@newagepublishers.com
- **Guwahati** Hemsen Complex, Mohd. Shah Road, Paltan Bazar, Near Starline Hotel, Guwahati-781 008,
  Tel.: (0361) 2513881, **Telefax:** 2543669, **E-mail: guwahati@newagepublishers.com**
- **Hyderabad** 105, 1st Floor, Madhiray Kaveri Tower, 3-2-19, Azam Jahi Road, Near Kumar Theater, Nimboliadda Kachiguda, Hyderabad-500 027, **Tel.:** (040) 24652456, **Telefax:** 24652457
  E-mail: hyderabad@newagepublishers.com
- **Kolkata** RDB Chambers (Formerly Lotus Cinema) 106A, 1st Floor, S N Banerjee Road, Kolkata-700 014
  Tel.: (033) 22273773, **Telefax:** 22275247, **E-mail: kolkata@newagepublishers.com**
- **Lucknow** 16-A, Jopling Road, Lucknow-226 001, **Tel.:** (0522) 2209578, 4045297, **Telefax:** 2204098
  E-mail: lucknow@newagepublishers.com
- **Mumbai** 142C, Victor House, Ground Floor, N.M. Joshi Marg, Lower Parel, Mumbai-400 013
  Tel.: (022) 24927869, **Telefax:** 24915415, **E-mail: mumbai@newagepublishers.com**
- **New Delhi** 22, Golden House, Daryaganj, New Delhi-110 002, **Tel.:** (011) 23262368, 23262370, **Telefax:** 43551305
  E-mail: sales@newagepublishers.com

ISBN: 978-81-224-1540-7

MRP: 299.00

C-17-01-7373

Printed in India at Print 'O' Pack, Delhi.

NEW AGE INTERNATIONAL (P) LIMITED, PUBLISHERS
7/30 A, Daryaganj, New Delhi-110002
Visit us at **www.newagepublishers.com**
(CIN: U74899DL1966PTC004618)

Dedicated
To
Our Parents

Preface

The authors have the priviledge of teaching several basic courses like Power Plant Engineering, Heat Power Engineering, Thermodynamics, Environmental Aspects of Thermal Systems, Opto-Electronics and Optical Communications, Non-Conventional Energy Systems, Renewable Energy Sources, at the final year under graduate level and post graduate level and also guiding several project and thesis work based on the above subjects. Apart from this core activity, the authors acted as coordinators and guest faculty members of various workshops, entrepreneurship training programmes, organized by B.E. College (Deemed University), West Bengal Renewable Energy Development Agency (WBREDA), Department of Science and Technology, Govt. of India, and some NGOs like FOSET, PBVM, AGRAGATI, Rotary Clubs etc.

In course of the aforesaid activities of the authors, it has been strongly felt that there are many books covering only some selected topics instead of the broad spectrum coverage of renewable energy sources and systems in existing demand by students of different Universities, Engineering Institutions, Polytechnics, and the working personnel in the relevant field. This necessity has prompted the authors to take an appropriate attempt in organizing this text book.

This book contains 11 (eleven) chapters, out of which Chapter 1 introduces the reader with various forms of Renewable Energy Sources. Chapter 2 has been devoted to the understanding of various forms of solar radiation, their terminologies, inter-relations, and the instruments employed for their measurement. Since solar energy is undoubtedly the father figure of various renewable energy sources, greater emphasis has been placed on the related components, their design and performance analysis, some integrated examples and economics relating to both solar thermal and solar photovoltaic conversion systems. The aforesaid aspects have been covered in details with several installed site examples (particularly on the Indian scenario) through Chapters 3, 4, and 5 respectively. Chapter 6 has highlighted all possible resources contributing to energy derived from biomass and biogas. Detailed technical features of biomass gasifiers and rice-husk gasifiers have been appended to project how this renewable energy source can be regarded as a true replacement technology in the current Indian energy scenario. Chapter 7 deals principally with the various components and figures of merit relating to wind energy conversion systems. A detailed discussion of a small hydro electric system along with the relevant design criteria has been presented in Chapter 8. A brief discussion has also been carried out on mini and micro hydel power systems which are being currently used by the Govt. of India vis-a-vis high hydro potential with low head for this country. Chapter 9 deals with the detailed discussion on the energy sources from the sea, e.g., ocean thermal energy, wave energy, tidal energy and their applications in producing electricity.

As a distinguishing feature of this text book, two additional chapters: Chapter 10 on hybrid energy systems and Chapter 11 on role of renewable energy with respect to environment have been systematically organized. The need for hybrid energy systems vis-a-vis stand alone and grid connected systems has been clearly established in Chapter 10. Various hybrid energy systems actually installed in India and abroad have been technically specified in details to bring out the essential difference with an integrated energy system. Chapter 11 has audited carefully the actual estimate of the impact of all possible conventional and non-conventional energy sources on the environment.

Authors would like to place on record their sincere thanks to the reviewers to provide them the effective guidelines in organizing the chapters in a manner appropriate with recent context. Sincere thanks are also due to M/s. New Age International (P) Ltd., for taking initiative in publishing this book with meticulous care. The authors are particularly grateful to Mr. T.K. Chatterjee for his painstaking efforts in communicating with head quarters for making the project successful. In this connection, sincere thanks are also due to Sri Rajat Mukhopadhyay of E. & T.C. Dept. for his constant help in drafting and correcting the manuscripts for this book.

The authors would very much appreciate criticisms and suggestions for improvement from the readers. It will be gratefully acknowledged and considered duly before the next edition.

D. Mukherjee and S. Chakrabarti

Contents

1

Introduction

INTRODUCTION

In a span of time situated about one hundred to one hundred and fifty years ago, people would have completed their daily routine work without moving more than ten kilometers from where they lived. The reason is obvious as the daily needs like cooking, lighting, eating, heating, etc. were largely met from resources within this radius. In today's era of information technology, the earth has been reduced to a global village where people from one remote corner exchange necessary information with friends, relatives, business partner situated several thousand kilometers across the globe. In fact, the three great challenges facing mankind in the new millennium are preservation of peace, eradication of poverty and conservation of our environment. Undoubtedly, each of the aforesaid goals is intertwined with questions of sources of energy, their demand and supply, access and harnessing.

Truly the measure of development in any society in these days is synonymous with quantum of energy consumption. This is why energy has been recognized as a critical input parameter for national development. Modern day energy demands are still met largely from fossil fuels like coal, oil and natural gas. Besides coal, natural gas, electricity from hydro and nuclear energy are considered to be commercial energy sources. Table 1.1 offers the primary commercial energy supply break-up for the years 1981–82, 1991–92, and 1997–98. It is seen that supply of commercial energy sources has grown from 97.1 million tons of oil equivalent (Mtoe) in 1981–82, 192.20 Mtoe in 1991–92 to 245.45 Mtoe in 1997–98. From this table, it is also evident that the % share in total supply of coal from 1981 to 1998 has remained unaltered while the % share in total supply of oil has declined during this period. During the same period, there has been an increase in % share of supply of natural gases.

It is a fact that most developing countries are passing through the initial stages of industrialization. Naturally, their energy consumption is growing at a greater rate in comparison to the developed countries. During 1990–2000, conventional energy consumption of the Asia Pacific countries has increased by 32.5 percent compared with 11 percent growth of the world consumption. In spite of the higher rate of growth of energy consumption, the present per capita energy consumption in the developing countries is much lower in comparison to the developed countries. It is also a fact that the energy input to the economy of developing countries is still at a very low level and a large section of the increasing population in many of these countries does not have access to commercial energy and electricity. It is also expected that their demand for energy will continue to grow at high rates for a number of decades to come. Therefore, a priority of developing countries is to ensure that supply of energy does not become a constraint to their economic growth.

Table 1.1: Primary Commercial Energy Supplies

Energy source	*Energy supply (MOTE)*[1]	*Share in total supply(%)*	*Energy supply (MOTE)*[1]	*Share in total supply(%)*	*Energy supply (MOTE)*[1]	*Share in total supply(%)*
	1981–82		*1991–92*		*1997–98*	
Coal	56.0	61.33	115.30	59.99	150.54	61.33
Oil	34.0	37.23	54.30	28.25	65.17	26.55
Natural Gas	1.3	1.42	16.00	8.32	22.63	9.22
Primary[2] Electricity	5.8	2.5	6.60	3.44	7.13	2.90
Total	97.1	100.00	192.20	100.00	245.47	100.00

[1] Million tones of oil equivalent
[2] Hydro and Nuclear power

Source: TERI energy data directory (1999–2000)

1.1 HUMAN DEVELOPMENT INDEX (HDI)

This parameter has come into existence through the United Nation Development Program (UNDP) in 1990. It is one way of measuring how well the countries and specially the developing countries are meeting not just the economics but also the social needs of their people. Typical value of this parameter lies between 0 and 1 and the variables that affect HDI are—

(i) Per capita income for evaluating satisfaction derived from the essential products and services and to focus on employment level.
(ii) Life expectancy, a social indicator to measure progress made in health and nutrition, infant mortality, etc.
(iii) Educational levels attained which can be studied taking into account adult literacy achievements and the mean schooling years.

It is worthwhile to note that HDI ranks countries in relation to one another on a uniform scale. According to World Development Report 1999, India was ranked 132 among 174 countries with the top five positions awarded to Canada, Norway, USA, Japan and Belgium. India, at the moment falls in the medium category with 93 other countries where 35 other countries being ranked in the low category group. UNDP has proposed a number of suggestions to improve the HDI of developing nations. Out of these one basis can be the per capita energy consumption for illumination. As a result of interactive survey work carried out by various agencies and NGOs in the rural belt, the aforesaid basis has shifted to per capita illumination level per unit of energy consumed.

There is little doubt that a direct correlation exists between the level of economic development of any nation and their consumption of energy. In fact, a new way of measuring HDI by combining indicators of life expectancy, educational attainment and prosperity into a composite HDI value is indicated through tables 1.1.1 and 1.1.2, where the tables are dedicated to a set of industrial and developing nations respectively. It can be easily seen that most of the developing countries consume less energies compared to industrialized countries and have been designated with low HDI values.

1.2 ENERGY ALTERNATIVE

Energy has been recognized to be one of the most important inputs for national development. If India is to achieve 7 to 8% growth in the economy, then it would need commensurate inputs of different energy resources; most of these will be in the form of fossil fuels and electricity. Before

Table 1.1.1: Human Development Index (HDI) for Selected Industrial Countries

Countries	*GDP/capita (PPP/$)*	*Energy use per capita (Kg. oil equivalent)*	*HDI*
Canada	21,916	7854	0.960
USA	26,977	7819	0.943
Sweden	19,297	5723	0.936
Germany	20,370	4128	0.925
France	21,176	4042	0.946

Table 1.1.2: HDI for Selected Developing Countries

Countries	*GDP/capita (PPP/$)*	*Energy use per capita (Kg. oil equivalent)*	*HDI*
Iran	5480	1505	0.758
Korea	4058	1129	0.766
China	2935	664	0.650
Srilanka	3408	97	0.716
India	1422	248	0.453

the first oil crisis/shock of 1973, energy meant only electrical energy. But over the past 25 years, the energy sources cover the whole spectrum, inclusive of fossil fuels, renewables and non-conventional sources. The depletion of fossil fuels is occurring at a faster rate due to enhancing gap between the demand and production of fossil fuels. It is therefore evident that we will have to search for the use of alternative energy sources for sustaining the energy requirement by human being. The various alternative energy sources are solar energy, nuclear energy, tar sands and oil shale, tidal energy, geothermal energy, hydrozen energy, fuel cells, magneto hydro-dynamic generator, thermoionic converter and thermo electric power. The solar energy can be used directly through solar thermal and solar photovoltaic routes. The indirect means of the use of solar energy include the hydro energy, wind energy, bio-mass energy, wave energy and ocean thermal energy. Among the above mentioned energy sources, solar energy, hydro energy, wind energy, bio-mass energy, wave energy, ocean thermal energy and tidal energy are considered to be renewable energy sources. Others are known as non-renewables. In this chapter no discussions on any of the renewable energy sources are being carried out as these have already been discussed in detail in different chapters of this book.

Among the non-renewable energy sources mentioned earlier, the potential of tar sands and oil shale, hydrozen energy, fuel cells, magneto hydro-dynamic generators, thermoionic converter, thermo electric power and geothermal energy as alternative energy sources with respect to conventional sources of energy is considered to be very little. Keeping in view the practical feasible contribution in the energy scenario, the only remaining option as an alternative for conventional energy sources is the option of nuclear energy sources. Therefore, a brief discussion on nuclear energy sources and its applications are being made in the following sub section.

1.2.1 Nuclear Energy

It is known from the modern theories of atomic structures that matter consists of minute particles known as atoms. These atoms contain enormous concentration of binding energy. From the heavier unstable atoms such as U^{235}, Th^{232}, Pu^{239}, U^{238}, U^{233} or very light atoms like hydrogen and heavy isotopes of hydrogen (deuterium D and tritium T) the enormous energy can be produced through some nuclear reaction processes. To generate the nuclear energy the two nuclear reactions, fission and fusion are used. The above mentioned reaction processes are being discussed in the following subsections.

1.2.2 Nuclear Fission

In fission, a heavy nucleus is splitted into two or more lighter nuclei. This enables to release enormous amount of heat energy which can be used in different ways for the mankind and their development. It is estimated that the complete fission of one Kg. of U^{235} releases the amount of heat energy which can be obtained by burning 4500 tonnes of high grade coal or 2200 tonnes of oil.

Naturally occurring Uranium contains three isotopes, U^{234}, U^{235} and U^{238}. The relative percentages of these isotopes are as U^{234} – 0.006 percent, U^{235} – 0.711 percent and U^{238} – 99.283 percent. Among these isotopes, U^{235} is the fissile material which can undergo spontaneous fission under the bombardment by slow neutrons. Fig. 1.2.2.1. shows the fission reaction of U^{235}. In this reaction, when slow neutrons are bombarded on the U^{235} nucleous, neutrons and fission products are produced along with large amount of generated energy as heat (8.2×10^7 kJ per gm of U^{235}). Moderator slows down the neutrons. The slow neutrons are used to bombard the U^{235} nucleus again, thereby sustaining a controlled chain reaction. A reactor which uses U^{235} only as the fuel and releases energy, is called burrier reactor.

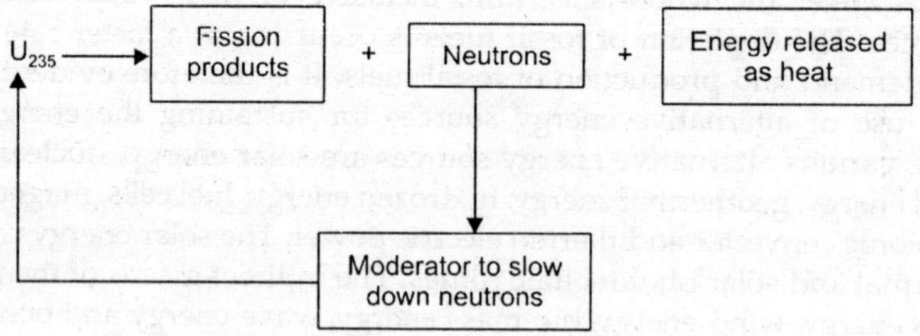

Fig. 1.2.2.1: Fission Reaction of U_{235}

Due to bombardment of neutrons on the nucleus of an atom of U^{235}, a neutron is captured by a nucleus of it and U^{236} isotope is formed. This isoptope is highly unstable in nature and gets splitted roughly into two equal parts with the total energy release of around 200 MeV. The fission fragments, neutrons and electromagnetic radiation or gamma rays are formed. Most of the total energy released by the fission process as kinetic energy is absorbed by fission products. As the fission fragments collide, it is converted to heat. The decay of the particles occur mainly by beta and gamma radiation with emission of some neutrons. Fig. 1.2.2.2 shows the schematic diagram for the fission of U^{235}. A typical reaction for U^{235} fission is represented by the following eqn. 1.2.2.1

$$ {}_{92}U^{235} + {}_{0}N^{1} \rightarrow {}_{56}Ba^{137} + {}_{56}Kr^{97} + 2\,{}_{0}N^{1} + \gamma \qquad ...(1.2.2.1)$$

In this reaction, Barium, Krypton are produced as fission fragments along with the generation of two or three neutrons and release of energy.

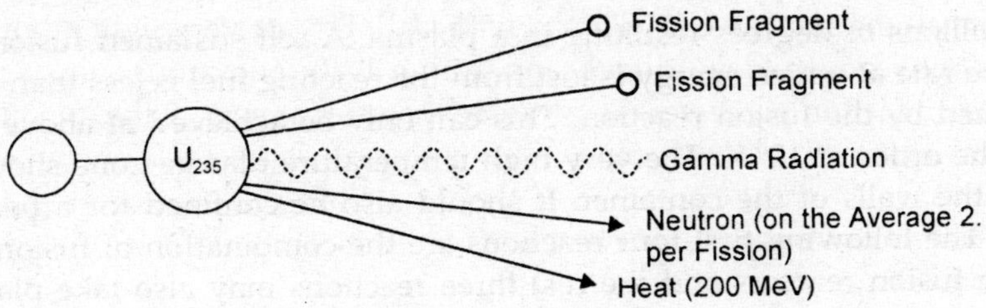

Fig. 1.2.2.2: Fission of U^{235}

Naturally occurring U^{238} or Th^{232} are not fissile materials. They are the fertile materials, i.e. they can be converted to fissile materials as Pu^{239} or U^{233} with the help of neutron bombardment. In these cases, the neutrons generated by the fission reactions, help in converting fertile materials to fissile one and also in sustaining the fission reaction for the formed fissile materials. These reactions are called the breeder reactions when the production of fissile materials are more than the consumption. The reactor in which the mentioned reaction takes place, is called breeder reactor. Breeder reactions for U^{238} and Th^{232} are shown in Fig. 1.2.2.3. In the context of large thorium deposits in the monazite beach sands in Kerala, utilizing Th^{232} to U^{233} breeder cycle in India is of great importance. It is estimated that India has reasonably assured supplies of 300,000 tonnes of monazite containing 7 to 9 percent of thorium.

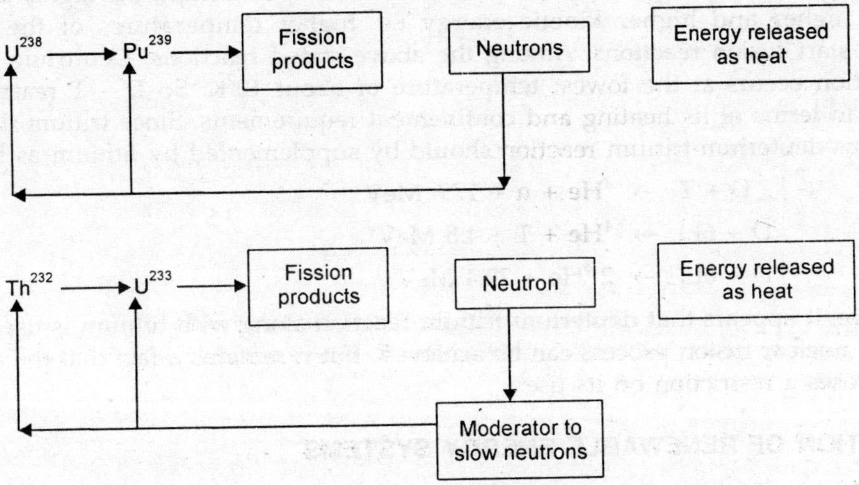

Fig. 1.2.2.3: Breeder Reactions for U^{238} and Th^{232}

1.2.3 Nuclear Fusion

Nuclear fusion is a process by which joining of very light atoms occur and thereby release of energy takes place. By the fusion process, two light nuclei, i.e. nuclei of low mass, combines and a heavier nucleus is produced. The artificially fusion process for hydrozen requires on an average billions of years for completion. But the fusion process for deuterium-deuterium needs a fraction of second.

For fusion process, the positively charged nuclei are to be accelerated to obtain high kinetic energies in overcoming electrical repulsive forces. This can be done by raising their temperature to hundred of millions of degrees resulting in a plasma. A self sustained fusion reaction can be attained when the rate at which energy is lost from the reacting fuel is less than the rate at which energy is generated by the fusion reaction. This can only be achieved at above a certain critical temperature of the order of 10^8 K. The very high temperature plasma zone should be prevented from contacting the walls of the container. It should also be confined for a period of time at a minium density. The following first four reactions are the combination of fusion reactions which are often used in fusion reactors and the rest three reactions may also take place.

(1) $D + D \rightarrow T + p + 3.25$ MeV ...(1.2.3.1)

(2) $D + D \rightarrow {}^3He + n + 4$ MeV ...(1.2.3.2)

(3) $D + T \rightarrow {}^4He + n + 17.6$ MeV ...(1.2.3.3)

(4) $D + 3\ He_2 \rightarrow {}^4He + p + 18.3$ MeV ...(1.2.3.4)

(5) $T + 3\ He_2 \rightarrow {}^4He + D + 14.31$ MeV ...(1.2.3.5)

(6) $T + 3\ He_2 \rightarrow {}^4He + p + n + 12.08$ MeV ...(1.2.3.6)

(7) $T + T \rightarrow {}^4He + 2n + 11.3$ MeV ...(1.2.3.7)

In all the equations, n, p, D, T represent neutron, proton, deuterium and tritium respectively.

The fusion process can take place when two nuclei approach each other within distances of the order of nuclear dimensions i.e. 10^{-12} to 10^{-13} cm or cross section in the order of 10^{-24} to 10^{-26} cm^2. Since positively charged nuclei repel each other more strongly, the highly charged nuclei fuels need higher and higher kinetic energy i.e. higher temperatures of the order of 10^8 to 10^9 K to start fusion reactions. Among the above stated reactions, Deuttrium – Tritium i.e. D – T reaction occurs at the lowest temperature of about 10^7K. So D – T reaction is having importance in terms of its heating and confinement requirements. Since tritium does not occur naturally, this deuterium-tritium reaction should by supplemented by lithium as follows,

$$D + T \rightarrow {}^4He + n + 17.6 \text{ MeV} \quad ...(1.2.3.8)$$

$$D + 6Li \rightarrow {}^4He + T + 4.8 \text{ MeV} \quad ...(1.2.3.9)$$

$$D + 6Li \rightarrow 2\,{}^4He + 22.4 \text{ MeV} \quad ...(1.2.3.10)$$

Therefore, it appears that deuterium-tritium reaction along with lithium is used first so that a controlled nuclear fusion process can be achieved. But it remains a fact that the availability of lithium imposes a restriction on its use.

1.3 ADOPTION OF RENEWABLE ENERGY SYSTEMS

It can now be clearly appreciated why the various forms of renewable energies from the total spectrum of alternative energy sources requires cultivation with serious emphasis. This is indeed the current practice in several developing countries out of which India has one of the largest renewable energy programme in the world. At the moment, the installed capacity of renewable energy systems is nearly 2000 MW, accounting for hardly 2 to 2.5% of the total installed capacity in the country. The Indian renewable energy programme is presently administered by the Ministry of Non-conventional Energy Sources (MNES) under which one state nodal agency exists in each one of almost all the states. This ministry has announced a number of important policy initiatives out of which the most significant one is to increase the share of renewable energy sources contributing to grid power to about 20% in the next two decades. This naturally calls for significant barrier removal activities in terms of the infrastructureal facilities, spreading awarness and educating consumers. India is a vast country with diversities in weather, culture and energy consumption pattern in people's daily lives. The efforts undertaken by MNES through state nodal agencies, various societies, schools and other institutions have been largely responsible for renewable energy systems making an important contribution to the renewable energy scenario in the country. As the concerned ministry has embarked on an ambitious plan to incorporate about 10% share of the renewables equivalent to 10,000 MW in the power generating capacity of the country by the year 2012, an actual assessment of the scenario targeted during the Xth plan period is presented in Table 1.3.1.

Table 1.3.1 is obviously indicative of a massive deployment of a whole range of product/system technologies, which may ultimately enable the attainment of economics of scale as well. Some of the proposed physical targets for renewable energy (RE) utilization during 10th plan period is focused in Table 1.3.2.

Table 1.3.1: Proposed Capacity to Power Generation by Renewables by 2012

Technology resource	*Capacity (in MW)*
Wind power	2000
Small Hydro power	800
Bio-mass power co generation	1000
Bio-mass gasification	122.5
Waste to energy	150.0
Solar photovoltaic (SPV)	15
Solar thermal power	140
Total	4227 MW (approx.)

Table 1.3.2: Proposed Physical Targets for Renewable Energy Utilization during 10th Plan

Resources	*Plan target*
Solar water heating systems (SWHS)	1 million sq.m. collector area
Solar air heating systems	10,000 sq.m. collector area
Solar cookers	3.60 lacs
Solar buildings	100 Nos.
SPV pumps	15,000 Nos.
SPV generators	30,000 Nos.
Wind pumps	2,000 Nos.
Hybrid systems	
Solar Aditya shops	1,000 Nos.
Energy parks	360 Nos.

- MNES draft 10th Five year plan (2002–2007) and Annual plan (2002–2003) October 2001.

1.3.1 High Initial Cost

The initial cost of installation of renewable energy technologies is higher than the cost of the devices that use conventional energy sources. While the difference in some cases is of the order of 50%, in other cases this could be as high as 500%. In almost all sectors of renewable energy technologies, the running cost of these systems is almost negligible compared to the running cost incurred in a typical life cycle of conventional energy sources. Unavoidably, the most significant part of capital cost arises from the storage devices, e.g., a battery bank which is an integral part of stand-alone or hybrid PV systems.

1.3.2 Limited Availability

The number of manufacturers of renewable energy devices and systems are very limited. As a result, it is not possible for an interested consumer to choose from a range of products as in the case of household consumer items like lights, fans, TV sets, etc. Retail outlets for sale of renewable energy items are also very few. One retail outlet for sale of renewable energy devices in India is 'Aditya'. In fact, the success of entrepreneurship awareness drive in this area will largely depend on the opening of more such retail outlets throughout the country.

1.3.3 Credibility

Manufacture of renewable energy devices is largely limited to the small and medium industry sectors. The larger industrial houses have deployed a marked reluctance to enter the field of manufacturing of these systems. Of course, the only exception to this trend is in the case of Solar PV systems. Besides the aforesaid status of credibility in the industrial sector, the consumers are also having inadequate information on the costs and performance analysis of these technologies. With the very rapid and recent strides of Information Technology (IT) based services, the consumer scenario is likely to turn over a new leaf very soon.

Technically qualified manpower and servicing agents are not available for providing quick and cost effective services in this sector particularly in the remotely located rural areas. This is why several people have still very little faith in the long term life and operation of these devices.

1.3.4 Financial Incentives

All manpower development efforts in this area are presently supplemented through instutional/ financial schemes of IREDA i.e. Indian Renewable Energy Development Agency and other than financial institutions, private investment, to a certain extent has been possible in this sector through fiscal incentives such as tax holidays, depreciation etc.

1.4 SPECIFIC APPLICATION AREAS OF RENEWABLE ENERGY OPTION

It is important to appreciate the conditions under which a technology can be labelled as a mature one. Any technology has to pass through qualifying lab tests and pilot scale testing under appropriate field conditions so as to become financially feasible projects which can be submitted to a bank. As a result of continued efforts for at least last twenty five years (1976–2001), a number of renewable energy systems have matured and a current list of them along with specific application areas are presented in section 1.4.1. to 1.4.5. In course of the last decade, a full fledged infrastructure has been installed and a manufacturing base established for these renewable energy systems including Balance of System (BOS) components for PV systems, device grade purity silicon. This is strongly supported by microelectronics of PV cells and a prevailing Indian standards for testing/ certification of Solar Water Heating Systems (SWHS), Solar cookers, Solar driers, Solar cells and various types of gasifiers.

1.4.1 Solar Thermal

(a) Solar water heating systems of capacity from 100 litres/day to 100,000 litres/day.
(b) Roof integrated solar air heating systems up to 400 m.sq. for plantation crops, spices and pulses.

(c) Solar passive systems for space heating in cold climates.
(d) Solar cookers for domestic and community cooking upto 10,000 persons.
(e) Solar distillation for labs, railways, automobile servicing stations, primary health centers and light houses.

1.4.2 Solar PV

(a) Roof top lighting systems for homes, schools and community centers (upto 5 KW).
(b) Pumping systems for supplying water to agriculture and community use from wells upto 100 m deep.
(c) 100 KW peak PV power being operated with grid interfacing or stand-alone with battery back-up.

1.4.3 Bio-mass

(a) Plants yielding upto 85 cu.metres/day with feedstock derived from animals, human being, kitchen and vegetable markets, slaughter house wastes and mixed wastes including garbage.
(b) Bio-mass plants upto 135 cubic metres/day from distillery spent wash and sugar press mud.
(c) Refuse derived fuel from Municipal Solid Wastes (MSW) upto 200 tons/day.
(d) Bio-mass gasifiers upto 500 KW based on wood, woody wastes and powdery bio-mass such as saw dust and rice husk.

1.4.4 Wind Power

(a) Multi bladed wind pumps with rotor diameter up to 7.5 metres and well depths upto 100 metres.
(b) Wind chargers upto 4 KW.
(c) Wind generators based on overseas technology with considerable indigenisation upto 0.5 MW for a single machine with configurations possible upto 10 MW.
(d) Wind generators up to 200 KW in a single machine of indigenous design (BHEL).

1.4.5 Passive and Low Energy Architecture (PLEA)

(a) Roof fabric cooling systems for multi-storied offices, hotels & hospitals.
(b) Passive draft evaporative cooling systems for commercial buildings & labs.
(c) Roof pond with moveable shades for houses for hot dry and hot humid climates.
(d) Low energy building material with embodied energy per unit floor area at one third that of the existing buildings without reduction of comfort.

The energy scenario of 20th century was dominated by fossil-fuel based technologies. The fossil fuels are going to be depleted soon and the energy scenario of 21st century will be dominated by the aforesaid components of Renewable Energy Systems each of which will take care about the preservation of the environment. In fact, most of todays' significant renewable energy components are gifted with (3E) dimensions as

(i) They generate additional energy.
(ii) They generate additional employment in the form of entrepreneurship ventures.
(iii) They care for a better environment.

REFERENCES

1. H.P. Garg, Appraisal of Indian Renewable Energy Programme, IREDA News, vol. 12, no. 2, pp. 31-37, April-June 2001.
2. R. Natarajan, Role of Renewable Energy in the XXI Century, IREDA News, vol. 10, no. 4, pp. 21-35, Oct-Dec 1999.
3. S. Rajagopal, Renewable Energy for Rural Development, IREDA News, vol. 10, no. 4, pp. 37-43, Oct-Dec 1999.
4. E.A.S. Sharma, Power Scenario in India, Int. Workshop on growth and sustainability, Chennai, 24th June 1998.
5. H.P. Garg, Issues in National Renewable Energy Programme, Proc. of 25th NREC-2001, Osmania University, Hyderabad, pp. 34-41, Dec 2001.
6. D. Yogi Goswami, New And Emerging Developments in Solar Energy, Proc. of 25th NREC-2001, Osmania University, Hyderabad, pp. 32-33, Dec 2001.
7. J. Madiath, Scaling up Renewable Rural Technologies, SESI Journal, 10(2): pp. 117-123, 2000.
8. C.S. Sinha, Cost of Renewable Electricity: Case of Wind Power generation in India, SESI Journal, vol. 6, no. 1, pp. 39-57, June 1996.
9. H.S. Mukunda, 10% of Energy from Renewables—Is this an achievable goal or an improbable slogan? Proc. of 25th NREC-2001, Osmania University, Hyderabad, pp. 273-279, Dec. 2001.
10. S.C. Bhattacharya and S. Kumar, Renewable Energy Technologies in Asia: A Review of Current Status and related issues, Proc. of Int. Seminar on Decentralised Energy: Options and Management, Kolkata, pp. 173-180, Feb. 22-24, 2002.
11. Baharuddin Bin Yatim, Malaysia: Application of Renewable Energy, ASEAN Energy Bulletin, vol. 3, no. 2, p. 11, June 1999.
12. K.N. Amulya Reddy, et al., Energy after Rio: Prospects and Challenges, UNDP, 1997.
13. Dr. (Mrs.) S. Majumdar, et al., Energy efficient illumination—An indicator of HDI in Rural life, Proc. of Int. Seminar on Decentralised Energy: Options and Management, Kolkata, pp. 186-192, Feb. 22-24, 2002.
14. D.P. Sengupta, Decentralised Power Generation: One of the essential features for System improvement, Proc. of the International Seminar on Decentralised Energy: Options and Management, Kolkata, p. 185, Feb. 22-24, 2002.
15. D. Subhakar, Development of Renewable Energy Sources and Technologies in Kerala, Proc. of the International Seminar on Decentralised Energy: Options and Management, Kolkata, pp. 197-206, Feb. 22-24, 2002.
16. Yuan Zhenbong, Research and Development on Renewable Energy in China—Project in National Hi-Tech D&R Program, Proc. of the International Seminar on Decentralised Energy: Options and Management, Kolkata, pp. 05-18, Feb. 22-24, 2002.
17. MNES Draft 10th V year plan (2002-2007) and Annual Plan (2002-2003), Oct 2001.
18. T. Persson, A Renewable Energy Future for India, Proc. of 24th NREC-2000, IIT Mumbai, pp. 349-353, 2000.
19. B. Bandyopadhyay, Solar Thermal Technologies in India—A brief Assessment, Proc. of 24th NREC-2000, Mumbai, pp. 140-145, Dec 2000.

2

Solar Radiation and Its Measurement

INTRODUCTION

In this Chapter an attempt has been made to discuss the extra-terrestrial radiation, its spectral distribution and the terrestrial radiation first. The various angles used in analyzing the radiation are defined then. The various instruments utilized in measuring solar radiation have been described and the associated measurement techniques have been discussed. Equations for assessing solar radiation at a location have been presented.

2.1 EXTRA-TERRESTRIAL SOLAR RADIATION AND TERRESTRIAL SOLAR RADIATION

The sun, a large sphere of very hot gases is having diameter of 1.39×10^6 km, while that of the earth is having diameter of 1.27×10^4 km. It is at a very large distance from the earth's surface. The mean distance between the sun and earth is 1.5×10^8 km. Due to this large distance, it subtends an angle of only 32 minutes at the earth's surface. Thus, the earth receives almost parallel beam radiation from the sun. Sun's brightness varies from its center to its edge. However, for engineering calculations, it is customary to assume that the brightness all over the solar disc is uniform.

The rate at which the solar energy reaches at the top of the atmosphere is called the solar constant I_{sc}. This is the rate at which energy is received from the sun on a unit area perpendicular to the rays of the sun, at the mean distance of the earth from the sun. Since the sun's distance varies throughout the year, rate of arrival of solar radiation varies accordingly. The value of the so called solar constant is average one. The actual values vary upto about 3 percent in either direction. For most practical purposes, the variation is not at all important one.

The National Aeronautics and Space Administration's (NASA) standard value for the solar constant in three common units is as follows:

1353 W/m^2

116.5 langleys (Cal/cm^2) per hour or

1165 kcal/m^2 – hour (1 langely is equal to 1 cal/cm^2 of solar radiation received in one day)

429.2 Btu per sq. ft. per hour

The radiation coming from the sun is essentially equivalent to blackbody radiation coming from a black surface at 5762 K for a solar constant of 1353 W/m^2.

The distance between the sun and the earth varies a little throughout the year. Due to this variation, the extra-terrestrial flux also varies. The extra-terrestrial flux on any day can be calculated from the equation:

$$I'_{sc} = I_{sc} [1 + 0.033 \cos (360n/365)] \quad ...(2.1.1)$$

Where n is the day of the year.

The earth is closest to the sun in the summer and furthest away in the winter.

Measurements of the spectral distribution of extra-terrestrial solar radiation are made and recorded by M.P. Thekaekara. The spectral distribution of extra-terrestrial solar radiation is shown in Fig. 2.1.1. The curve depicts that the spectral first increases sharply with wavelength and passes through the maximum value of 2074 W/m^2 –um at a wavelength of 0.48 um and then decreases asymptotically to zero. 99 percent of the sun's radiation is obtained upto a wavelength of 4 um.

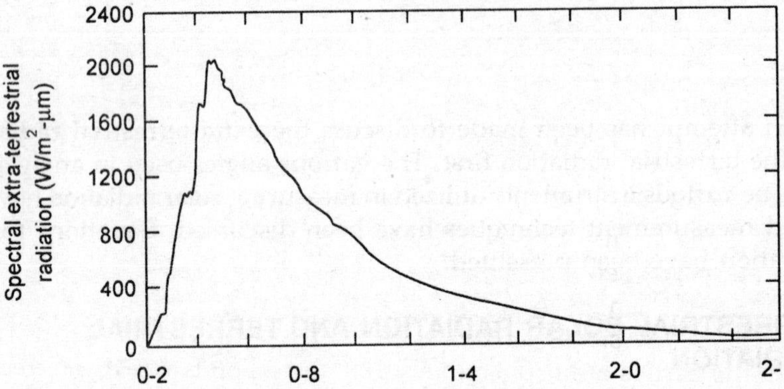

Fig. 2.1.1: Spectral Distribution of Extra-terrestrial Solar Radiation from NASA SP-8005 (1971).

The solar energy that arrives at the earth's atmosphere, consists of about 8 percent ultraviolet radiation (wave length, of less than 0.39 um), 46 percent visible light (0.39 to 0.78 um), 46 percent infra-red radiation (wave length of more than 0.78 um).

When the solar radiation passes through the earth's atmosphere (Fig. 2.1.2), it gets subjected to absorption and scattering mechanisms before reaching the earth's surface. Mainly because of presence of ozone and water vapour in the atmosphere, and to a lesser extent the other gases, e.g., CO_2, NO_2, O_2, CH_4 and CO etc. and particulate matter, absorb the solar radiation. This leads to an increase in the internal energy of the atmosphere. On the other hand, all gaseous molecules as well as particulate matter in the atmosphere assist in scattering the solar radiation. Some scattered radiation goes back into space and some reaches the earth's surface.

Moreover, the earth's atmosphere may be of without clouds or with clouds. In the former case, the sky is cloudless everywhere, while in the latter, the sky is partly or fully covered by clouds. For both types of atmosphere the absorption and scattering mechanisms, as stated earlier, occur. It is fact that less attenuation takes place in a cloudless sky and consequently earth's surface receives maximum radiation under the said condition.

Solar radiation received at the earth's surface without change in direction is called beam or direct radiation. The radiation received at the earth's surface from all parts of the sky's hemisphere after being scattered in the atmosphere is called diffuse radiation. The sum of the beam and diffuse radiation is referred to as total or global radiation. This is the total solar radiation received at any point on the earth's surface. In general sense, it is also referred to as the insolation at that point.

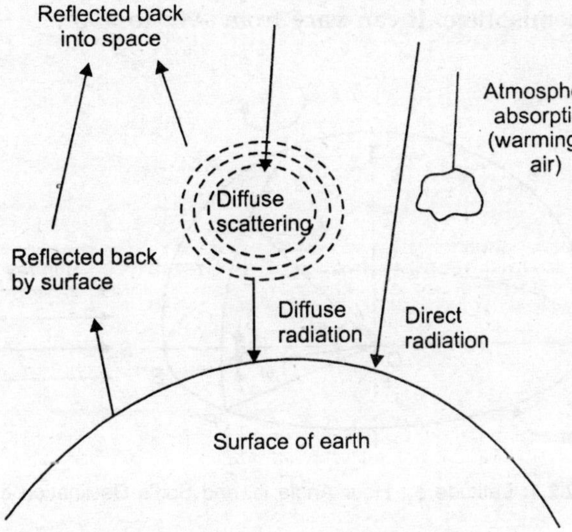

Fig. 2.1.2: Direct, Diffuse and Total Radiation.

A term called the air mass (m) is often used to assess the distance traveled by beam radiation through the atmosphere before reaching a location on the earth's surface. The air mass (m) is the ratio of the path of the sun's rays through the atmosphere to the length of the path when the sun is directly overhead, i.e., sun is at the zenith.

Just above the earth's atmosphere, $m = 0$. It corresponds to extra-terrestrial radiation. When the sun is at zenith, i.e., directly overhead, $m = 1$. At the zenith angle θ_z, the angle subtended by the zenith and the line of sight to the sun is 60°, then $m = 2$.

When $m > 3$, $m = \sec \theta_z$.

2.2 SOLAR RADIATION GEOMETRY

The following angles are often used in analyzing the solar radiation:

ϕ = latitude of location
δ = declination
w = hour angle
β = slope
α = altitude angle
θ_z = zenith angle
γ = surface azimuth angle
γ_s = solar azimuth angle

If θ, the incident angle, is the angle between incident beam radiation I_{bn} and the normal to a plane surface, then the equivalent radiation intensity or flux falling normal to the surface is given by $I_{bn} \cos\theta$. Let us first define all the aforesaid angles.

The latitude ϕ of a point or location is the angle made by the radial line joining the location to the center of the earth with the projection of the line on the equatorial plane. In accordance with Fig. 2.2.1, it is the angle between the line *OP* and *OP'*. Here *P* represents location on the earth surface and *O* represents the center of the earth. By convention, the latitude is measured as positive for the northern hemisphere. It can vary from –90° to +90°.

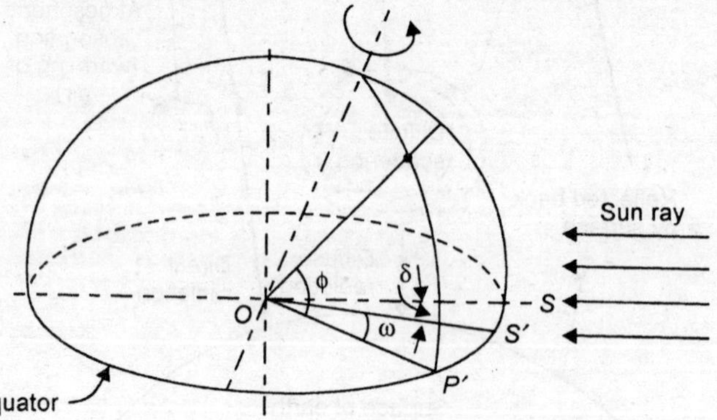

Fig. 2.2.1: Latitude ϕ_1, Hour Angle ω, and Sun's Declination δ.

The declination is the angular distance of the sun's rays north (or south) of the equator. It is the angle made by the line joining the centers of the sun and the earth with its projection on the equatorial plane. This arises by virtue of the fact of tilt and it varies from a maximum value of +23.45° on June 21 to a minimum value of –23.45 on December 21. Declination is zero on the two equinox days of March 22 and September 22. The Latin word "Equinox" means equal nights. The nights are equal when the declination of the sun is zero. The declination may be calculated from the following simple relationship given by Cooper (1969).

$$\delta \text{ (in degrees)} = 23.45 \sin [(360/365)(284 + n)] \quad \text{...(2.2.1)}$$

where n is the day of the year, [e.g. for June 21, 1996, $n = 31 + 29 + 31 + 30 + 31 + 21 = 173$]

The variation of sun's declination is shown in Fig. 2.2.2.

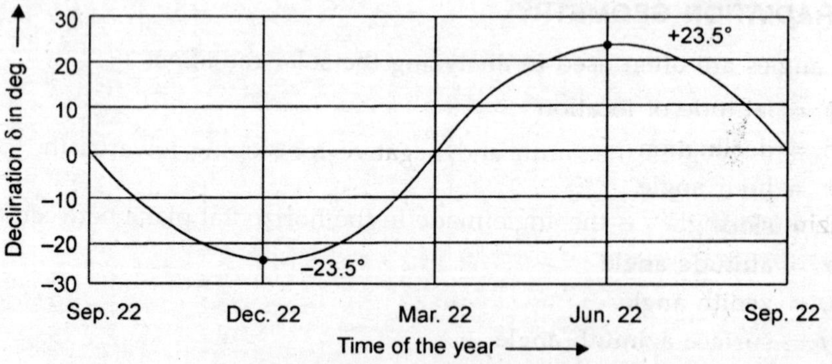

Fig. 2.2.2: Variation of Sun's Declination.

The hour angle w is the angle through which the earth must turn to bring the meridian at a point directly in line with the sun's rays shown in Fig. 2.2.1. It is equivalent to 15° per hour. It also varies from –180° to +180°. It is measured from noon based on the local solar time (LST) or local apparent time (LAT), being positive in the morning and negative in the afternoon.

Altitude angle α (solar altitude) is the vertical angle between the projection of the sun's rays on the horizontal plane and direction of sun's rays (passing through the point). This is shown in Fig. 2.2.3.

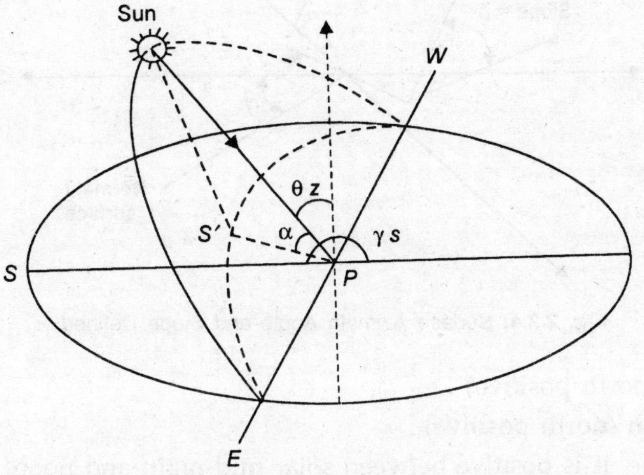

Fig. 2.2.3: Sun's Zenith, Altitude and Azimuth Angles.

Zenith angle θ_z is a vertical angle between sun's rays and a line perpendicular to the horizontal plane through the point (Refer Fig. 2.2.3).

$$\theta_z = (\pi/2) - \alpha$$

solar azimuth angle γ_s is the horizontal angle measured from north to the horizontal projection of the sun's rays. The angle is positive when measured west wise.

In terms of ϕ (latitude angle), δ (declination), and w (hour angle), altitude angle α, zenith angle θ_z and solar azimuth angles can be presented by followings:

$$\cos \theta_z = \cos \phi \cos w \cos \delta + \sin \phi \sin \delta \quad \text{...(2.2.2)}$$

$$\cos \gamma_s = \sec \alpha (\cos \phi \sin \delta - \cos \delta \sin \phi \cos w) \quad \text{...(2.2.3)}$$

$$\text{and} \quad \sin \gamma_s = \sec \alpha \cos \delta \sin w \quad \text{...(2.2.4)}$$

The slope β is the angle made by the plane surface with the horizontal. It is considered to be positive for sloping towards the south and negative for sloping towards the north. (Refer Fig. 2.2.4).

The surface azimuth angle γ is the angle made in the horizontal plane between the line due south and the projection of the normal to the surface on the horizontal plane. It can vary from –180° to +180°. The angle is considered to be positive if the normal is east of south and negative if normal is west of south.

The relation between θ and other angles can be given by the equation:

$$\begin{aligned}\cos \theta = {} & \sin \phi (\sin \delta \cos \beta + \cos \delta \cos \gamma \sin w \sin \beta) \\ & + \cos \phi (\cos \delta \cos w \cos \beta - \sin \delta \cos \gamma \sin \beta) \\ & + \cos \delta \sin \gamma \sin w \sin \beta \quad \text{...(2.2.5)}\end{aligned}$$

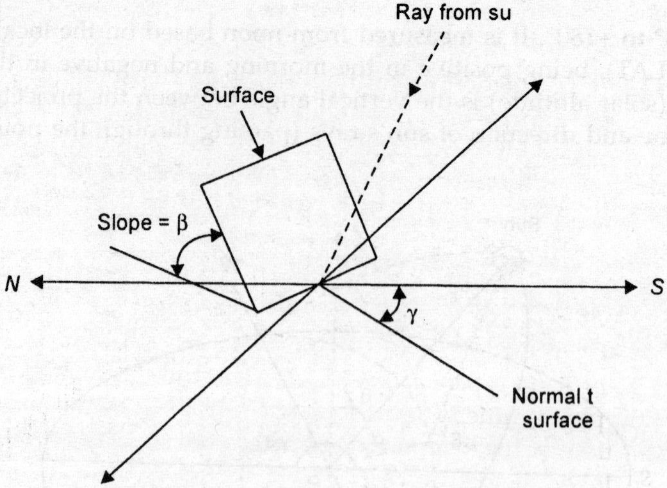

Fig. 2.2.4: Surface Azimuth Angle and Slope Defined.

where ϕ = latitude (north positive)

δ = declination (north positive)

w = hour angle, it is positive between solar mid-night and noon, otherwise negative.

For different conditions the equation (2.2.5) can be simplified as follows:

For vertical surfaces, $\beta = 90°$, equation (2.2.5) becomes

$$\cos\theta = \sin\phi\cos\delta\cos\gamma\cos w - \cos\phi\sin\delta\cos\gamma + \cos\delta\sin\gamma\sin w \quad ...(2.2.6)$$

For horizontal surfaces, $\beta = 0°$, $\theta = \theta_z$

$$\cos\theta_z = \sin\phi\sin\delta + \cos\phi\cos\delta\cos w \quad ...(2.2.7)$$

$$= \sin\alpha$$

i.e.,
$$\cos\theta = \cos\theta_z = \sin\alpha \quad ...(2.2.8)$$

For surface facing due south $\gamma = 0$,

$$\cos\theta_T = \sin\phi(\sin\delta\cos\beta + \cos\delta\cos w\sin\beta) + \cos\phi(\cos\delta\cos w\cos\beta - \sin\delta\sin\beta)$$

(θ_T denotes the incident angle for a tilted surface)

$$= \sin\delta\sin(\phi - \beta) + \cos\delta\cos w\cos(\phi - \beta) \quad ...(2.2.9)$$

For vertical surfaces facing due south, $\beta = 90°$, $\gamma = 0$

$$\cos\theta_z = \sin\phi\cos\delta\cos w - \cos\phi\sin\delta \quad ...(2.2.10)$$

Sunrise or Sunset Hour Angle (ws) and Day Length: The hour angle corresponding to sunrise or sunset on a horizontal surface is called sunrise or sunset hour angle (w_s). At the time of sunrise or sunset, the zenith angle, θ_z is 90°.

Substituting this in equation (2.2.7), we obtain sunrise or sunset hour angle (w_s) as:

$$\cos w_s = -\tan\phi\tan\delta, \quad ...(2.2.11)$$

Positive or negative value for w_s may be obtained. The positive value corresponds to sunrise and the negative to sunset. Since 15° of the hour angle is equivalent to 1 hour, the corresponding day length (in hours) is given by

$$t_d = (2/15) \cos^{-1} (-\tan \phi \tan \delta) \quad ...(2.2.12)$$

So the day length (t_d) is a function of latitude and solar declination.

For tilted surface, the sunrise or sun set hour angle (w_{st}) will be lesser than the value obtained by equation, if the day under consideration lies between March 21 and September 22. This can be obtained by substituting $\theta = 90°$ in equation (2.2.5) or one of its simple versions. Thus for an inclined surface facing south, substituting $\theta = 90°$, in equation (2.2.9), we obtain

$$\omega_{st} = \cos^{-1} [-\tan (\phi - \beta) \tan \delta] \quad ...(2.2.13)$$

The corresponding day length in hours is

$$t_d = (2/15) \cos^{-1} [-\tan (\phi - \beta) \tan \delta] \quad ...(2.2.14)$$

Local solar time (LST) or Local apparent time (LAT): The time estimated by the equations (2.2.5) – (2.2.14) is the local apparent time or local solar time. It is not same as that of local clock time. To assess the local solar time or local apparent time from the standard time observed on a clock, two corrections are to be done. The first correction appears due to the difference between the longitude of a location and the meridian on which the standard time is based. The correction is having a magnitude of 4 minutes for every degree difference in longitude. The second correction is called the equation of time correction. This correction arises due to the fact that the earth's orbit and rate of rotation are subject to small perturbations.

The equation of time correction is based on experimental observations and plotted in Fig. 2.2.5. Thus, Local solar time (LST) or Local apparent time (LAT) = standard time + 4 (standard time longitude – longitude of location) + (Equation of time correction) ...(2.2.15)

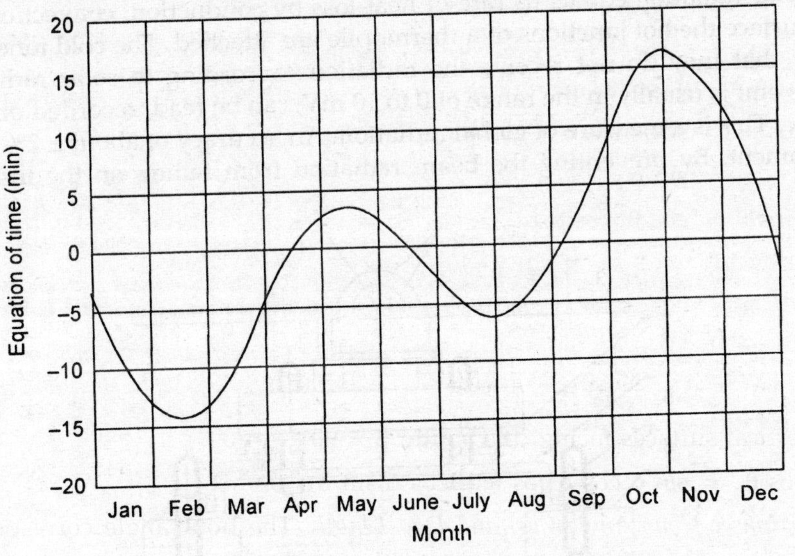

Fig. 2.2.5: Graph of Equation of Time Correction

The negative sign in the first correction is applicable for the eastern hemisphere, while positive sign is applicable for western hemisphere.

Illustrative Example: Calculate the local apparent time (LAT) corresponding to 13 - 30 (IST) on July 16, 1998 at Delhi (28°35′ N, 77°12′ E). The equation of time correction on July 16 is (–6) minutes. Indian standard time (IST) is the local civil time corresponding to 82.50° E longitude. Also calculate the declination.

Solution: The local apparent time (LAT) = IST – 4 (Standard time longitude – longitude of location) + (Equation of time correction) = $1330^h - 4\ (82°30' - 77°12') - 6'$

$= 1330^h - 4\ (82.5 - 77.2)$ minutes – 6 minutes

$= 1330^h - 15.2$ minutes $= 13^h\ 30' - 15'12''$

$= 13^h\ 14'48''$ **Ans.**

Declination (from Cooper's equation) , i.e., = 23.45 sin [(360/365) (284 + n)]

Here n = day of the year = 31 + 28 + 31 + 30 + 31 + 30 + 16 = 197 on July 16.

$\therefore\ \delta = 23.45 \sin [(360/365)\ (284 + 197)] = -23.45 \sin 22° = -8.78°$.

2.3 SOLAR RADIATION MEASUREMENTS

Day by day the use of solar appliances is increasing. In predicting the performance of the appliances, the accurate solar insolation data is needed. For this, measurements of solar radiation with the help of some instruments are required.

In the area of measurements in relating to solar radiation the following instruments are mostly used.

2.3.1 Pyranometer

It is an instrument by which either global or diffuse radiation over a hemispherical field of view can be measured. Basically the pyranometer (shown in Fig. 2.3.1) is having a 'black' surface which gets heated up when exposed to solar radiation. Its temperature gets increased until its rate of heat gain by solar radiation equals its rate of heat loss by conduction, convection and radiation. On the black surface the hot junctions of a thermopile are attached. The cold junctions are placed in such a way that they do not receive the radiation. According to solar radiation an emf is generated. This emf is usually in the range of 0 to 10 mV, can be read, recorded or integrated over a period of time. This is a measure of global radiation. An accuracy of about ± 2% can be obtained with the instrument. By preventing the beam radiation from falling on the instrument sensor,

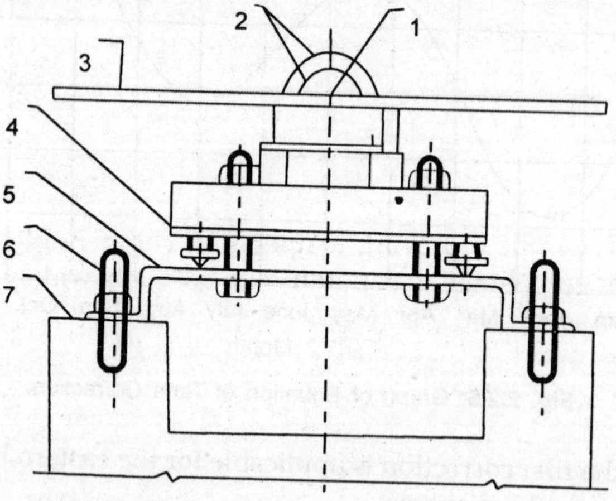

1. Black surface, 2. Glass domes, 3. Guard Plate
4. Three levelling screws, 5. Mounting plate,
6. Grouted bolts, 7. Platform

Fig. 2.3.1.1: Pyranometer for Measuring Global Radiation

with the help of a shading ring, the diffuse radiation is measured. The measurements with the help of pyranometer are the most common.

2.3.2 Pyrheliometer

Pyrheliometer measures only beam radiation. Diffuse radiation can not be measured with this instrument. In this instrument, a sensor disc is located at the base of a tube whose axis is aligned with the direction of the sun's rays. So, the diffuse radiation is essentially blocked from the sensor surface.

The most common current practice to measure direct solar rediation, is to use the thermo electric transducers. But low cost photovoltaic transducers are also becoming popular now-a-days. There are mainly three types of pyrheliometers commonly used for normal incident beam radiation. They are as follows,

(i) The Angstrom pyrheliometer
(ii) The Abbot silver-disc pyrheliometer
(iii) Eppley pyrheliometer

For all the cases, the receiving surface must be placed in such a manner so that it should be normal to direct solar rays.

In Angstrom Pyrheliometer, two manganin strips of size 20 × 2 × 0.1 mm each are used. One is shaded and another is exposed directly to solar radiation. The blackened shaded strip is electrically heated to the same temperature which is obtained by the exposed strip. Under steady state condition, the energy used for heating should be equal to the solar energy absorbed by the exposed strip. On the back of each strip, thermocouples are connected in opposition through a sensitive galvanometer. By this, equality of temperature for both the strips is tested. The energy of direct radiation (H_D) is calculated by the formula as:

$$H_D = ki^2$$

where i = heating current in amperes

k is a dimension and instrument const = $R/(Wa)$

where R is the resistance per unit length of the absorbing strip (Ω/cm).

W is the mean width of the absorbing strip and

a is the absorbtion coefficient of the absorbing strip.

For the case of Abbot silver-disc pyrheliometer a blackened silver disc is positioned at the lower end of a tube with diaphragms to control the limit of whole aperture upto 5.7°. This pyrheliometer is mainly used in calibrating pyranometers. In the case of Eppley pyrheliometer, there is a bismuth silver thermopile of having temperature compensated 15 junctions connected in series. The limiting diaphragms limits the aperature to 5.7°. It is widely used in USA and many other countries in the world.

2.3.3 Sunshine Recorder

With the help of a sunshine recorder (shown in Fig. 2.3.3.1) the duration of bright sunshine in a day is measured. The sun's rays are focused by a glass sphere to a point on a card strip held in a groove in a spherical bowl mounted concentrically with the sphere. During the bright sunshine in a day, the image formed moves along the strip. This image is so intense that it burns a spot on the card strip. Thus, a burnt trace whose length is proportional to the duration of sunshine is obtained on the strip.

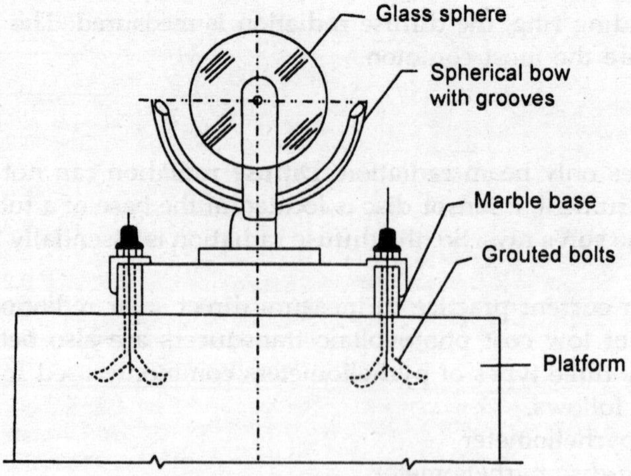

Fig. 2.3.3.1: Sunshine Recorder

2.4 ASSESSMENT OF AVERAGE DAILY GLOBAL SOLAR RADIATION

According to Angstrom (1924), the monthly average of the daily global radiation on a horizontal surface can be related to the amount of sunshine by a simple linear relation of the form:

$$H/H'_0 = a' + b'(\bar{n}/N) \qquad \text{...(2.4.1)}$$

where, H = monthly average of the daily global radiation on a horizontal surface at a location (W/m^2 – day).

H'_0 = monthly average of the daily global radiation on a horizontal surface at the same location on a clear day (W/m^2 – day).

$\bar{n}$ = monthly average of the sunshine hours per day at the location (h).

N = monthly average of the maximum possible sunshine hours per day at the location, i.e., the day length on a horizontal surface (h).

a', b' = arbitrary constants (Fritz 1951, suggested that $a' = 0.35$ and $b' = 0.61$).

Because of difficulties in taking decision on a clear day, page (1961) suggested a better from of equation 2.4.1 as,

$$H/H_0 = a + b(\bar{n}/N) \qquad \text{...(2.4.2)}$$

where, H_0 = monthly average of the daily extra-terrestrial radiation which would fall on a horizontal surface at the location under consideration.

a, b = modified constants depending upon the location. These constants can be obtained by carrying out regression analysis.

Table 2.4.1 gives the values of constants a and b for some Indian cities and towns.

The day length, i.e., N can be obtained by $N = (2/15)\cos^{-1}(-\tan\phi\tan\delta)$ ...(2.4.3)

H_0 can be obtained from charts or by following empirical relation.

$$H_0 = (24/\pi) I_{sc} [\{1 + 0.033\cos(360n/365)\} \times \{\cos\phi\cos\delta\sin w_s + (2\pi w_s/360)\sin\phi\sin\delta\}] \qquad \text{...(2.4.4)}$$

Table 2.4.1: Constants *a* and *b* for Some Indian Cities and Towns

Location	*a*	*b*	*Location*	*a*	*b*
Ahmedabad	0.28	0.48	Mangalore	0.27	0.43
Bangalore	0.18	0.64	Minicoy	0.26	0.39
Baroda	0.28	0.48	Nagpur	0.27	0.50
Bhavnagar	0.28	0.47	New Delhi	0.25	0.57
Bhopal	0.27	0.50	Pune	0.31	0.43
Calcutta	0.28	0.42	Roorkee	0.25	0.56
Goa	0.30	0.48	Shillong	0.22	0.57
Jodhpur	0.33	0.46	Srinagar	0.35	0.40
Kodaikanal	0.32	0.55	Trivandrum	0.37	0.39
Madras	0.30	0.44	Vishakhaaptnam	0.28	0.47

where, I_{sc} = solar constant per hour = 1353 W/m^2 or 4871 KJ/m^2 – h in SI units and 1165 Kcal/m^2 – hr in MKS units.

n = day of the year

ω_s = sunrise hour angle.

The declination δ can be obtained by Cooper Equation (2.2.1). The sunrise hour angle can be obtained from equation (2.2.11), i.e. $w_s = \cos^{-1}(-\tan\phi\tan\delta)$.

Illustrative Example: Calculate the average value of solar radiation on a horizontal surface for June 19, at the latitude of 10° N. The constants *a* and *b* are as 0.30 and 0.51 respectively. The average sunshine hours per day are 9.1.

Solution: Declination = 23.45 sin [(360/365) (284 + *n*)]

$= 23.45 \sin[(360/365)(284 + 170)] = 23.45 \sin 86 = 23.4°$

Sunrise hour angle $w_s = \cos^{-1}(-\tan\phi\tan\delta) = \cos^{-1}(-\tan 10° \tan 23.43°)$

Or $w_s = 94.4° = 1.646$ radian.

Now $H_0 = (24/\pi)I_{sc}[1 + 0.033\cos(360n/365)\{\cos\phi\cos\delta\sin w_s + (2\pi w_s/360)\sin\phi\sin\delta\}]$

$= (24/\pi)I_{sc}[\{1 + 0.033\cos(360 \times 170)/365\}$

$\{\cos 10 \cos 23.4 \sin 94.4 + \{(2\pi \times 1.646)/360\} \sin 10 \sin 23.4)\}]$

$= (24/3.14)\, I_{sc}\, [0.967\,(0.899 + 0.00192)] = (24/3.14)\, I_{sc}\, 0.871$

$= 6.658 \times 1.353$

$= 9\text{kW/m}^2 - \text{day}.$

Now, the day length $N = (2/15)\cos^{-1}(-\tan\phi\tan\delta) = 12.58$ h.

∴ Here, $\bar{n} = 9.1$ h. ∴ $H_0 = 9000$ w/m^2 – day.

$H = H_0(0.3 + 0.51 \times \bar{n}/N) = 6020$ w/m^2 – day.

2.5 SOLAR RADIATION ON TILTED SURFACES

Most of the solar collectors (e.g. flat plate collectors) are tilted to an angle to the horizontal. It therefore becomes necessary to convert data for a hourly radiation (measured or estimated) on a horizontal surface to radiation on a tilted surface. This radiation is the sum of the beam and diffuse radiation falling directly on the surface and the radiation reflected on to the surface from the surroundings.

Beam Radiation: Usually the tilted surface faces due south i.e. $\gamma = 0$, for this case,

$$\cos\theta = \sin\delta\sin(\phi-\beta) + \cos\delta\cos w\cos(\phi-\beta)$$

while for a horizontal surface ($\theta = \theta_z$),

$$\cos\theta_z = \sin\phi\sin\delta + \cos\phi\cos w\cos\delta$$

The ratio of the beam radiation falling on a tilted surface to that falling on a horizontal surface (Fig. 2.5.1) is called the tilt factor for beam radiation and is given by

$$R_b = H_T/H = H_n\cos\theta_T/H_n\cos\theta_z = \cos\theta_y/\cos\theta_z$$
$$= \sin\delta\sin(\phi-\beta) + \cos\delta\cos w\cos(\phi-\beta)/\sin\phi\sin\delta + \cos\phi\cos w\cos\delta \quad \text{...(2.5.1)}$$

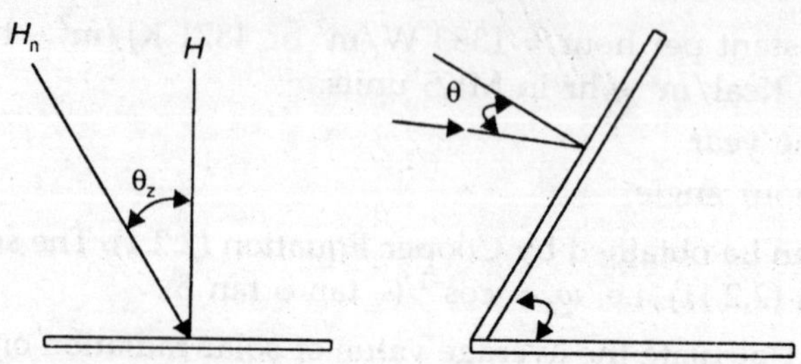

Fig. 2.5.1: Radiation on Horizontal and Tilted Surface

Diffuse Radiation: The tilt factor R_d for diffuse radiation is the ratio of the diffuse radiation flux falling on the tilted surface to that falling on a horizontal surface.

Assuming that the sky is an isotropic source of diffuse radiation, R_d is given by equation

$$R_d = (1+\cos\beta)/2 \quad \text{...(2.5.2)}$$

Because a surface, tilted at a slope of β from the horizontal, encounters $(1 + \cos\beta)/2$ as the radiation shape factor for a tilted surface with respect to the sky. This tilt factor depends on the distribution of the diffuse radiation over the sky and on the portion of the sky dome seen by the tilted surface.

Reflected Radiation: As $(1 + \cos\beta)/2$ is the radiation shape factor for a tilted surface with respect to the sky, so $(1 - \cos\beta)/2$ is the radiation shape factor for the tilted surface with respect to the surrounding ground. If it is assumed that the reflection of the beam and diffuse radiations falling on the ground is diffuse and isotropic, and those surroundings have a diffuse reflectance of ζ for solar radiation, the tilt factor for reflected radiation is given by

$$R_r = \zeta(1-\cos\beta)/2 \quad \text{...(2.5.3)}$$

Total Radiation: Total radiation flux falling on a tilted surface at any instant is given by

$$H_T = H_b R_b + H_d R_d + (H_b + H_d) R_r$$
$$= H_b R_b + H_d(1+\cos\beta)/2 + (H_b + H_d)(1-\cos\beta)/2\,\zeta \quad \text{...(2.5.4)}$$

The value of diffuse reflectivity ζ is not well-known for most situations. According to Liu and Jordan the value of ζ is around 0.2 when there is no snow and around 0.7 when there is a snow cover. For Indian conditions, a typical value around 0.2 is generally expected with surfaces of concrete or glass, being used. Since in India the value of the tilt *b* would rarely exceed 30°, the reflected radiation term does not very often contribute much to the total.

2.6 PERSONAL COMPUTER (PC) BASED MEASUREMENT OF SOLAR RADIATION

It has been pointed out that radiation measurements can be carried out by a variety of instruments like Radiometer, Pyrheliometer, Pyranometer, Pyroradiometer, etc. With all such instruments, the sensor output has to be callibrated or manipulated to obtain a particular insolation value. Thus, if such data are to be collected at short and regular intervals, a person has to be earmarked for the job only.

Data Acquisition Systems (DAS): These systems can eliminate the presence of a person to measure the radiation data. As PC's are now widely available in all laboratories such PC's can be utilized for the purpose of using suitable software and hardware interfacing. PC's have some empty slots in their motherboard for use by any user. Several types of modular matching cards are available commercially for fixing them into these slots. One such option is the Analog to Digital Conversion (ADC) card which serves as the vital interface between the sensor or instrument and the PC. The ADC receives analog data from the sensor, digitizes it and passes it on to the PC. Now, this data can be processed in the PC using suitable software. A block diagram of a complete measuring system is shown in Fig. 2.6.1, where the radiation incident on the Pyronometer is converted into thermoelectric emf. This emf is then fed into one of the channels of ADC card fixed with the PC.

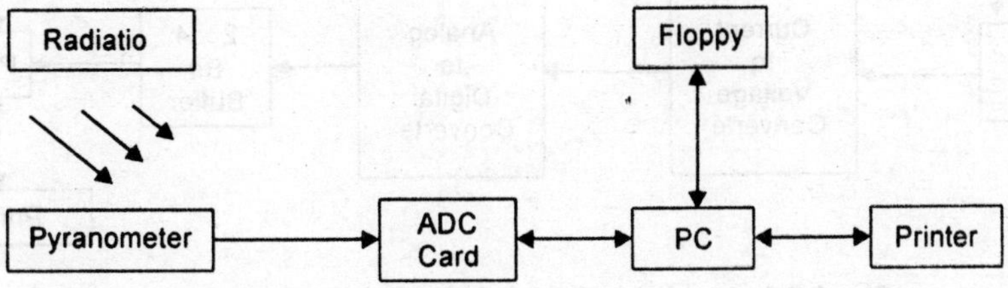

Fig. 2.6.1: Block Diagram of the Radiation Measuring System

Here the instantaneous voltage values in the digital form is stored in the Programmable Peripheral Interface (PPI) 8255. This memory can be accessed by one simple program written in basic language and another to process the data and print out the flux.

An alternative low-cost on-line system has been proposed by Mukherjee and Ray (NSEC-98), which avoids the use of common instruments like Pyranometers and Pyrheliometers and uses a 8-bit A/D converter instead of commercially available PC add on cards. The sensor used is a standard 4 inch diameter single crystal silicon solar cell in the photo-conducting mode. This device delivers a short circuit current upto 96mA corresponding to an incident insolation of 120m W/cm^2, as noted by a standardized Watt meter. This current is converted to a D.C. voltage in the range of 0 to 5 volts by designing a current to voltage converter as shown in Fig. 2.6.2. It is to be noted that OP-AMP 1 and transistor T_1 are used to convert the current to voltage while OP-AMP 2 makes this voltage compatible with the ADC-0809 which uses a clock frequency of 20 KHz. Almost linear variation of d.c. voltage (upto 4.8 volts) is observed against radiation. The 8-bit digital data from the A/D converter moves through a 4-bit buffer (IC74125) to avoid loading effect with the PC. Interfacing, in this case, has been performed through an LPT1 port, by using some of its standard terminals.

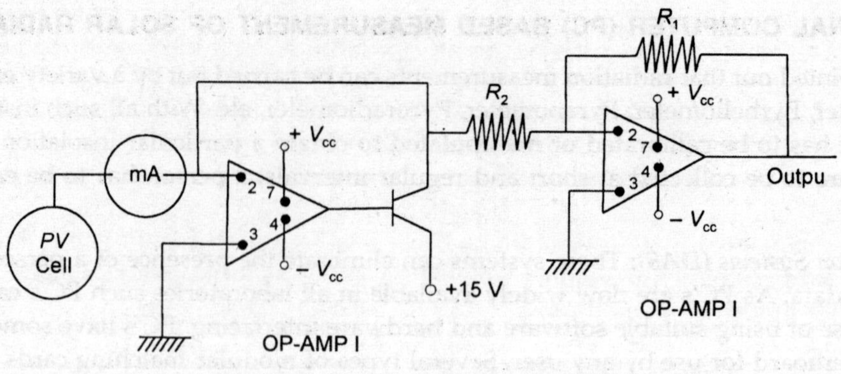

Fig. 2.6.2: Current to Voltage Converter.

A block diagram of the alternative low cost system is presented in Fig. 2.6.3.

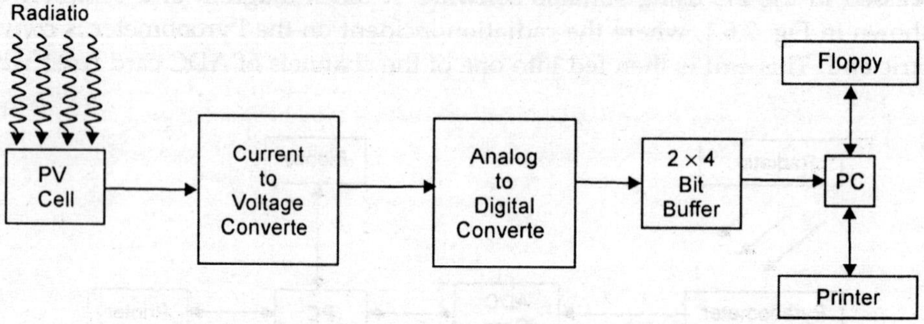

Fig. 2.6.3: Alternative On-line Measurement of Radiation.

EXERCISE

1. Calculate the day length on April 1 and November 1 for a south facing surface tilted at an angle of 40° and located at Bombay (19°07′N, 72°51′E).
2. Calculate the sunset hour angle and day length at a location latitude of 40°N, on June 21.
3. Calculate the daily global radiation falling on a horizontal surface at Shillong (25°34′N, 91°56′E) during the month of March. The average sunshine hours per day are 9.2. The constants a and b are 0.22 and 0.57 respectively.
4. What do you mean by solar constant and air mass ratio ? With a diagram, mention the related angles used in solar radiation geometry.
5. State the difference between a pyranometer and a pyrheliometer. Describe the principle of operation of an Angstrom type pyrheliometer.
6. Suggest a schematic arrangement for measuring online solar radiation with the aid of a personal computer.

SUMMARY

This chapter includes a detailed discussion on extra-terrestrial and terrestrial solar radiation. The meaning of solar constant has been explained in a lucid manner along with some typical data. The beam radiation and diffused radiation have been explained. The important parameter air mass has also been explained. The different terms e.g. latitude, altitude, zenith angle, sunset hour angle, local solar time and local apparent time etc. used to understand the solar radiation geometry have been discussed in detail. The instruments used relating to the solar radiation measurements have been described elaborately. Mathematical formulation for the assessment of average daily global solar radiation on the horizontal surface and the tilted surface have been presented. Personal computer based on-line solar radiation measurement system, which is getting importance day by day, has been explained in a very simple way. In this regard, details of hardware and interface ports have been presented in a manner suitable for initiating typical project work in a laboratory lacking in high grade infrastructure.

REFERENCES

1. A. Mani, Handbook of Solar Radiation Data for India, Allied Publishers, New Delhi, 1980.
2. A. Mani and O. Chacko, Solar Radiation—Climate of India, Solar Energy, 14, p. 139, 1973.
3. Y. Kemmoku, Daily Insolation Forecasting Using a Multi-stage Neural Networks, Solar Energy, 66 (3), pp. 193-199, 1999.
4. M. Iqbal, An Introduction to Solar Radiation, Academic Press, Canada, 1983.
5. S.P. Sukhatme, Solar Energy—Principles of Thermal Collection and Storage, Tata Mc-Graw Hill, 1996.
6. J.S. Hsieh, Solar Energy Engineering, Prentice Hall, New Jersy, 1986.
7. G.N. Tiwari and S. Suneja, Solar Thermal Engineering Systems, Narosa Publishing House, New Delhi, 1997.
8. B.S. Magal, Solar Power Engineering, Tata Mc-Graw Hill, 2000.
9. F. Kreith and J. Krieder, Principles of Solar Engineering, Washington, Hemisphere Publishing Corporation, 1978.
10. APCTT, Solar Energy, Asian-Pacific Centre for Technology Transfer, VATIS Update, 2 (4), 1994.
11. NSCT, Solar Energy: Promise and Challenge, National Committee on Science and Technology, New Delhi, April, 1974.
12. P.V. Narasimha Rao, Solar Power can solve our energy problems, A Pamphlet, Directorate of Advertising and Visual Publicity, Ministry of Information and Broadcasting, New Delhi, pp. 8-10, Feb. 1993.
13. M.P. Thekaekara and A.J. Drummond, Standard Values for the Solar Constant and its Spectral Components, Nat. Phy. Scu., 229, p. 6, 1971.
14. P.I. Cooper, The Absorption of Solar Radiation in Solar Stills, Solar Energy, 12, p. 3, 1969.
15. J.E. Braun and J.C. Mitchell, Solar Geometry for Fixed and Tracking Surfaces, Solar Energy, 31, p. 439, 1983.
16. N. Robinson, Solar Radiation, Elsevier Publ. Comp., New York, 1966.
17. A. Misra and N.S. Hukmani, Weather-data and Ambient Design conditions for Indian Stations (WeD Co Database), SESI Journal, 12(1), pp. 25-35, 2002.
18. N. Halouani, et al., Calculation of Monthly Average Global Solar Radiation on Horizontal Surfaces using Daily Hours of Bright Sunshine, Solar Energy, 50 (3), pp. 247-258, 1993.
19. J.M. Gorden and T.A. Reddy, Time Series Analysis of Daily Horizontal Solar Radiation, Solar Energy, 41, p. 215, 1988.
20. J.M. Gorden and T.A. Reddy, Time Series Analysis of Hourly Global Horizontal Solar Radiation, Solar Energy, 41, p. 423, 1988.
21. V.A. Graham and K.G.T. Hollands, A Method to Generate Synthetic Hourly Solar Radiation Globally, Solar Energy, 44, p. 333, 1990.
22. J.A. Davies and D.C. Mckay, Estimating Solar Irradiance and Components, Solar Energy, 29, p. 55, 1982.

23. J.E. Hay and J.A. Davis, Calculation of the Solar Radiation Incident on An Inclined Surface, Proc. of the 1st Canadian Solar Radiation Workshop, pp. 59-72, 1980.
24. D.T. Reindl, et al., Evaluation of Hourly Tilted Surface Radiation Models, Solar Energy, 45 (1), pp. 9-17, 1990.
25. H. Suehrcke and P.G. McCormick, The Distribution of Average Instantaneous Terrestrial Solar Radiation over the Day, Solar Energy, 42 (4), pp. 303-310, 1989.
26. S. Chakrabarti, et al., Prediction of Monthly Average Daily Global Solar Radiation for Calcutta-Howrah Region of West Bengal, Proc. of 24th NSEC-98, University of Roorkee, pp. 220-225, Dec. 1998.
27. M.A.C. Chendo and A.A.L. Maduekwe, Hourly Global and Diffuse Radiation of Lagos, Nigeria–Correlation with Some Atmospheric Parameters, Solar Energy, 52 (3), pp. 247-251, 1994.
28. K.K. Gopinathan, Estimation of Hourly Global and Diffuse Solar Radiation from Hourly Sunshine Duration, Solar Energy, 48 (1), pp. 3-5, 1992.
29. A. Mani and O. Chacko, Attenuation of Solar Radiation in the Atmosphere, Solar Energy, 24, pp. 347-349, 1980.
30. V.V. Satyamurthy and P.K. Lahiri, Estimation of Symmetric and Asymmetric Hourly Global and Diffuse Radiation from Daily Values, Solar Energy, 48 (1), pp. 7-14, 1992.
31. H.C. Hottel, A Simple Model for Estimating the Transmittance of Direct Solar Radiation Through Clean Atmospheres, Solar Energy, 18 (1), pp. 129-134, 1976.
32. J. Chandrasekharan and S.Kumar, Hourly Diffuse Fraction Correlation at a Tropical Location, Solar Energy, 53 (6), pp. 505-510, 1994.
33. L.T. Wong and W.K. Chow, Solar Radiation Models, Applied Energy, 69, pp. 191-224, 2001.
34. R.M.E., Diamant, Electronic Heat Meters, The Heating and Ventillation Engineering, 54 (625), p.p 6-8, 1980.
35. D. Mukherjee and A. Ray, Design of a Simple Data Acquisition System for On-line Recording of Solar Radiation, Proc. of NSEC-98, University of Roorkee, p.p.226-230, Dec'1998.
36. M.B. Pillai and J. Nagaraju, Thermal Energy Meter for Solar Water Heating Systems, SESI Journal, 8(1), pp. 37-47, 1998.

3

Solar Thermal Collectors

INTRODUCTION

A solar thermal collector is a device where solar energy is collected basically by absorbing the energy with the help of an absorber and then the absorbed energy is transferred to an working fluid passing in contact with it. Basically there are two types of collectors. One is flat plate collector where no optical concentration is allowed to occur. In this case, the collector area is the same as the absorber area. This collector is simple in design and has no moving parts. Maintenance requirement is less. Both beam and diffused solar radiation are utilized in this collector. With the help of it, the temperature of the working fluid can be raised to range from 40°C to 100°C. The other one is concentrating type collector where the area intercepting the solar radiation is number of times greater than the absorbed area. It is required to raise the temperature of the working fluid higher than 100°C. Mirrors and lenses are used to concentrate the Sun's rays on the absorber. The efficiency of the concentrating type collectors is higher than that of non-concentrating type. The flat plate collectors are classified into two types depending on the type of heat transfer fluids used. Liquid heating flat plate collectors are one type where water or non-aqueous non-freezing aqueous solutions are used as heat transfer fluids. The other type is air or gas heating collector which will be discussed in subsection 3.3.

3.1 LIQUID FLAT PLATE COLLECTOR

A schematic diagram of liquid flat plate collector is shown in Fig. 3.1.1.

The solar radiation first passes through one or more transparent covers (usually made of glass) and falls on the absorber plate. The absorbed radiation is partly transferred to the working heat transfer fluids passing through the tubes. The tubes are fixed to the absorber plate or are integrated with it. Rest energy is lost by convection and re-radiation to the surroundings from the top covers, and by conduction through the back and the edges. The convection and re-radiation losses are reduced by using the transparent covers and the conduction losses are reduced by using thermal insulation on the back and the edges. It uses both beam and diffuse components of solar radiation. It is simple in design and it needs little maintenance. But its collection efficiency is comparatively low because it does not use optical concentration leading to more heat loss. Typically a metal plate ranging in thickness from 0.2 mm to 1 mm is used as absorber. The most common metal is copper. But other metals are also being used effectively. In the last few years, plastic material is used for the absorber plate as it is light in weight, easy to manufacture and cost is less. A coating is placed on the surface of the absorber plate to enhance its absorptivity. Commonly black paint or black chrome is considered as the absorber coatings.

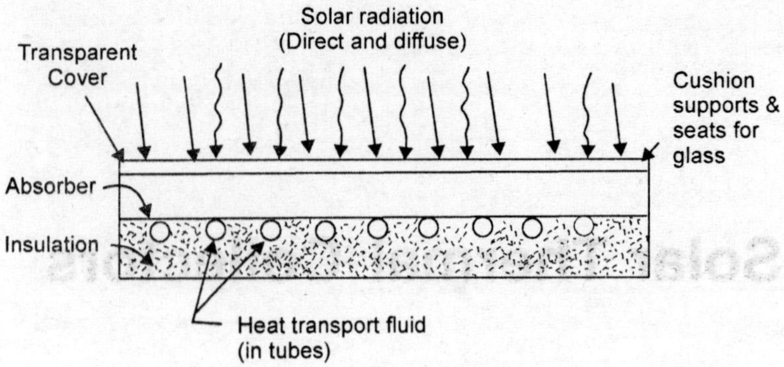

Fig. 3.1.1: Schematic Diagram of Flat Plate Collector

The tubes, through which the heat transfer fluid flows, are also made of metal or plastics ranging in diameter from 1 to 1.5 cm. The tubes are attached to the bottom of the absorber plate by welding, soldering or, brazing, clamping or pressure bonding. In some designs, the tubes are bonded on the top or are in-line and integral with the absorber plate. The pitch of the tubes ranges from 5 to 12 cm. The header pipes through which heat transfer fluid reaches the tubes of the collector and goes out of the collector, are made of the same metal as the tubes. The headers are of larger diameters ranging from 2 to 2.5 cm. The various types of configuration of heat transfer fluid passage and their method of bonding are shown in Fig. 3.1.2(a), (b), (c), (d), (e) and (f). Fig. 3.1.2(a) and (b) and (c) show two ways which are widely used throughout the world, Fig. 3.1.2(d) and (e) show method of bonding or clamping of tubes to flat or corrugated sheet. Fig. 3.1.2(f) highlights the "tube in strip" or roll bond design, where the tubes are formed in the sheet leading to good thermal bond between the sheet and the tubes.

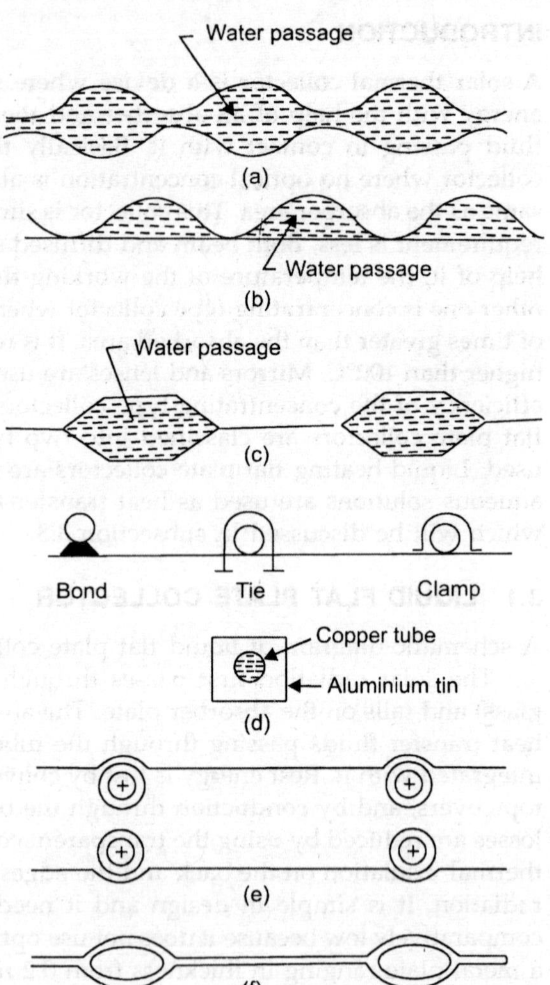

Fig. 3.1.2: Configuration of Heat Transfer Fluid Passages and their Bondings

The bottom and sides are usually insulated by mineral wool, rock wool or glass wool with a covering of aluminium foil to prevent heat losses from the back and edges of the collector. The thickness of the insulation ranges from 2.5 to 10 cm.

For the transparent covers, the plain or toughened glass of 4 to 5 mm thickness is commonly used. Usually 1 or 2 cover with spacings ranging from 1.5 to 3 cm. is used.

The collector box contains the whole assembly. It may be made of aluminium, steel sheet or fibre glass. It can be tilted at a suitable angle.

The commercially available collectors are of around 2m × 1m size, with the length kept along sloping direction. The followings are the typical salient data for a flat plate collector of a solar water heater of 100 lpd capacity (taken from BHEL manual).

Overall size: 2.114 m × 1.04 m
Absorber panel: Copper – Copper type
Absorber surface: Selectively coated with black chrome
Glazing: 4 mm thick toughened glass
Collector housing material: Aluminium extruded section
Inter connecting flange: Copper
Insulation: Rock Wool pad.

3.1.1 Transmissivity of the Cover System

Transmissivity of the cover system of a collector can be assessed with adequate accuracy if reflection – refraction and absorption are considered separately. It is given by $\tau = \tau_\rho \, \tau_\alpha$

where, τ = transmissivity of the glass cover system. It can be defined as the ratio of the solar radiation coming through after reflection at the glass-air interfaces and absorption in the glass to the radiation incident on the glass cover system,

τ_ρ = transmissivity obtained by considering only reflection and refraction

and τ_α = transmissivity obtained by considering only absorption.

Transmissivity for Beam Radiation: Transmissivity based on Reflection – Refraction at interface ($\tau\rho$):

According to Snell's law, $$\frac{\sin\theta_1}{\sin\theta_2} = \frac{n_2}{n_1} \quad ...(3.1.1.1.)$$

(Refer Fig. 3.1.1.1.)

where θ_1 = Angle of incidence, θ_2 = Angle of refraction,

n_1, n_2 = Refractive indices of the two media.

The reflectivity ρ which is nothing but I_ρ/I_0, where I_ρ is reflected beam intensity and I_0 is incident beam radiation and is related to the angle of incidence and refractions by:

$$\rho = \frac{1}{2}(\rho_1 + \rho_{11}) \quad ...(3.1.2)$$

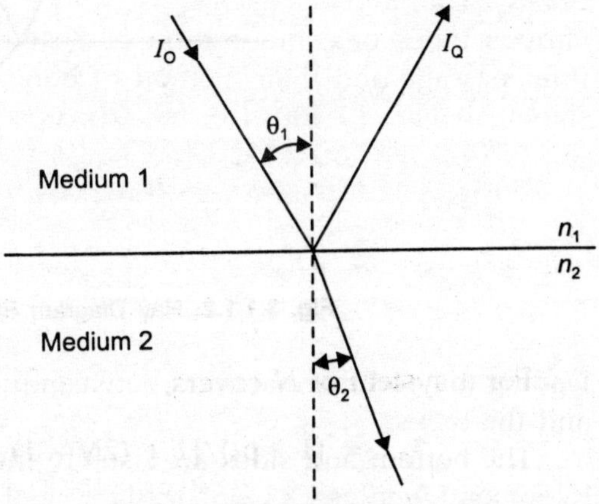

Fig. 3.1.1.1: Angle of Incidence and Reflection in Two Media

where, $$\rho_1 = \frac{\sin^2(\theta_2-\theta_1)}{\sin^2(\theta_2+\theta_1)} \text{ and } \rho_{11} = \frac{\tan^2(\theta_2-\theta_1)}{\tan^2(\theta_2+\theta_1)}$$

ρ_1 and ρ_{11} are the reflectivities of the two components of polarization.

So $$\rho = I_\rho/I_0 = \frac{1}{2}[\sin^2(\theta_2-\theta_1)/\sin^2(\theta_2+\theta_1) + \tan^2(\theta_2-\theta_1)/\tan^2(\theta_2+\theta_1)]$$

For normal incidence (i.e., $\theta_1 = \theta°$), it can be shown

$$\rho = \rho_1 = \rho_{11} = [(n_1 - n_2)/(n_1 + n_2)]^2 \qquad ...(3.1.1.3)$$

The transmissivity τ_ρ can be given by an expression similar to that for ρ

$$\text{Thus } \tau_\rho = \frac{1}{2}\left(\tau_{\rho 1} + \tau_{\rho 11}\right) \qquad ...(3.1.1.4)$$

Considering the transmission through a single cover where reflection – refraction takes place alone, (shown in Fig. 3.1.1.2) one of the components of polarization of a beam incident on the cover can be considered. As there are two interfaces, multiple reflections and refractions will take place.

$$\text{Hence, } \tau_{\rho_1} = (1 - \rho_1)^2 + \rho_1^{\,2}(1 - \rho_1)^2 + \rho_1^{\,4}(1 - \rho) + ...$$

$$= (1 - \rho_1)^2 [1 + \rho_1^{\,2} + \rho_1^{\,4} + ...] = (1 - \rho_1)^2/(1 - \rho_1^{\,2})$$

$$= (1 - \rho_1)/(1 + \rho_1) \qquad ...(3.1.1.5)$$

Similarly,

$$\tau_{\rho_{11}} = (1 - \rho_{11})/(1 + \rho_{11}) \qquad ...(3.1.1.6)$$

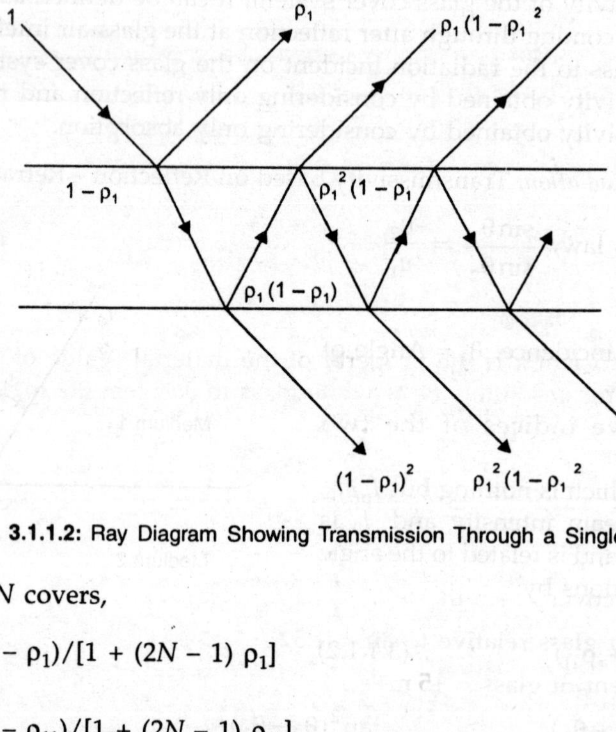

Fig. 3.1.1.2: Ray Diagram Showing Transmission Through a Single Cover

For a system of N covers,

$$\tau_{\rho_1} = (1 - \rho_1)/[1 + (2N - 1)\,\rho_1] \qquad ...(3.1.1.7)$$

$$\text{and} \quad \tau_{\rho_{11}} = (1 - \rho_{11})/[1 + (2N - 1)\,\rho_{11}]$$

Transmissivity based on Absorption (τ_α): According to Bauger's Law, the attenuation due to absorption is proportional to the local intensity. If a beam of intensity I_0 incident normally on a transparent cover of thickness δ and emerging with an intensity I_1 (shown in Fig. 3.1.1.3), then from Bouger's Law, $dI = -KI\,dx$

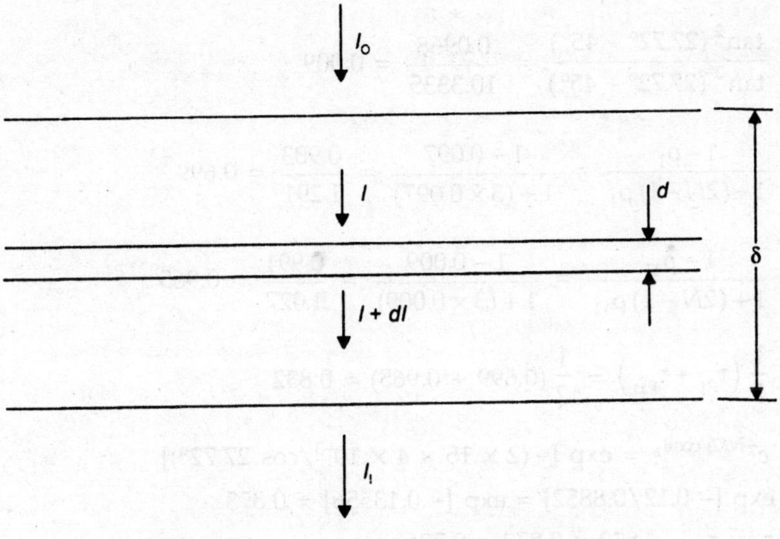

Fig. 3.1.1.3: Absorption in a Transparent Cover

Where K is a constant of proportionality known as the extinction coefficient. It is assumed to have a value independent of wavelength.

Integrating over the length traversed by the beam, we have

$$\tau_\alpha = I_1/I_0 = e^{-k\delta} \quad ...(3.1.1.8)$$

when the beam is incident at an angle θ_1, the path traversed through the cover would be $(\delta/\cos\theta_2)$, where θ_2 is the angle of refraction.

So, the eqn. 3.1.1.8 becomes $\tau_\alpha = e^{-k\delta/\cos\theta_2}$ ...(3.1.1.9)

For N covers, $\tau_\alpha = e^{-Nk\delta/\cos\theta_2}$ ...(3.1.1.10)

The extinction coefficient k is the property of the material. Value of k for glass varies from about 0.01/cm (1 m^{-1}) for absolutely clear white glass to 0.32/cm (32 m^{-1}) for poor quality glass with greenish cast of the edges.

Example 3.1.1.1: For a glass cover system, estimate τ_ρ, τ_α and τ for the following data.

Angle of incidence = 45°

Number of covers = 2

Thickness of each cover = 4 mm

Refractive index of glass relative to air = 1.52

Extinction coefficient of glass = 15 m^{-1}

Solution:

Here $\theta_1 = 45°$, $\sin\theta_1/\sin\theta_2 = 1.52$

$$\therefore \quad \theta_2 = \sin^{-1}[\sin 45°/1.52] = 27.72°$$

$$\rho_1 = \frac{\sin^2(27.72° - 45°)}{\sin^2(27.72° + 45°)} = \frac{0.0882}{0.9118} = 0.097$$

$$\rho_{11} = \frac{\tan^2(27.72^\circ - 45^\circ)}{\tan^2(27.72^\circ + 45^\circ)} = \frac{0.0968}{10.3335} = 0.009$$

$$\tau_{\rho_1} = \frac{1-\rho_1}{1+(2N-1)\rho_1} = \frac{1-0.097}{1+(3\times 0.097)} = \frac{0.903}{1.291} = 0.699$$

$$\tau_{\rho_{11}} = \frac{1-\rho_{11}}{1+(2N-1)\rho_{11}} = \frac{1-0.009}{1+(3\times 0.009)} = \frac{0.991}{1.027} = 0.965$$

$$\therefore\ \tau_\rho = \frac{1}{2}\left(\tau_{\rho_1} + \tau_{\rho_{11}}\right) = \frac{1}{2}(0.699 + 0.965) = 0.832$$

$$\text{Now, } \tau_\alpha = e^{-NK\delta/\cos\theta_2} = \exp\left[-(2 \times 15 \times 4 \times 10^{-3}/\cos 27.72^\circ)\right]$$

$$= \exp[-0.12/0.8852] = \exp[-0.13556] = 0.873$$

$$\therefore\ \tau = \tau_\rho \times \tau_\alpha = 0.832 \times 0.873 = 0.726$$

Transmissivity for Diffuse Radiation: Since, the diffuse radiation comes from many directions, the estimation of the transmissivity of a cover system under diffuse radiation is difficult. Usually it is assumed that the diffuse radiation is equivalent to beam radiation coming at an angle of incidence of 60°.

Transmittance—Absorptance Product: The transmittance-absorptance product is denoted by the symbol (τ_α). Frequently it is also referred as transmissivity—absorptivity product. An appropriate subscript (*b* or *d*) is added in the symbol (τ_α) to indicate the type of radiation. Here, τ is the transmissivity of the glass cover system and α is the absorptivity of the absorber plate.

It is defined as the ratio of insolation absorbed by the absorber plate to the insolation incident on the cover system.

When a fraction of radiation τ falls on the absorber plate after transmitting through the cover system, a part of radiation is absorbed and a part gets reflected diffusely. Among the reflected part, a portion gets transmitted through the cover system and a portion gets reflected back to the absorber plate. The process of absorption and reflection at the absorber plate surface is shown in Fig. 3.1.1.4. The process continues indefinitely. The quantities involved in the process becomes gradually smaller.

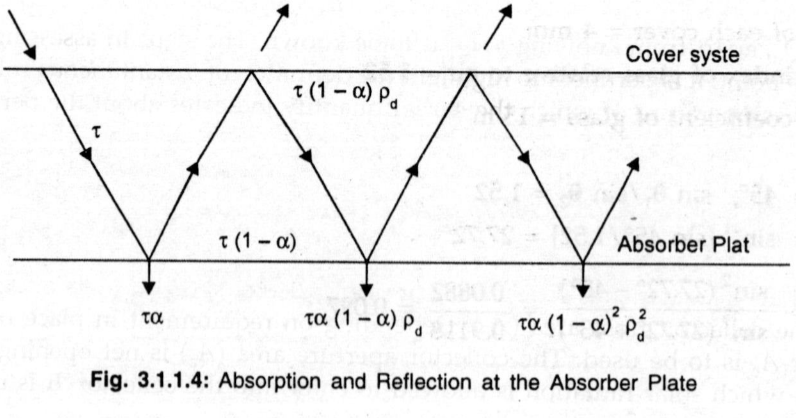

Fig. 3.1.1.4: Absorption and Reflection at the Absorber Plate

Thus the net fraction absorbed (τ_α) can be given by

$$(\tau_\alpha)_{net} = [\tau_\alpha + \tau_\alpha (1-\alpha)\rho_d + \tau_\alpha (1-\alpha)^2 \rho_d^2 + ...]$$

$$= \tau_\alpha [1 + (1-\alpha)\rho_d + (1-\alpha)^2 \rho_d^2 + ...]$$

$$= \tau_\alpha / [1 - (1-\alpha)\rho_d] \qquad ...(3.1.1.11)$$

In this case, ρ_d denotes the diffuse reflectivity of the cover system, and it can be assessed by evaluating the value $\tau_\alpha (1 - \tau_\rho)$ for the cover system for an incidence angle of 60°.

3.1.2 Performance Analysis

As discussed in sec. 2.5, the total solar radiation flux falling on a tilted surface at any instant can be given by

$$I_T = I_b R_b + I_d R_d + (I_b + I_d) R_r \qquad ...(3.1.2.1)$$

Where I_T = total radiation flux, I_b = beam radiation flux, I_d = diffused radiation flux and the value of R_b is same as given in equation (2.5.1) for a tilted surface with $\gamma = 0$. R_d and R_r can be estimated by the equations (2.5.2) and (2.5.3). These equations are valid for any tilted surface with a slope β.

For the steady state condition, the energy balance condition on the absorber plate of a liquid flat plate collector, in which the tubes are considered to be bonded on the under-side of the absorber plate can be written as $Q_u = A_p S - Q_1$ ...(3.1.2.2)

where, Q_u = Useful heat gain, i.e. the rate of heat transfer to the working fluid, (watt).

S = Solar flux absorbed by the absorber plate, (W/m^2).

A_p = Area of the absorber plate, m^2

Q_1 = Rate of heat loss by convection and re-radiation from the top, and by conduction and convection from the bottom and sides, (watt).

The term S can be estimated by—

$$S = I_b R_b (\tau_\alpha)_b + [I_d R_d (I_b + I_d) R_r] (\tau_\alpha)_d \qquad ...(3.1.2.3)$$

Where $(\tau_\alpha)_b$ is the transmissivity-absorptivity product for beam radiation falling on the collector and $(\tau_\alpha)_d$ is the transmissivity-absorptivity product for diffuse radiation falling on the collector. The radiation flux absorbed by the absorber (i.e., S) can also be expressed as

$$S = I_T (\tau_\alpha)_e \qquad ...(3.1.2.4)$$

where $(\tau_\alpha)_e$ is the effective transmissivity-absorptivity product of cover system for beam and diffuse radiation.

To assess Q_u, apart from knowing S, Q_1 is to be known. The steps in assessing Q_1 is taken up later on. At this point, it is pertinent to highlight the definition of instantaneous collector efficiency and stagnation temperature of which the earlier quantity indicates about the performance of the collector and the later is used as an indicator for comparing designs of different collectors.

The instantaneous collector efficiency is defined as the ratio of useful heat gain and radiation incident on the collector. This is given by

$$n_i = Q_u / A_p I_T \qquad ...(3.1.2.5)$$

Sometimes the instantaneous collector efficiency is calculated on the basis of collector aperature area (A_a) or the collector gross area (A_c). Depending on requirement in place of A_p in the eqn. (3.1.2.4), A_a or A_c is to be used. The collector aperture area (A_a) is net opening in the topmost cover through which solar radiation is allowed to enter into the collector. It is usually about 10 to 15 percent more than absorber plate area. The collector gross area (A_c) is the area of the topmost cover including the frame. It is usually about 15 to 20 percent more than absorber plate area.

When the collector is used without flowing liquid through the collector, its useful gain and efficiency become zero. At this condition, the absorber plate attains the highest temperature such that $A_p S = Q_l$ gets satisfied. This temperature is sometimes referred to as the stagnation temperature. It helps in comparing different collector designs and in choosing proper materials for the collector.

3.1.3 Heat Losses and Heat Loss Coefficients

The heat lost from the collector can be given by

$$Q_l = U_l A_p (T_p - T_a) \quad \text{...(3.1.3.1)}$$

Where, T_p = Average temperature of the absorber plate.

T_a = Temparature of surrounding air.

U_l = Overall heat loss coefficient

A_p = Area of the absorber plate

It is evident that heat losses take place from the top, bottom and sides. So the heat lost from the collector can also be written as

$$Q_l = Q_t + Q_b + Q_s$$

Where, Q_t = Rate of heat loss from the top

Q_b = Rate of heat loss from the bottom

Q_s = Rate of heat loss from the sides

Each heat loss can also be determined with respect to their respective heat loss coefficient by the following equations.

$$Q_t = U_t A_p (T_p - T_a) \quad \text{...(3.1.3.2)}$$

$$Q_b = U_b A_p (T_p - T_a) \quad \text{...(3.1.3.3)}$$

$$Q_s = U_s A_p (T_p - T_a) \quad \text{...(3.1.3.4)}$$

From these above said equations, the overall heat loss coefficient is given by—

$$U_l = U_t + U_b + U_s \quad \text{...(3.1.3.5)}$$

The typical range of values for U_l varies from 2 to 10 W/m^2-K.

Among the losses, top loss dominates the others. For a single cover system, the typical top loss becomes about 85 percent. Therefore, in practice, measures are taken to minimize the top loss as far as possible. The top loss can usually be reduced by (i) Using multi cover systems, (ii) Using selective coating on the absorber plate, and (iii) maintaining vacuum in the spaces between covers and between cover and absorber plate.

3.1.3.1 Top Loss Coefficient (U_t)

The heat loss from the top occurs due to convection and re-radiation losses from the absorber plate in the upward direction. Transparent covers and absorber plate may be considered as a system of infinite parallel surfaces where the heat flow is steady and occurring in one-direction. It can be conveniently assumed that the temperature drop across the covers is negligible. The insolation absorbed by the covers and the outgoing large wavelength's re-radiation loss from it may be neglected. The aforesaid assumptions are very much appropriate for the transparent cover made of glass.

According to a schematic diagram for a two-cover system as shown in Fig. 3.1.3.1.1, the steady state heat transfer due to convection and re-radiation will take place between absorber plate and the bottom cover, between bottom cover and top cover, and between top cover and the surroundings. For each case the heat transfer will be equal. Hence

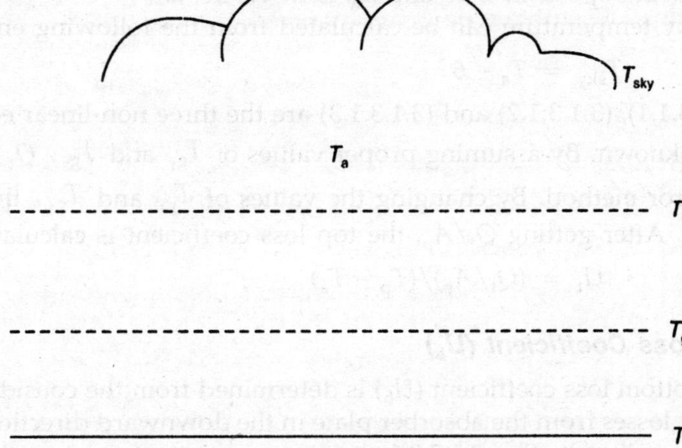

Fig. 3.1.3.1.1: Calculation of the Top Loss Coefficient

$$\frac{Q_t}{A_p} = h_{p-g_2}(T_p - T_{g_2}) + \frac{\sigma\left(T_p^4 - T_{g_2}^4\right)}{\left[1/\epsilon_p + 1/\epsilon_g - 1\right]} \qquad ...(3.1.3.1.1)$$

$$= h_{g_2-g_1}\left(T_{g_2-} - T_{g_1}\right) + \frac{\sigma\left(T_{g_2}^4 - T_{g_1}^4\right)}{\left[1/\epsilon_g + 1/\epsilon_g - 1\right]} \qquad ...(3.1.3.1.2)$$

$$= h_w\left(T_{g_1} - T_a\right) + \sigma\,\epsilon_g\left(T_{g_1}^4 - T_{sky}^4\right) \qquad ...(3.1.3.1.3)$$

where, h_{p-g_2} = Convective heat transfer coefficient between the absorber plate and the bottom cover, W/m^2-K

$h_{g_2-g_1}$ = Convective heat transfer coefficient between the bottom and top covers, W/m^2-K

h_w = Convective heat transfer coefficient between the top cover and the surrounding air, W/m^2-K

T_{g_2}, T_{g_1} = Temperatures attained by bottom and top covers respectively, K

T_{sky} = Effective temperature of the sky with which radiative heat exchange takes place, K

ϵ_p = Emissivity of the absorber plate for long wavelength radiation.

ϵ_g = Emissivity of the covers for long wavelength radiation.

Many researchers have developed various correlations regarding the heat transfer coefficient between the surfaces and the top cover.

For example, the empirical correlation for the heat transfer coefficient at the top cover suggested by McAdams is given by

$$h_w = 5.7 + 3.8\, V_\alpha \qquad ...(3.1.3.1.4)$$

in which V_α is the wind speed in m/s and h_w is in W/m^2-k.

The effective sky temperature can be calculated from the following empirical relation

$$T_{sky} = T_a - 6 \qquad ...(3.1.3.1.5)$$

Equations (3.1.3.1.1), (3.1.3.1.2) and (3.1.3.1.3) are the three non-linear equations in which Q_t, T_{g_2} and T_{g_1} are unknown. By assuming proper values of T_{g_1} and T_{g_2}, Q_t is calculated with the help of trial and error method. By changing the values of T_{g_1} and T_{g_2}, the optimum value for Q_t/A_p is calculated. After getting Q_t/A_p, the top loss coefficient is calculated by

$$U_t = (Q_t/A_p)/(T_p - T_a) \qquad ...(3.1.3.1.6)$$

3.1.3.2 Bottom Loss Coefficient (U_b)

In a collector, the bottom loss coefficient (U_b) is determined from the consideration of conduction and convection heat losses from the absorber plate in the downward direction through the bottom of the collector. According to Fig. 3.1.3.2.1, it is appropriate to assume that in most cases the insulation thickness provided is such that thermal resistance associated with conduction dominates. Therefore, the convective resistance at the bottom surface of the collector casing can be neglected. For the steady and one dimensional heat flow, the bottom loss coefficient is given by

$$U_b = K_b/L_b \qquad ...(3.1.3.2)$$

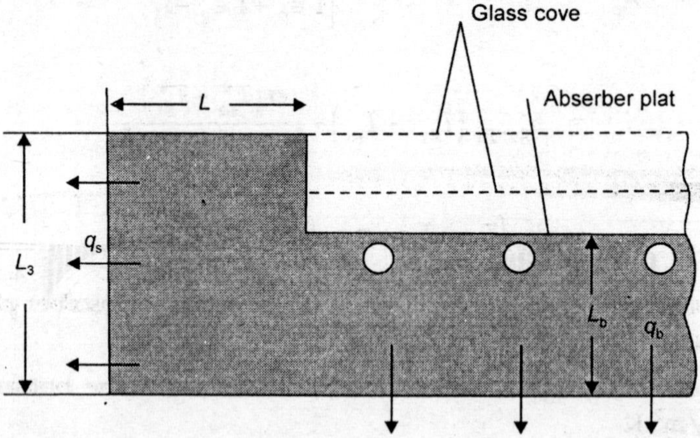

Fig. 3.1.3.2.1: Bottom and Side Losses from a Flat-plate Collector

Where K_b = thermal conductivity of the insulation

L_b = thickness of the insulation

3.1.3.3 Side Loss Coefficient (U_s)

The area across which the heat flows sideways is given by

$$A_s = 2\,(L_1 + L_2)\,L_3 \qquad \text{...(3.1.3.3.2)}$$

where L_1, L_2 = Dimensions of the absorber plate.

L_3 = Height of the collector casing.

The temperature drop across which the heat flow occurs varies from $(T_p - T_a)$ at the absorber plate level to zero both at the top and bottom. Therefore it can be assumed that the average temperature drop across the side insulation is $(T_p - T_a)/2$. If the thickness of insulation is L_s, for one dimensional and steady heat flow condition like the earlier case, the heat loss through the sides is given by

$$Q_s = 2\,L_3\,(L_1 + L_2)\,K_s \cdot (T_p - T_a)/2\,L_s \qquad \text{...(3.1.3.3.3)}$$

Thus, from eqn. (3.1.3.4)

$$U_s = [(L_1 + L_2\,)\,L_3\,K_s]/A_p \cdot L_s \qquad \text{...(3.1.3.3.4)}$$

$$= U_s' \cdot (A_s/2)/A_p \qquad \text{...(3.1.3.3.5)}$$

where, $U_s' = K_s/L_s$ ...(3.1.3.3.6)

(The eqn. 3.1.3.3.6) is like eqn. (3.1.3.2.1)

where $A_p = L_1 \times L_2$

3.1.4 Thermal Analysis of Flat Plate Collector

Let there be n fluid tubes in a collector having absorber plate of length L_1 and width L_2. The pitch of the tubes is $W\ (= L_2/n)$ and D_i, D_0 and S be the inside, outside diameters of the tubes and thickness of the absorber plate respectively

If a section of the absorber plate with two adjacent tubes (shown in Fig. 3.1.4.1) is considered, the absorber plate temperature distribution in the x-direction is also shown in the same figure.

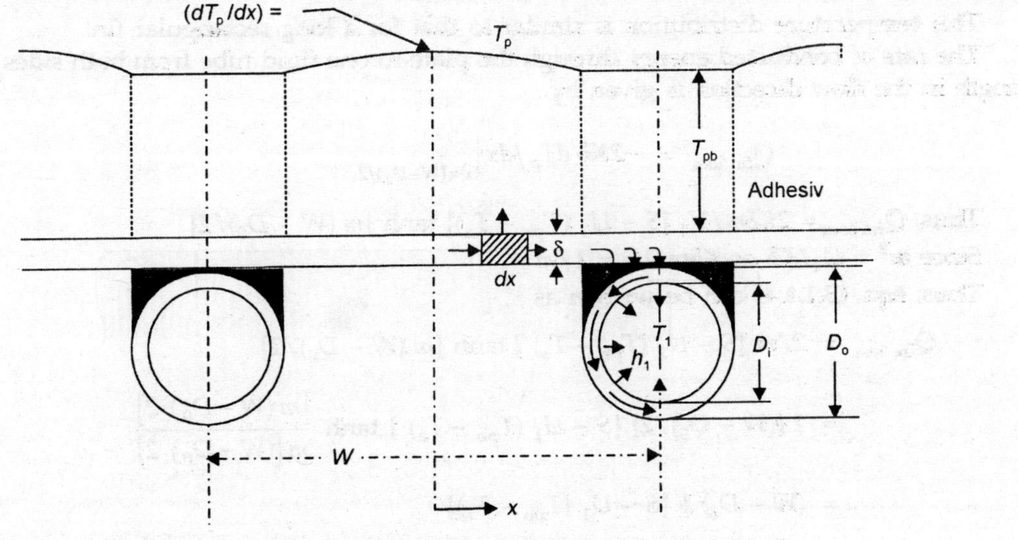

Fig. 3.1.4.1: Temperature Distribution in an Absorber Plate in a Direction at Right Angles to the Fluid Flow

The same distribution is also assumed prevailing between any two tubes. It is also assumed that the temperature of the sheet above the fluid tube i.e. above the bond is constant (T_{pb}), while in between the tubes, temperature will pass through a maximum value. For elemental region Δx and unit length in the flow direction, assuming the negligible heat conduction in the plate in that direction, energy balance (shown in Fig. 3.1.4.2) gives as

$$S\,\Delta x - U_1\,\Delta x\,(T_p - T_a) + [-K\,\delta\,(dT/dx)]_x - [-K\delta\,(dT/dx)]_{x+dx} = 0 \qquad ...(3.1.4.1)$$

The simplification yields

$$d^2T_p/dx^2 = (U_1/K\delta)\,[T_p - T_a - S/U_1] \qquad ...(3.1.4.2)$$

where, U_1 = Overall heat loss coefficient

K = Thermal conductivity of the plate

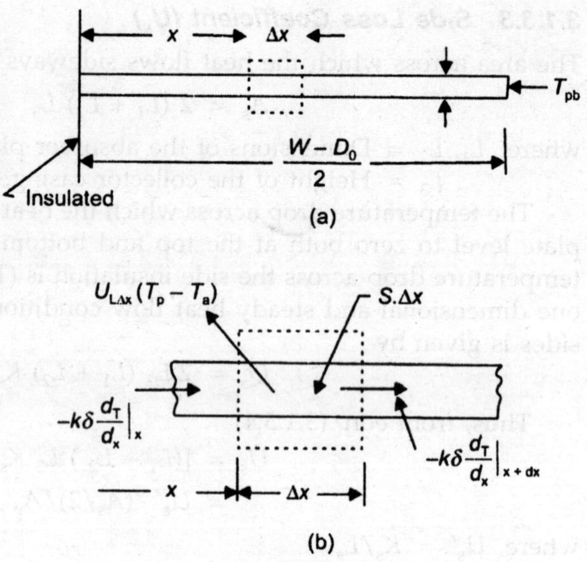

Fig. 3.1.4.2: Energy Balance Conditions in a Flate Plate Collector

Equation (3.1.4.2) can be solved by using the following boundary conditions.
At $x = 0$, $dT_p/dx = 0$ and at $x = (W - D_0)/2$, $T_p = T_{pb}$
This yields the solution

$$\frac{T_p - T_a - S/U_1}{T_{pb} - T_a - S/U_1} = \frac{\cosh mx}{\cosh\left[m\,(W - D_0)/2\right]} \qquad ...(3.1.4.3)$$

where $m^2 = U_1/K\delta$

This temperature distribution is similar to that for a long rectangular fin.

The rate of conducted energy through the plate to one fluid tube from both sides per unit of length in the flow direction is given by

$$Q_{\text{fin loss}} = -2K\delta\,dT_p/dx\Big|_{x=(W-D_0)/2}$$

Thus, $Q_{\text{fin loss}} = 2K\delta m/U_1\,[S - U_1\,(T_{pb} - T_a)]\,\tanh\,[m\,(W - D_0)/2]$ ...(3.1.4.4)

Since $m^2 = U_1/K\delta$ or $K\delta m/U_1 = 1/m$

Thus, eqn. (3.1.4.4) can be written as

$$Q_{\text{fin loss}} = 2/m\,[S - U_1\,(T_{pb} - T_a)\,]\,\tanh\,[m\,(W - D_0)/2]$$

$$= 2\,[(W - D_0)/2]\,[S - U_1\,(T_{pb} - T_a)\,]\,\tanh\frac{\left[m\,(W - D_0)/2\right]}{m\left[(W - D_0)/2\right]}$$

$$= (W - D_0)\,F\,[S - U_1\,(T_{pb} - T_a)] \qquad ...(3.1.4.5)$$

where $F = \dfrac{\tanh\left[m\,(W - D_0)/2\right]}{m\left[(W - D_0)/2\right]}$ and it is usually known as the fin efficiency.

The rate at which energy is absorbed in the tube just where the fluid flows in i.e. the energy collected above the tube region is given by

$$Q_{tube} = D_0 [S - U_1 (T_{pb} - T_a)] \qquad ...(3.1.4.6)$$

Thus, the total energy gain of the collector tube per unit length in the flow direction can be written as

$$Q_u = Q_{fin\ loss} + Q_{tube} = [(W - D_0) F + D_0] [S - U_1 (T_{pb} - T_a)] \qquad ...(3.1.4.7)$$

The flow of heat from the plate to the fluid occurs through three thermal resistances in the path. They are expressed as

(i) Resistance due to the adhesive used for attaching the tubes to the absorber plate, i.e., due to bonding material used between plate and the tube.
(ii) Resistance due to tube wall thickness.
(iii) The resistance due to the heat transfer coefficient at the inner surface of the tube.

$$\text{Hence, } Q_u = \frac{T_{pb} - T_f}{1/C_a + 1/D_1 h_f + 1/C_w} \qquad ...(3.1.4.8)$$

where, C_a = the conductance of the adhesive.
h_f = the local film heat transfer coefficient on the inside surface of the tube.
C_w = the conductance of the tube wall.
and T_f = local fluid temperature.

The conductance of the bond can be calculated as, $C_a = K_a\ L_a/\delta_a$ ...(3.1.4.9)

where, K_a = thermal conductivity of the adhesive material.
δ_a = average thickness of the adhesive.
L_a = length of adhesive or bond.

The bond conductance is important in describing collector performance. To protect from significant loss, proper bonding between the plate and tube is to be ensured so that it should have good metal to metal contact by providing bond resistance less than about 0.33 m – °C/W.

From equations (3.1.4.7) and (3.1.4.8) eliminating T_{pa}, the useful heat gain is given by—

$$Q_u = \frac{S - U_1 (T_f - T_a)}{U_1 \left[1/\left(U_1\{(W - D_0) F + D_0\}\right) + 1/C_a + 1/\pi D_1 h_f + 1/C_w\right]} \qquad ...(3.1.4.10)$$

Now Q_u also be expressed as

$$Q_u = W F^1 [S - U_1 (T_f - T_a)] \qquad ...(3.1.4.11)$$

$$\text{where } F^1 = \frac{1/U_1}{W [1/(U_1 (W - D_0) F + D_0)] + 1/C_a + 1/C_w + 1/\pi D_1 h_f} \qquad ...(3.1.4.12)$$

It represents the actual useful heat gain per tube per unit length to the gain which would occur if the collector absorber plate were at the temperature T_f. It can also be represented by the ratio of thermal resistance between absorber plate and ambient air to the thermal resistance between fluid and ambient air. In practice, the thermal resistance of the tube wall is neglected with respect to the other resistances.

The one dimensional analysis can also be carried out along the direction of the fluid flow in linking the useful heat gain rate with the fluid inlet temperature. In obtaining the temperature distribution in the fluid flow direction, the application of first law of thermodynamics on a control volume, having an elementary length dy of one tube, yields (refer fig. 3.1.4.3): Rate of change of enthalpy of the fluid flowing through the control volume = Rate of heat transfer to fluid inside the control volume.

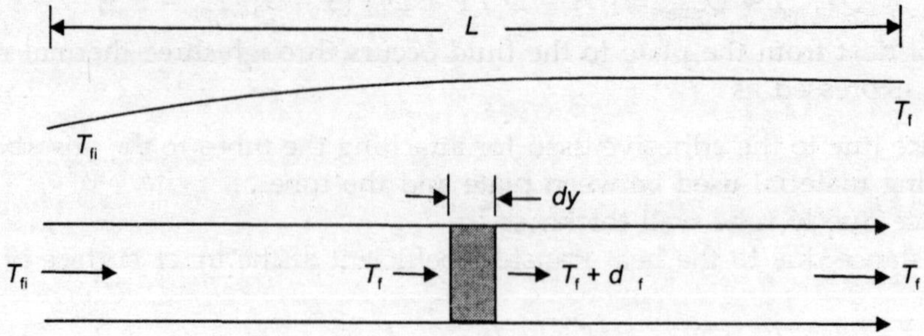

Fig. 3.1.4.3: Variation of Fluid Temperature in Flow Direction

Thus $(m/n)\, C_p\, d\,T_f = d\,Q_u = W F^1 [S - U_1 (T_f - T_a)]\, dy$

Where m/n = mass flow rate through one tube

dQ_u = useful heat gain in one tube

So, $dT_f/dy = W F^1 U_1/(m/n) \cdot C_p [(S/U_1 + T_a) - T_f]$ ...(3.1.4.13)

Integrating the eqn. (3.1.4.13) and using the boundary condition as, at $y = 0$, $T_f = T_{fi}$ Where T_{fi} is the fluid inlet temperature, the temperature distribution is given by—

$$\frac{T_f - T_a - S/U_1}{T_{fi} - T_a - S/U_1} = e^{-(U_1 L_2 F^1/m C_p)} \quad ...(3.1.4.14)$$

At $y = L_1$, $T_f = T_{fo}$, where T_{fo} the fluid outlet temperature, is obtained by substituting $T_f = T_{fo}$ and $y = L_1$ in eqn. 3.1.4.14

Thus,

$$\frac{T_{fo} - T_a - S/U_1}{T_{fi} - T_a - S/U_1} = e^{-(F^1 U_1 L_1 L_2)/mC_p} \quad ...(3.1.4.15)$$

or,

$$\frac{T_{fo} - T_a - S/U_1}{T_{fi} - T_a - S/U_1} - 1 = e^{-(F^1 U_1 L_1 L_2)/mC_p} - 1$$

or,

$$\frac{T_{fo} - T_{fi}}{(S/U_1 + T_a) - T_{fi}} = 1 - e^{-(F^1 U_1 A_p)/m C_p} \quad ...(3.1.4.16)$$

The useful heat gain rate for the collector can be written as—

$$Q_u = m\, C_p\, (T_{fo} - T_{fi}) = m\, C_p\, [(S/U_1) + T_a - T_{fi}] \left[1 - e^{-(F^1 U_1 A_P)/m C_p}\right]$$

$$= (m\, C_p/U_1)\, [S - U_1 (T_{fi} - T_a)] \left[1 - e^{-(F^1 U_1 A_P)/m C_p}\right]$$

$$= F_R \cdot A_p \cdot [S - U_1 (T_{fi} - T_a)] = F_R \cdot A_p\, [I_T (\tau\alpha)_e - U_1 (T_{fi} - T_a)] \quad ...(3.1.4.17)$$

where, $F_R = (m\,C_p/U_1\,A_p)\left[1-e^{-(F^1 U_1 A_P)/m\,C_p}\right]$ ...(3.1.4.18)

The term F_R is called the collector heat removal factor. It represents the ratio of actual rate of heat transfer to the working fluid to the heat transfer rate if the entire absorber plate was at the temperature T_{fi} everywhere. It is considered as an important design parameter. Its value ranges between 0 and 1. Usually inlet fluid temperature remains known. Therefore, eqn. (3.1.4.17) offers convenience in calculating the useful energy gain.

The useful heat gain rate can be assessed by any one of the following equations.

$$Q_u = m\,C_p\,(T_{fo} - T_{fi}) = A_p\,[I_T\,(\tau\alpha)_e - U_1\,(T_p - T_a)]$$

$$= A_p F^1\,[I_T\,(\tau\alpha)_e - U_1\,(T_f - T_a)] = A_p\,F_R\,[I_T\,(\tau\alpha)_e - U_1\,(T_{fi} - T_a)]$$

3.2 ANALYSIS FOR OTHER TYPES OF LIQUID FLAT PLATE COLLECTORS

Apart from conventional type, having the geometry of a flat absorber plate with tubes bonded below, there are two other types, one having tubes bonded on the top of the absorber plate and the other having tubes in line (refer Fig. 3.2.1). Method of analysis for both the types is same as the case considered earlier. But the different expression, for the collector efficiency, F' will be obtained for these two types, neglecting the thermal resistance of the tube wall, the expression for collector efficiency in the case of tubes bonded above, is given by

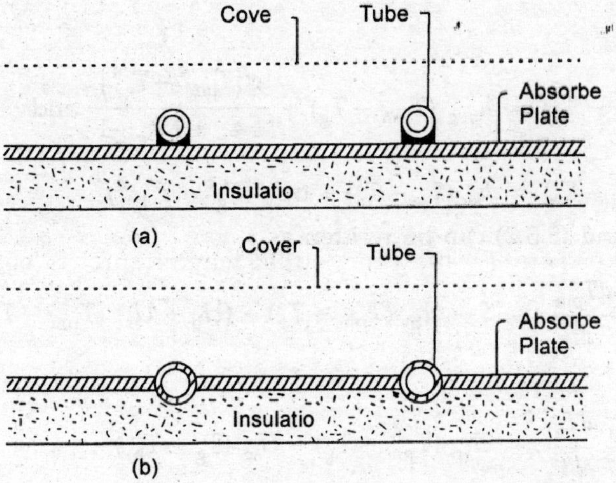

Fig. 3.2.1: Plate and Tube Flat-plate Collector (a) Tubes Bonded above Absorber Plate (b) Tubes in-line with Absorber Plate

$$F^1 = \frac{1}{W U_1\left[1/\{1/(U_1[(W-D_0)\,F)]+1/C_a\}^{-1}+U_1 D_0]+1/\pi D_i\,h_f\right]} \qquad ...(3.2.1)$$

The collector efficiency for the case of tubes having in line with the absorber plate is expressed by

$$F^1 = \frac{1}{W U_1\left[1/(U_1\,\{(W-D_0)\,F+D_0)\}+1/\pi D_i\,h_f\right]} \qquad ...(3.2.2)$$

3.3 THERMAL ANALYSIS UNDER TRANSIENT CONDITION

For a single cover collector, let us assume that the absorber plate and tubes, the bottom and side insulation, and the water in the tubes have a heat capacity $(mC)_p$ and are at the same temperature T_{pm}, while the cover has a heat capacity $(mC)_g$ and is at a temperature T_g. Since there will be no useful heat gain during the transient period, the following energy balance equations can be obtained for the absorber plate and the cover.

For the absorber plate:

$$\frac{(mC)_p}{A_p}\frac{dT_{pm}}{dt} = S - [h_{p-g}(T_{pm} - T_g) + \sigma(T_{pm}^4 - T_g^4)/(1/\epsilon_p + 1/\epsilon_g - 1)] - (U_b + U_s)(T_{pm} - T_a) \quad \text{.....(3.3.1)}$$

For the cover:

$$\frac{(mC)_g}{A_p}\frac{dT_g}{dt} = [h_{p-g}(T_{pm} - T_g) + \sigma(T_{pm}^4 - T_g^4)/(1/\epsilon_p + 1/\epsilon_g - 1)] - [h_w(T_g - T_a) + \sigma\epsilon_g(T_g^4 - T_{sky}^4)] \quad \text{...(3.3.2)}$$

Combining the convective and radiative losses from the top, the following overall Coefficients are defined as

$$U_{tP}(T_{pm} - T_g) = h_{p-g}(T_{pm} - T_g) + \frac{\sigma(T_{pm}^4 - T_g^4)}{1/\epsilon_p + 1/\epsilon_g - 1} \quad \text{and}$$

$$U_{tg}(T_g - T_a) = h_w(T_g - T_a) + \sigma\epsilon_g(T_g^4 - T_{sky}^4)$$

The eqns. (3.3.1) and (3.3.2) can be written as

$$\frac{(mC)_p}{A_p}\frac{dT_{pm}}{dt} = S - U_{tp}(T_{pm} - T_g) - (U_b + U_s)(T_{pm} - T_a) \quad \text{...(3.3.3)}$$

$$\frac{(mC)_g}{A_p}\frac{dT_g}{dt} = Ut_p(T_{pm} - T_g) - U_{tg}(T_g - T_a) \quad \text{...(3.3.4)}$$

In the steady state, $U_{tg}(T_g - T_a) = U_t(T_{pm} - T_a)$ ...(3.3.5)

If it is assumed that this relation holds during transient state also and T_a is constant, the differentiation of eqn. (3.3.5) yields

$$U_{tg}(dT_g/dt) = U_t(dT_{pm}/dt) \quad \text{...(3.3.6)}$$

Eqn. (3.3.4) becomes

$$\frac{(mC)_g}{A_p}\frac{U_t}{U_{tg}}\frac{dT_{pm}}{dt} = U_{tp}(T_{pm} - T_g) - U_{tg}(T_g - T_a) \quad \text{...(3.3.7)}$$

Addition of eqns. (3.3.3) and (3.3.7) gives

$$\frac{1}{A_p}\left[(mC)_p + (mC)_g \cdot \frac{U_t}{U_{tg}}\right]\frac{dT_{pm}}{dt} = S - U_l\,(T_{pm} - T_a) \qquad \text{...(3.3.8)}$$

The effective heat capacity of the collector is denoted by

$$(mC)_e = (mC)_p + (mC)_g \frac{U_t}{U_{tg}} \qquad \text{...(3.3.9)}$$

At $t = 0$, $T_{pm} = T_{pi}$, the integration of the eqn. (3.3.8) over a time interval t, gives

$$\frac{S - U_l\,(T_{pm} - T_a)}{S - U_l\,(T_{pi} - T_a)} = e^{-\left[A_p U_l t/(mC)_e\right]} \qquad \text{...(3.3.10)}$$

This is valid if the time interval of the integration is considered reasonably small and S and T_a are constants.

3.4 SOME SALIENT FEATURES OF SUB COMPONENTS FOR A FLAT PLATE COLLECTOR

(a) *Absorber:* It is a black solar energy absorbing surface or plate which absorbs solar energy and converts it into heat. Copper, Aluminium, Mild Steel, Stainless Steel and some plastics are commonly used as the absorber materials.

(b) *Absorber Coating:* Coating on the absorber plate is used for converting solar incident energy into heat. Black paint or selective coatings are commonly used. The coating should have the following criterion : it should be perfect absorber of solar radiation, perfect reflector of thermal radiation, poor emitter of thermal radiation. It should have high resistance to temperature and humidity variation and high adhesive nature. It should not evaporate at the operating temperature and pressure.

(c) *Collector Glazing:* It is transparent cover placed over the absorber to reduce heat losses from the collector to the environment. It should be able to transmit maximum short wave length solar radiation and minimum long wave length thermal radiation. It should have dimensional stability, low thermal conductivity, long term durability under exposure to sun and weather, low absorptivity for solar spectrum and high absorptivity for thermal spectrum. Iron content in the glass should be as low as possible.

(d) *Collector Insulation:* It reduces the conduction losses from the bottom and sides of the collector. It should not deteriorate at high temperature. It should withstand temperature variation from of about 30°C to 200°C. It should have high structural stability and fire resistance. It should not settle or compact in time. Moisture should have no effect on insulation.

(e) *Collector Housing:* It holds absorber plate, glazing, insulation in position and protects them from surroundings. It should withstand loading and handling stresses, corrosion and weathering effect. There should not be any stress on the cover at high temperature.

The life cycle expectancy of the collector depends on subcomponent materials, their properties and quality of working fluid used. This can be evaluated nicely from Table 3.4.1.

Table 3.4.1: Life Cycle Expectancy of Collector

	Item	*Material*		
		Grade-I 15 Yrs. and above	*Grade-II* 10 Yrs. and above	*Grade-III* 5 Yrs. or less
1.	Collector Box	Anodized Aluminium	Aluminium	Mild Steel, Mild Steel
2.	Absorber Plate	Copper	Copper	Galvanized Steel or Copper
	Riser & Headers	Copper	Copper	– do –
3.	Coating	Selective black chrome coating $\alpha = 0.96$, $\epsilon = 0.2$	Selective black chrome coating $\alpha = 0.96$, $\epsilon < 0.2$	Black paint
4.	Gasketing	Silicon rubber	EPDM	Neoprene or rubber
5.	Cover Glass	Tempered/Low iron float glass	Toughened window/float glass	Window Glass
6.	Insulation	Polyurethene or Isocyanurates	High density fiber glass/polyurethane	Low density mineral wool
7.	Sealants	Silicone	Silicone	Nil
8.	Piping	All copper	Galvanized Iron B class	Galvanized Ironc A class
9.	Screws, bolts and nuts	Stainless Steel or non-corrosive materials	Stainless Steel or non-corrosive materials	Zinc coated/ Nickel plated mild steel

3.5 CONCENTRATING COLLECTORS

Concentrating Collector is a device by which solar energy is collected with high intensity of solar radiation on the absorbing surface. It is a whole system mainly comprising of concentrator and receiver. The concentrator is the optical system in the form of reflecting mirrors or refracting lenses.

In the term receiver, the energy absorbing surface, transparent cover and other accessories are included. In this system, the radiation falls on a relatively large area and this radiation is focused on to a receiver having considerably smaller area. Depending on the magnitude of energy concentration, fluid temperature can be enhanced even more than 500°C. These concentrators are used in the medium temperature or high temperature energy conversion cycles.

In the flat plate collectors, both the beam and diffuse radiations are encountered. But concentrating collectors concentrate only beam radiation coming from a specific direction. This happens because only a very small proportion of diffuse radiation matches the direction of beam radiation.

The optical system such as mirrors or lenses directs the radiation on-to an absorber of smaller area usually surrounded by a transparent cover. Apart from the losses which occur while the radiation is transmitted through the cover, the reflection or absorption losses in the mirrors or lenses due to geometrical imperfections in the optical system also take place in this case. A term called optical efficiency takes care of the combined effect of all aforesaid losses. The extra optical losses is compensated for by the fact that the flux incident on the absorber surface is concentrated on a smaller area. Due to this, the thermal loss terms do not dominate to the same extent as in a flat-plate collector. This leads to higher collection efficiency.

During use of concentrating collectors, tracking arrangement is generally used to follow or track the sun so that, for all time, the beam radiation can be focused on-to absorber surface. Depending on desired degree of concentration, the tracking arrangement may be simple or complicated.

3.5.1 Types of Concentrating Collector

The concentrating collectors are classified in two following groups:

(i) Focussing type,
(ii) Non-focussing type

In the focussing type collector, optical system focuses the solar radiation onto the absorber, while for non-focussing type collector, the solar radiation is allowed to fall on the absorber after using reflection in the mirror.

Focussing collectors may be of line focussing type or point focussing type. In the line focussing, the collector pipe, through which working fluid flows, may be considered as a line, while in the case of point focussing, the point to be considered will be a small volume through which fluid flows. Depending upon the geometries of the concentrators, the following types of concentrating collectors and their features are being discussed. The line focussing collectors usually offer temperatures between 150 and 400°C with concentration ratios between 10 and 80.

Mirror-strip reflectors: This is a line focussing collector in which a plane or slightly curved (concave) mirror strips are mounted on a flat base. The individual mirrors are placed in such angles that the reflected solar radiation can fall on to the same focal line where the collector pipe is placed (Fig. 3.5.1.1).

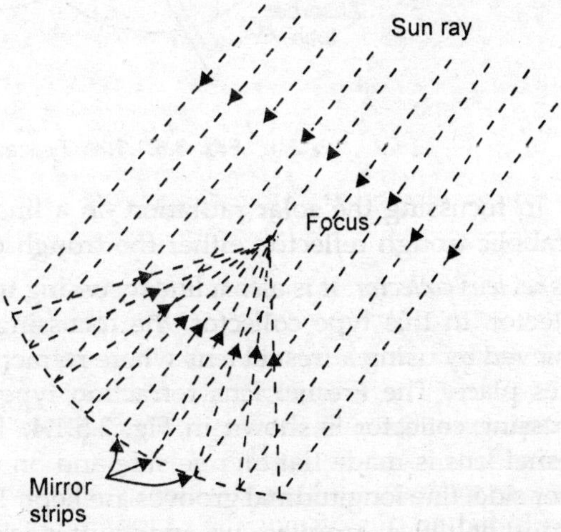

Fig. 3.5.1.1: Mirror-strip Solar Collector

With respect to the changes in the Sun's elevation in viewing of focussing of the reflected radiation onto a focal line, either collector pipe is to be moved continuously keeping mirror strips in fixed condition, or the angles of the mirrors are allowed to be adjusted continuously keeping collector pipe in fixed position.

Parabolic trough reflector: This type of collector collects the solar radiation coming from the particular direction over the area of the reflecting surface and concentrates at the focus of parabola (Fig. 3.5.1.2). When the reflector is manufactured in the form of a trough with parabolic cross-section, the solar radiation gets focused along a line. Parabolic trough reflectors are usually made of highly polished aluminium or silvered glass or of a film of aluminized plastic on a firm base. The reflector may be of continuous form or a number of line flat mirror strips can be placed on a parabolic base. Cylindrical parabolic concentrators (Fig. 3.5.1.3) are mostly used in practice. Here, the collector pipe is used as an absorber preferably with a selective coating and it is placed along the focus axis.

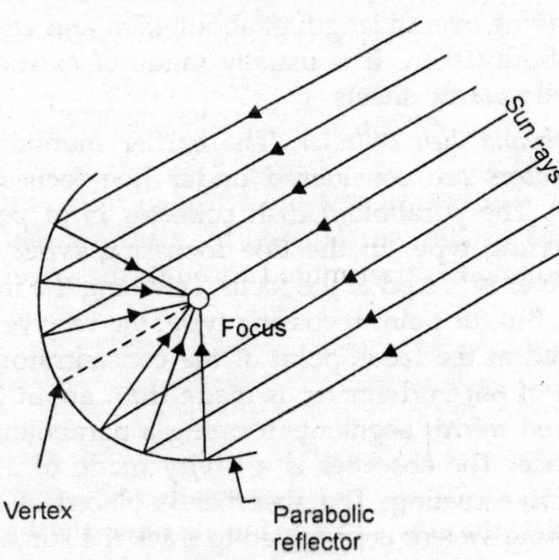

Fig. 3.5.1.2: Cross-section of Parabolic-trough Collector

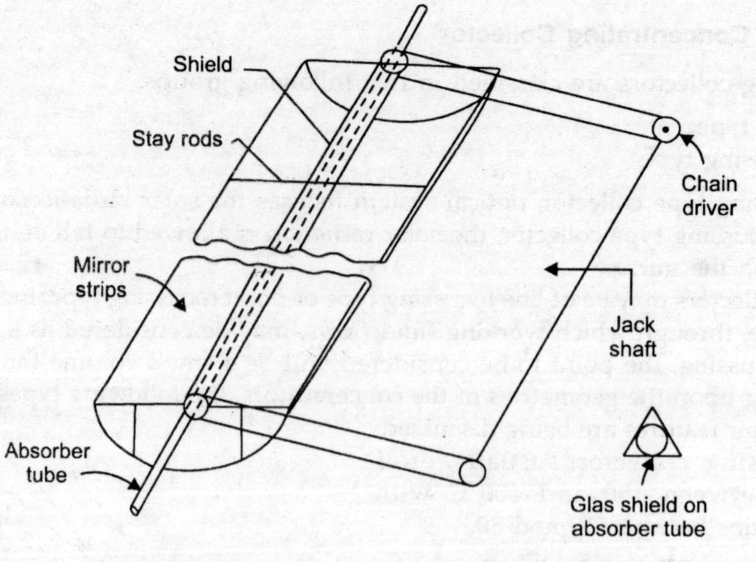

Fig. 3.5.1.3: A Typical Cylindrical Parabolic System

In focussing the solar radiation on a line, with respect to changes in suns elevation by the parabolic trough reflector, either the trough or collector pipe is to be rotated continuously.

Fresnel lens collector: It is also a line focussing type collector. In this type collector, the focussing is achieved by using a fresnel lens where refraction takes place. The Fresnel lens refraction type of focussing collector is shown in Fig. 3.5.1.4. The Fresnel lens is made flat on one side and on the other side, fine longitudinal grooves are kept. The angles of these grooves are made in such a manner that solar radiation can be focussed on a line. A typical Fresnel lens which is usually used, is having overall length of about 4.7m and width of about 0.05m. It is usually made of extruded acrylic plastic sheets.

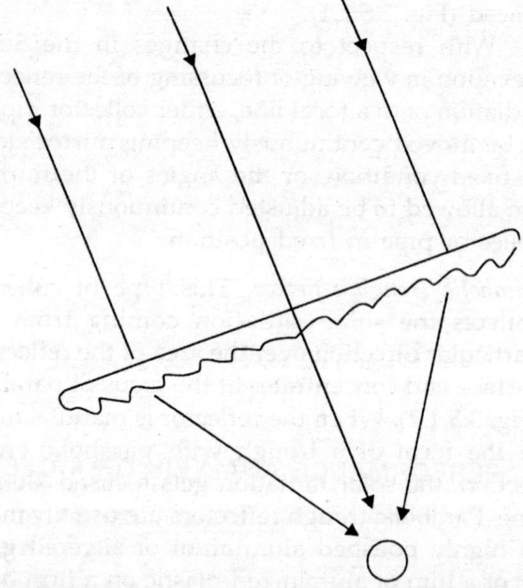

Fig. 3.5.1.4: Fresnel Lens Concentrating Collector

Paraboloid dish collector: The earlier mentioned collectors are considered under line focussing type. The paraboloid dish collector is of point focussing type. In the line focussing type, the receiver is placed at the focus, i.e. along the focal line. But for point focussing type, the receiver is placed at the focus point of the concentrator. A dish of 6m in diameter is made from about 200 curved mirror segments forming a paraboloidal surface. The absorber is a cavity made of a Zirconium-Copper alloy. Black chrome is used as selective coating. The absorber is placed at the focus (shown in Fig. 3.5.1.5). In this case, the tracking system is provided to track the sun automatically for all times. These collectors can have concentration ratios ranging from 100 to a few thousand. Around 2000°C temperature can be achieved by this collector.

Flat plate collectors with plane reflectors: This is a non-focussing type collector where flat plate collector facing south with reflectors attached is used. The reflectors are set at proper angles. Reflectors reflect the solar radiation to the absorber plate. Apart from the reflected radiation, absorber plate also receives normal radiation falling on it. The concentration ratios are normally ranging from 1 to 4. The attainable operating temperature is about 130 to 140°C. This type of collector is shown in Fig. 3.5.1.6.

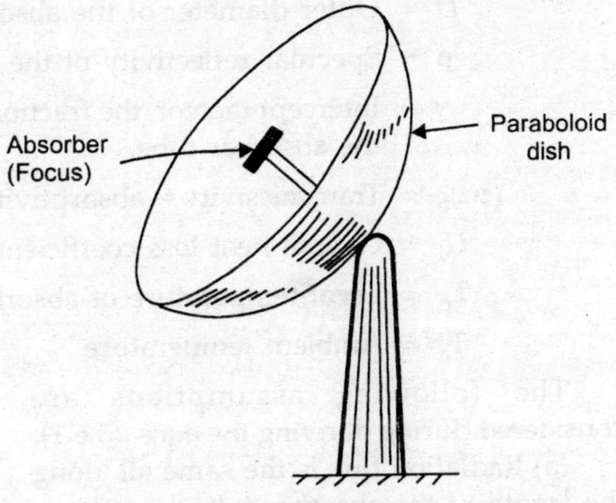

Fig. 3.5.1.5: Paraboloid Dish Collector

Compound parabolic collector: It is a non-focussing type collector (shown in Fig. 3.5.1.7). The concentrators are of curved segments which are parts of two parabolas. Solar radiation from many directions get reflected on to the absorber. It collects both direct and diffuse radiations on a small area. The concentration ratio for these type of collectors generally ranges from 3 to 10. This collector is having high acceptance angle and it requires occasional tracking.

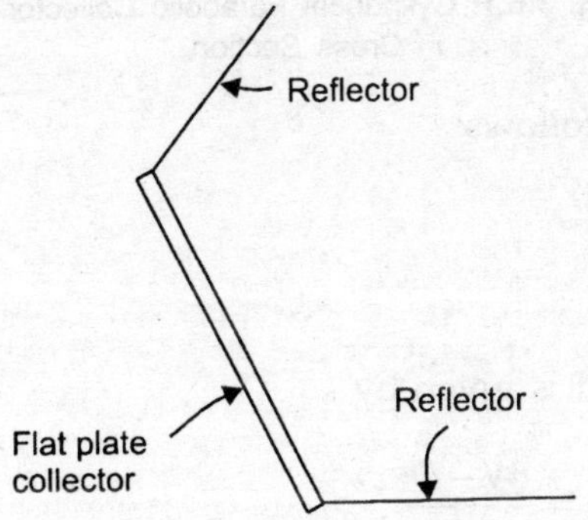

Fig. 3.5.1.6: Flat Plate Collector with Plane Reflector

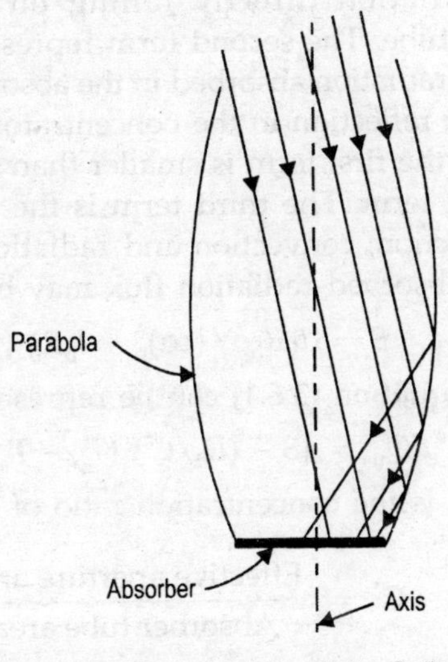

Fig. 3.5.1.7: Compound Parabolic Collector

3.6 PERFORMANCE ANALYSIS OF A CYLINDRICAL PARABOLIC COLLECTOR

The cross section of a cylindrical parabolic collector is shown in Fig. 3.6.1. With the help of energy balance, for an elemental length dx of the absorber tube (shown in Fig. 3.6.1.), at a distance x from the inlet, the following equation for the steady state condition can be obtained :

$$dQ_u = [I_b R_b D\,(\tau\alpha)_b + I_b R_b (W - D)\,\rho\gamma\,(\tau\alpha)_b - U_1 \pi D\,(T_p - T_a)]\,dx \qquad \text{...(3.6.1)}$$

Where, dQ_u = Useful heat gain rate for the length dx

I_b = Beam radiation flux

R_b = Beam radiation tilt factor

W = Aperture of the concentrator

D = Outer diameter of the absorber tube

ρ = Specular reflectivity of the concentrator surface

γ = Intercept facotor, the fraction of the specularly reflected radiation intercepted by the absorber tube.

$(\tau\alpha)_b$ = Transmissivity – absorptivity product for beam radiation

U_l = Overall heat loss coefficient

T_p = Local temperature of absorber tube

T_a = Ambient temperature

The following assumptions are considered during deriving the eqns. (3.6.1):

(a) Radiation flux is the same all along the length of the absorber tube.

(b) Negligible temperature drop across the absorber tube and the glass cover.

The first term on the right hand side in eqn. (3.6.1) indicates the absorbed incident beam radiation directly falling on the absorber tube. The second term represents the beam radiation absorbed in the absorber tube after reflection in the concentrator. In this case, the first term is smaller than that of second term. The third term is the loss by conduction, convection and radiation.

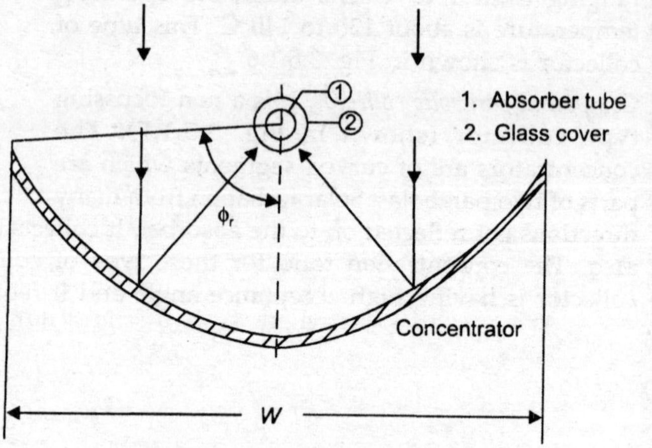

Fig. 3.6.1: Cylindrical Parabolic Collector in Cross Section

The absorbed radiation flux may be expressed as follows:

$$S = I_b R_b \rho\gamma (\tau\alpha)_b + I_b R_b (\tau\alpha)_b (D/(W - D)) \quad ...(3.6.2)$$

The equation (3.6.1) can be represented as,

$$dQ_u = [S - (U_l/C^*) (T_p - T_a)] (W - D) dx \quad ...(3.6.3)$$

where, C^* is the concentration ratio of the collector and is given by

$$C^* = \frac{\text{Effective aperture area}}{\text{Absorber tube area}} = \frac{(W-D)L}{\pi DL} = \frac{W-D}{\pi D} \quad ...(3.6.4)$$

In which L is the length of the concentrator.

The useful heat gain rate can also be expressed as—

$$dQ_u = h_f \pi D_i (T_p - T_f) dx \quad ...(3.6.5)$$

$$= m C_p dT_f \quad ...(3.6.6)$$

Where D_i = inner diameter of the absorber tube

h_f = heat transfer coefficient on the inner surface of the tube

T_f = local fluid temperature

m = mass flow rate of the working fluid flowing through the absorber tube

Combining eqns. (3.6.3) and (3.6.5) the useful heat gain can be written as—

$$dQ_u = F^1 [S - (U_l/C^*)(T_f - T_a)] (W - D)\, dx \quad ...(3.6.7)$$

where F^1 is the collector efficiency factor given by—

$$F^1 = \frac{1}{U_1 [1/U_1 + D/h_f D_i + D 1_n (D/D_i)/2k]} \quad ...(3.6.8)$$

In which k is the thermal conductivity of the tube material.

Combining eqns. (3.6.6) and (3.6.7) and putting the value of C^* we have the following differential equation

$$\frac{dT_f}{dx} = \frac{F^1 \pi D U_1}{m C_p} [C^*(S/U_l) - (T_f - T_a)] \quad ...(3.6.9)$$

Integrating the above equation with the boundary condition as at $x = 0$, $T_f = T_{fi}$ we have,

$$\frac{(C^*(S/U_1) + T_a) - T_f}{(C^*(S/U_1) + T_a) - T_{fi}} = \exp\{-F^1 \pi D U_l\, x/m\, C_p\} \quad ...(3.6.10)$$

Here T_{fi} indicates the temperature of the working fluid at the inlet of the absorber tube. At $x = L$, $T_f = T_{fo}$, where T_{fo} is the fluid outlet temperature yields,

$$\frac{T_{fo} - T_{fi}}{C^*(S/U_1) + T_a - T_{fi}} = 1 - \exp\{-F^1 \pi D U_l\, L/m\, C_p\} \quad ...(3.6.11)$$

Now, useful heat gain rate can be written as,

$$Q_u = m\, C_p\, (T_{fo} - T_{fi})$$

$$= m\, C_p\, [C^*(S/U_l) + T_a - T_{fi}]\, [1 - \exp\{-F^1 \pi D U_l\, L/m\, C_p\}] \quad ...(3.6.12)$$

The eqn. (3.1.2.12) can also be written as,

$$Q_u = F_R\, (W - D)\, L\, [S - (U_l/C^*)(T_{fi} - T_a)] \quad ...(3.6.13)$$

Where F_R is the heat removal factor and can be defined by

$$F_R = (m\, C_p/\pi D L U_l)\, [1 - \exp\{-F^1\, \pi D U_l\, L/m\, C_p\}] \quad ...(3.6.14)$$

The instantaneous collection efficiency n_i on the basis of beam radiation alone is given by

$$n_{ib} = \frac{Q_u}{I_b R_b\, WL} \quad ...(3.6.15)$$

In general, the instantaneous collection efficiency is given by—

$$n_i = \frac{Q_u}{(I_b R_b + I_d R_d)\, WL} \quad ...(3.6.16)$$

Here, the ground reflected radiation is neglected.

The heat loss coefficient U_l can be obtained from—

$$U_l = (1/h_{wind} + 1/h_r)^{-1} \quad ...(3.6.17)$$

Where, h_{wind} = film coefficient due to wind = 5.7 + 3.8 V, W/m^2 °C

In which, V is the wind velocity in meter/sec.

$$h_r = \text{radiation coefficient.}$$

The radiation coefficient can be determined as follows:

$$h_r (T_r - T_a) = \sigma \epsilon_r (T_r^4 - T_a^4)$$

where σ = Boltzman Plank's constant = 5.67×10^{-8} W/m² °C.

ϵ_r = emissivity of the surface

T_r = Temperature at the radiant surface

T_a = ambient temperature

So, $h_r = \sigma\epsilon_r (T_r + T_a)(T_r^2 + T_a^2)$

If it is assumed that T_r and T_a do not differ much and if $\bar{T} = (T_r + T_a)/2$ then we have,

$$h_r = 4\sigma\epsilon_r \bar{T}^3 \qquad \text{...(3.6.18)}$$

Example: A cylindrical parabolic concentrator is having 2.5m width and 9m length. The outside diameter of the absorber tube is 6.5. cm. The collector is used to heat a fluid whose temperature at the inlet of the absorber is 160°C and flow rate is 450 Kg/hr. The beam radiation falling on the collector is 700 W/m². The ambient temperature is 28°C. Estimate, (i) useful heat gain rate, (ii) exit fluid temperature and (iii) instantaneous collection efficiency based on beam radiation alone. The following fluid and optical properties may be used.

C_p = 1.256 KJ/Kg°C; ρ = 0.85; $(\tau\alpha)_b$ = 0.78

γ = 0.93; Collector efficiency factor (F') = 0.85;

Overall heat loss coefficient (U_l) = 7.0 W/m² °C.

Solution:

The given data are as follows—

I_bR_b = 700 W/m²; W = 2.5m; L = 9m; D = 0.065 m; m = 450/3600 = 0.125 Kg/sec.

T_{fi} = 150°C; T_a = 28°C; C_p = 1.256 KJ/Kg-K; ρ = 0.85; $(\tau\alpha)_b$ = 0.78; γ = 0.93;

F' = 0.85; U_l = 7.0 W/m² °C.

The absorbed radiation flux (S) is given by,

$$S = I_bR_b\rho\gamma(\tau\alpha)_b + I_bR_b(\tau\alpha)_b (D/W - D)$$

$$= 700 \times 0.85 \times 0.93 \times 0.78 + 700 \times 0.78\,[0.065/(2.5 - 0.065)] \text{ W/m}^2$$

$$= 431.611 + 0.027 = 431.6 \text{ W/m}^2$$

Heat removal factor F_R is $= \dfrac{mC_p}{\pi DLU_l}\,[1 - \exp\{-F'\pi DU_l\, L/mC_p\}]$

$$\frac{mC_p}{\pi DU_l L} = \frac{0.125 \times 1.256 \times 10^3}{3.14 \times 0.065 \times 7.0 \times 9} = 12.21$$

$$F_R = 12.21\,[1 - \exp\{-0.85/12.21\}] = 12.21\,[1 - \exp(-0.0696)]$$

$$= 12.21\,[1 - 0.9327] = 12.21 \times 0.06723 = 0.821$$

Concentration ratio (C^*) is given by

$$C^* = (W - D)/\pi D = (2.5 - 0.065)/(3.14 \times 0.065)$$
$$= 11.93$$

Useful heat gain rate

$$Q_u = F_R (W - D) L [S - (U_l/C^*) (T_{fi} - T_a)]$$
$$= 0.821 (2.5 - 0.065) 9 [431.6 - (7.0/11.93)/(150 - 28)]$$
$$= 0.821 \times 2.435 \times 9 [360.015] = 6477.5 \text{ W}$$
$$Q_u = m C_p (T_{fo} - T_{fi})$$

So, $$T_{fo} = Q_u/m C_p + T_{fi} = 6477.5/(0.125 \times 1.256 \times 10^3 + 150)$$
$$= 191.2°\text{C}.$$

$$n_{ib} = [6477.5/(700 \times 2.5 \times 9)] \times 100\% = 41\%.$$

EXERCISE

1. What do you mean by a solar thermal collector? With a neat labelled schematic diagram, discuss in brief the functioning of each components of a liquid flat plate collector.
2. How the transmissivity of the cover system of a flat plate collector is assessed? For a system of two glass covers, estimate τ_ρ, τ_α and τ when the angle of incidence is 40°. Assume that thickness of each cover is 4mm. Refractive index of glass relative to air is 1.5 with extinction coefficient of glass as 14m^{-1}.
3. Define the instantaneous collector efficiency. Mention the types of losses that take place in a flat plate collector. State in brief the methodology used in assessing the coefficients for each losses.
4. Mention the physical concepts of (i) collector efficiency factor and (ii) heat removal factor used for thermal analysis of a flat plate collector.
5. How a concentrating collector does differ from a flat plate collector? Classify the concentrating collectors. With schematic diagram, discuss in brief each collector.
6. What do you mean by a cylindrical parabolic collector? A cylindrical parabolic collector, having 3m width and 9.5m length, is used for generating hot water. The temperature of water at the inlet of absorber is 170°C. The flow rate of water inside the collector is 430 kg/hr. The outside diameter of the absorber is 7.0cm. If the beam radiation falling on the collector is 680W/m^2, calculate (i) the useful heat gain rate, (ii) the exit water temperature and (iii) the instantaneous collection efficiency based on beam radiation alone. Assume T_a = 35°C, C_p = 1.3 kJ/kg°C; ρ = 0.86, $(\tau\alpha)_b$ = 0.8, F^1 = 0.87, U_1 = 7.2 W/m² °C.

SUMMARY

Basic principle, classification and the application of solar thermal collector have been presented in a lucid manner. The liquid flat plate collector has been described and its principle of operation has also been discussed along with some typical data regarding the sub components of the collector. A typical salient data for the commercially available collector of a solar water heater of 100 litres per day, manufactured by BHEL, has been presented. The procedure for the assessment of the transmissivity for the cover system has been outlined. The meaning of Transmittance-Absorptance Product and its importance has been explained in simple manner. The procedure for the performance analysis of a solar collector has been elaborated. The typical losses relating to the solar collector and their assessment procedure have been narrated. Thermal analysis procedure for a flat plate collector has been presented in details. The important terms in relating to the solar collector, e.g. fin efficiency (F), collector efficiency (F') and the collector heat removal factor (F_R), have been discussed in detail. The analysis procedure for the other types of liquid flat plate collectors, apart from conventional one, has also been outlined. The procedure for the thermal analysis under transient condition has been presented. The salient features of the sub components for a flat plate collector have been discussed along with a Table on the life cycle expectancy of the collector. The meaning of concentrating collector, its importance and its classification have been presented. The typical mirror-strip reflector, parabolic trough reflector, Fresnel lens collector, paraboloid dish collector and compound parabolic collector etc. have been described elaborately. This chapter has also covered a detailed performance analysis procedure for a cylindrical parabolic collector.

REFERENCES

1. G.N. Tiwari and S. Suneja, Solar Thermal Engineering Systems, Narosa Publishing House, New Delhi, 1997.
2. S.P. Sukhatme, Solar Energy-Principles of Thermal Collection and Storage, Tata Mc-Graw Hill, 1996.
3. J.S. Hsieh, Solar Energy Engineering, Prentice Hall, New Jersey, 1986.
4. K.M. Mital, Non-conventional Energy Systems—Principles, Progress and Prospects, Wheeler Publishing, New Delhi, 1997.
5. Solar Flat Plate Collector—Specification, Part 2—Components, Indian Standard 12933:1992, Bureau of Indian Standards, 1992.
6. Manual for Solar Thermal Systems, IREDA Ltd., New Delhi, 1997.
7. Manual of BHEL for Solar Water Heating System.
8. W.A. McAdams, Heat Transmission, 3rd edn., Mc-Graw Hill, NY, 1954.
9. J.A. Duffie and W.A. Beckman, Solar Engineering of Thermal Processes, John Wiley, NY, 1980.
10. Y. Tripanagno Stopoulos, CPC Solar Collectors with Flat-Bifacial Absorbers, Solar Energy, 69 (3), pp. 191-203, 2000.
11. H.P. Garg and T.C. Kandpal, Laboratory Manual on Solar Thermal Experiments, Narosa Publishing House, New Delhi, 1999.
12. D.K. McDaniels, et al., Enhanced Solar Energy Collection using Reflector-Solar Thermal Collector Combinations, Solar Energy, 17, p. 277, 1975.
13. A. Rabi, Comparison of Solar Concentrators, Solar Energy, 18, pp. 93-111, 1976.
14. T. Beikircher, et al., Gas Heat Conduction in Evacuated Flat Plate Solar Collectors: Analysis and Reduction, J. Solar Energy Engg., 117, pp. 229-235, 1995.
15. W. Spirkl, et al., In Situ Characterisation of Solar Flat Plate Collectors under Intermittent Operation, Solar Energy, 61 (3), pp. 147-152, 1997.
16. J.H. Lee, et al., An Experimental and Theoretical Study on The Corrugated Water-Trickle Collector, Solar Energy, 38 (2), pp. 113-123, 1987.
17. E.H. Amer and J.K. Nayak, Experimental Evaluation of a Short-Term Dynamic Test Method for Flat Plate Solar Collectors under Indian Climatic Conditions, SESI Journal, 8 (2), pp. 85-96, 1998.
18. J.J. O'Gallagher, et al., A New Evacuated CPC Collector Tube, Solar Energy, 29, p. 575, 1982.
19. A. Steinfeld, Cone Compared to CPC as Secondary Concentrator in tandem with A Paraboloidal Dish Primary, Jl. of Solar Energy Engineering, 114, pp. 201-202, 1992.

20. N. Benz and T. Beikircher, High Efficiency Evacuated Flat Plate Solar Collector for Process Steam Production, Solar Energy, 65 (2), pp. 111-118, 1999.
21. J.P. Holman, Heat Transfer, Mc-Graw Hill, NY, 1976.
22. R. Siegel and J.R. Howell, Thermal Radiation Heat Transfer, Mc-Graw Hill, NY, 1972.
23. B.S. Magal, Solar Power Engineering, Tata Mc-Graw Hill, 2000.
24. H.V. Kumar, et al., Design, Construction and Performance Analysis of A Parabolic Trough Concentrator, Proc. of 26th NREC-2002, Coimbatore, pp. 49-52, Jan 2003.
25. P. Ray, Steering of Mirrors by means of Guys, SESI Journal, 4 (2), p. 73, 1994.
26. P. Ray, Steering of Mirrors by means of Guys-Part II, SESI Journal, 9 (2), pp. 77-85, 1999.
27. F. Kreith and J. Kreider, Principles of Solar Engineering, Washington: Hemisphere Publishing Corporation, 1978.
28. A. Senthil Kumar and M. Senthil Kumar, Thermodynamic Optimization of Solar Collector, Proc. of 26th NREC-2002, Coimbatore, India, pp. 43-48, Jan 2003.
29. A. Saito, Transient Response of Flat Plate Solar Collector for Periodic Solar Intensity Variation, Solar Energy, 32(1), pp. 17-23, 1984.
30. V.S. Gorantiwar and P.A. Podukhe, Design and Fabrication of Solar Paraboloid Concentrator Steam Generator, Proc. of 25th NREC-2001, Warangal, pp. 57-63, Dec 2001.
31. G.N. Tiwari, Simple and Transient Analysis of A Normal and Flat Plate Collector, Energy Conversion and Management, 26, pp. 145-146, 1985.
32. S.K. Bhardwaj, et al., Techno-economic Optimization of A Hybrid Collector Cum Storage System, Int. Jl. of Energy Research, 14, p. 705, 1990.
33. C.L. Gupta, Solar Collectors—Global Milestones and Directions for Indian R&D, Proc. of 24th NREC-2000, Mumbai, pp. 3-4, Dec 2000.
34. A.V. Narasimha Rao, et al., Performance of East/West Plane Booster Mirror, J. Energy Conversion and Management, vol. 35, no. 6, pp. 543-554, 1994.
35. R.K. Mazumder and M. Hussain, Optical and Thermal Performance Testing of A Discrete Mirror, Seasonally Adjusted, Linear Solar Concentrator with Tubular Absorber, Energy Conversion and Management, 31 (2), pp. 187-192, 1991.
36. A.V. Narasimha Rao, et al., Energy Contribution by Booster Mirrors, J. Energy Conversion and Management, 34, pp. 309-326, 1993.
37. Y. Tripanagno Stopaulos, Cost Effective Asymetric CPC Solar Collectors, Proc. of Int. Conference WREC-V, pt. II, pp. 628-631, Italy, 1998.
38. B. Bandyopadhyay, Solar Thermal Technologies in India—A Brief Assessment, Proc. of 24th NREC-2000, Mumbai, pp. 140-145, Dec 2000.

4

Solar Thermal Conversion Systems

INTRODUCTION

In this Chapter, attention will be focused on the systems where the solar energy can be converted to the thermal energy. Detail discussion and performance analysis have been carried out for the solar water heating system in which solar energy is utilized to generate hot water. Economic analysis for solar water heating systems has also been discussed to facilitate the students and working personnel, involved in this area. The methodologies followed in assessing the pay back period and extended period for free energy have also been included. Among these, solar air drier, solar cooker, solar distillation, solar pumping systems, solar cooling, solar ponds and solar thermal power plant have come under the purview of this chapter. For each case, the system configuration its operation and practical use have been discussed in detail. For all the cases the available current data have also been reported in this chapter.

4.1 SOLAR WATER HEATING SYSTEM (SWHS)

Solar thermal collector has been discussed in detail in sec. 3.1, wherein it is stated that in a solar collector, solar energy incident on the absorber panel coated with selective coating gets absorbed and the absorber panel gets heated. The generated heat is transferred to the working fluid (here it is water) passing through the tubes underneath the absorber panel. The water get heated up and its temperature will increase. It is delivered to the storage tank. The total system with solar collector, storage tank and pipelines is called solar water heating system (SWHS). The rise in water temperature depends on solar radiation, weather conditions and solar collector system efficiency. The system produces hot water for around 300 days in a year. It generates hot water maximum on clear sunny day, moderate on partially cloudy day and nil on rainy or heavily overcast day. The system can be installed at roof tops, building terrace, open ground.

A typical solar water heating system is shown in Fig. 4.1.1. The system consists of:

(i) Solar flat plate collector or set of collectors.
(ii) Storage tank
(iii) Make-up of overhead tank
(iv) Pipe lines and circulation system
(v) Auxiliary heating system
(vi) Control of the system

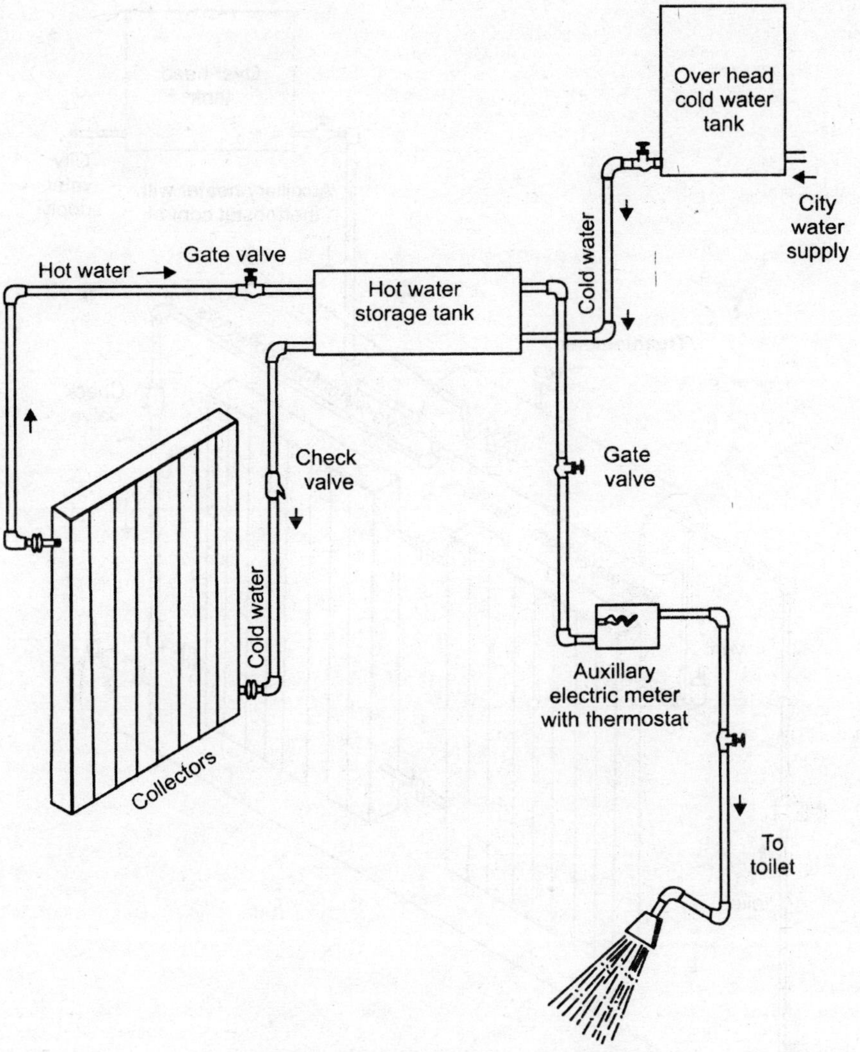

Fig. 4.1.1: Solar Water Heating System

Basically there are two types of systems, natural circulation or thermo-syphon system and forced circulation or pumped system.

4.1.1 Natural Circulation or Thermo-syphon Solar Water Heating System

A typical natural circulation system is shown in Fig. 4.1.1.1. In this system the water circulates through the system due to natural convection, i.e., due to density difference between the hot and cold water. When the water is heated in the collector, its density decreases and flows to the storage tank, placed atleast 50 cms. above the outlet of the collector. Whereas the colder water whose density is higher flows from the bottom of the storage tank to collector to replace the hot water discharged into the tank. Hot water is drawn from the top of tank as and when required and the cold water is fed to the tank from overhead or makeup tank. During the stronger radiation, circulation will be faster. When there will be no incident solar energy on the collector, there will

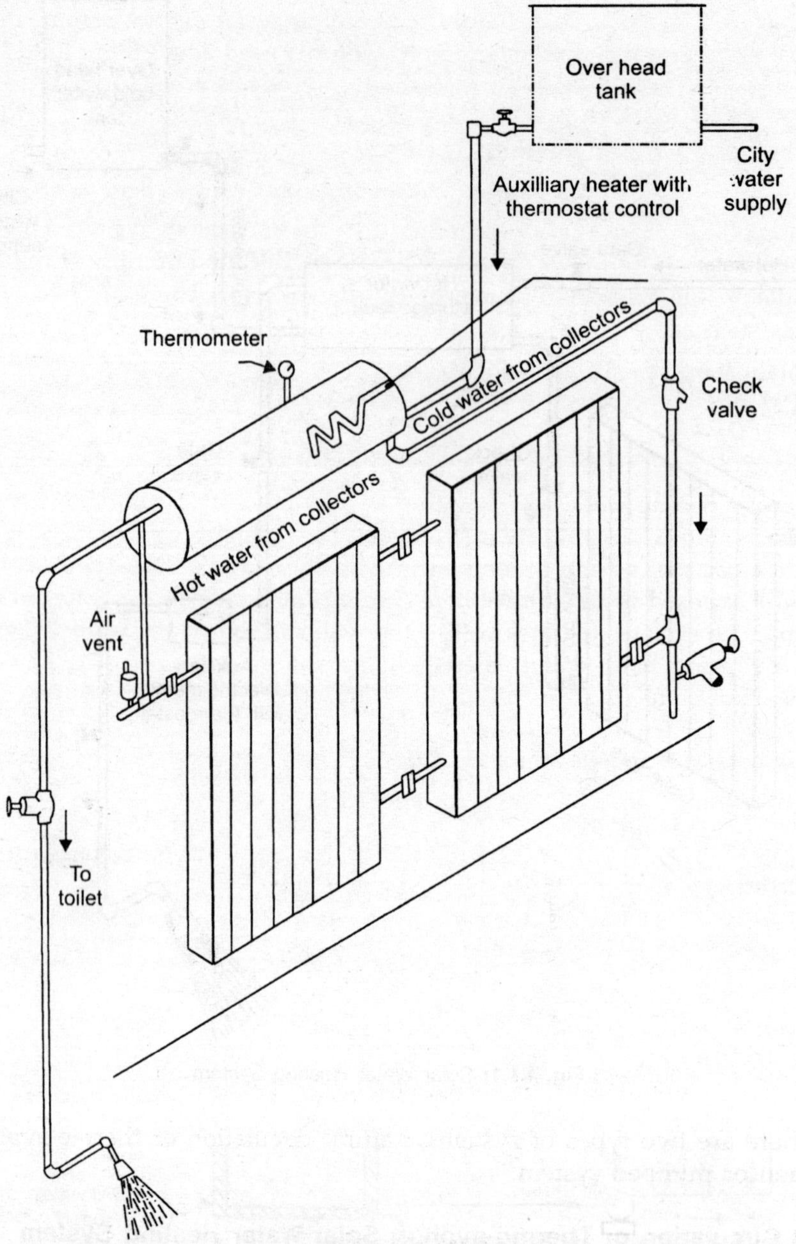

Fig. 4.1.1.1: Thermosyphon Type Solar Water Heater

be no circulation. This system is simple, less expensive and requires little maintenance. It is not, however, suitable in freezing cold water, because the water remaining in the collector gets freezed at very low ambient temperature, which results in bursting of the pipes. This can be avoided by draining the water in the collector when freezing temperature occurs. The circulation of water is low and heat removal rate is also low. This results in less efficiency of the system. This system is suitable for domestic purpose with relatively low capacity (e.g. 100 litres per day). For large systems , i.e., above 1000 liters per day natural circulation system is not recommended because of low efficiency. It is also not suitable where hard water is used as working fluid. The auxiliary heating system is used to provide heat during cloudy periods as a backup for the solar system.

4.1.2 Forced Circulation Water Heating System

In this system, the circulation of the working fluid through the circuit is achieved by using a pump. This system collects more energy compared to a thermosystem system. This happens due to higher flow rates, leading to higher efficiency as the higher the rate of heat removal from the collector, the greater is the absorption of energy by the collector. There are two types of forced circulation systems such as:

(i) Direct or open loop system
(ii) Indirect or closed loop system

4.1.2.1 Direct or Open Loop System

The schematic diagram of a direct or open loop forced circulation system is shown in Fig. 4.1.2.1.1. In this system, the water circulates and recirculates through the collectors, storage tank and associated pipe line until the maximum temperature is attained in a day. The water in the forced circulation system is exposed to the atmosphere through air space above water in the hot water tank or other point. The system operates well if the water is soft. Hard water is not suitable for these systems as by using hard water, there will be charge and corrosion of scaling on the collector pipes and storage tank. To maintain the required temperature of the water, sometimes auxiliary heating may be required. For this purpose, an electrical heater is provided in the hot water storage tank as a standby to the solar collector.

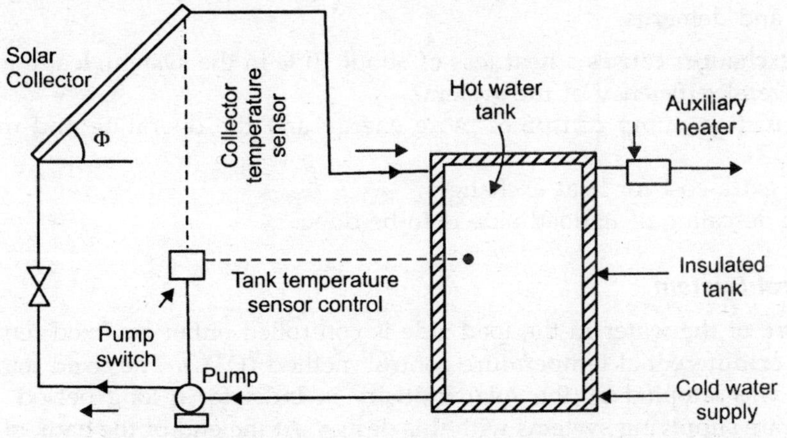

Fig. 4.1.2.1.1: Schematic of a forced Circulation Solar Water Heater

4.1.2.2 Indirect or Closed Loop System

Fig. 4.1.2.2.1. shows the diagram of a closed loop or indirect forced circulation system. When the quality of water is hard, the water should either be treated adequately to provide the water with acceptable level of softness, or an indirect system should be adopted with a suitable fluid for the primary system consisting of collectors and heat exchangers. The latter one is called closed loop system. A coil type heat exchanger is provided in secondary circuit, i.e., on the load side, so that

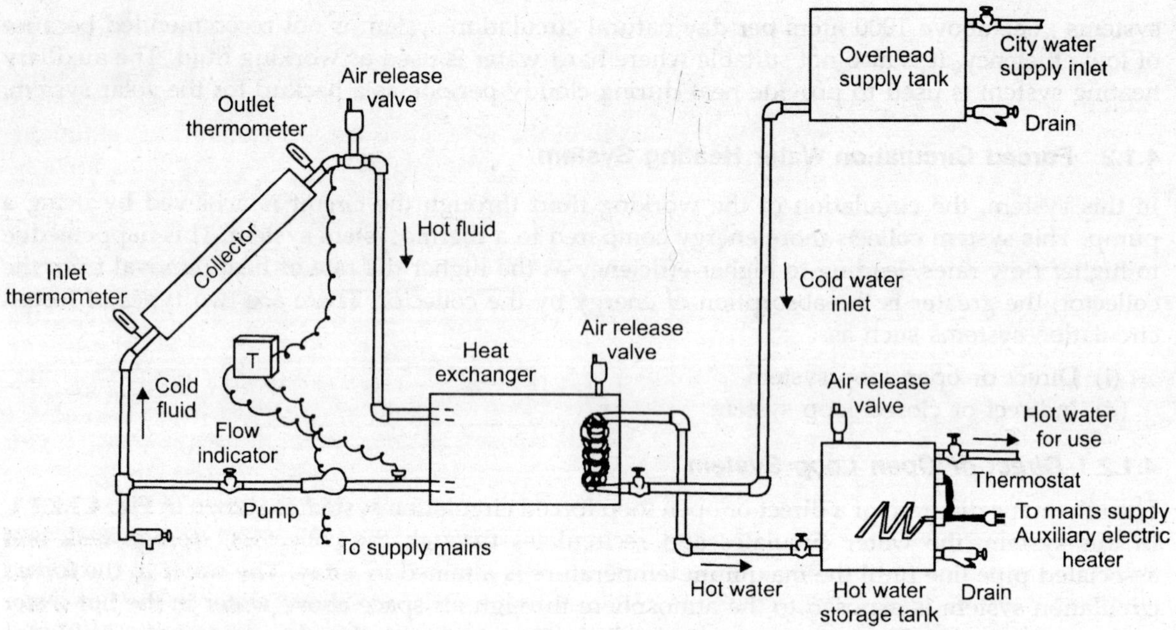

Fig. 4.1.2.2.1: Solar Hot Water System with Heat Exchanger 100 LPD

scales are not formed in the collector tubes to block the flow of water. The fluid in the primary circuit will not come in contact with the water in solar hot water storage tank. The primary heat transfer fluid should be of non-freezing type. By incorporating the heat exchanger, the followings are the merits and demerits.

(i) A heat exchanger causes a heat loss of about 10°C in the system, leading to the reduction to the overall efficiency of the system.

(ii) The circulating pump consumes more energy and the operating and maintenance costs enhance.

(iii) It offers extra cost for heat exchanger.

(iv) Periodic descaling of its load side is to be done.

4.1.2.3 Control System

The temperature of the water in the load side is controlled either by fixed temperature control method (FTC) or differential temperature control method (DTC). The fixed temperature control (FTC) method was adopted by the solar industry in India for a long period. Even now some manufacturers are supplying systems with this design. At the end of the bank of collectors on the delivery load side, a thermostat and an electro-hydraulic valve are fixed. The cold water circulates with the circuit until the desired temperature is reached. When the water in the collectors reaches the set temperature, the electro-hydraulic valve opens and releases the hot water to the load side. This type is cheap but not efficient. Because there should be a good temperature difference and an optimum flow in the collectors. The flow should be neither too slow nor too fast. When the flow is too slow, the absorber loses heat because of its high temperatures. For the too fast flow, time becomes insufficient to transfer heat form the absorber to the liquid. This method offers high thermal losses.

In contrary, the differential temperature control (DTC) method senses the set difference of temperature between outlet and inlet of the collector field. If the difference in temperatures becomes lower than the set value, the pump gets automatically started. As soon as the difference in temperatures becomes more than the set value, the generated signal automatically shuts off the pump. The electrical energy consumption by the pump reduces as the pump runs for less time period.

4.1.3 Two Tank System

In the earlier stated systems, a continuous source of cold water supply is provided to replace the hot water drawn from the storage tank. Particularly in the smaller size solar hot water system, it is noticed that cold water from the overhead tank flows into the bottom of the storage tank and hot water from the collector flows to the top of the hot storage tank. The hot water is also taken out from the top of the tank. As hot water is drawn, cold water flows in to fill up the space. There will be mixing up of cold and hot water. The overall temperature of water in the tank reduces gradually. To avoid their mixing up problem, two tank systems can be used. Instead of one tank, two hot water storage tanks are provided. One tank gets filled up with hot water coming out from the collectors after circulation, while the other tank is used for discharging hot water already collected previously to the users. The filling and discharge cycles of two tanks get alternated. This system is particularly recommended for large capacity system. It provides hot water with more or less constant temperature. In spite of the additional cost for the another tank, sophisticated instrumentation and control, this type of system is preferred as its efficiency is high and it becomes economical in long term aspect.

4.1.4 Installation of Solar Collectors

The placement of solar collector or collectors is an important aspect as far as the optimum solar energy utilization is concerned. It should be placed with proper orientation and tilt. Shading problem generally between 9 am and 4 pm should be properly tackled. Minimum distance between rows of collector should be correctly provided.

The collectors should be installed in such a manner that they face due south. Generally south facing surface is a potential collector location. However, a variation of fifteen degrees either way from due south is used as it results in little or no difference in the performance of the system. A typical collector orientation is shown in Fig. 4.1.4.1. Tilt angle of collector is another important criterion. Collector should be tilted in such a way that it will be at right angle to the sun's rays as it absorbs the most

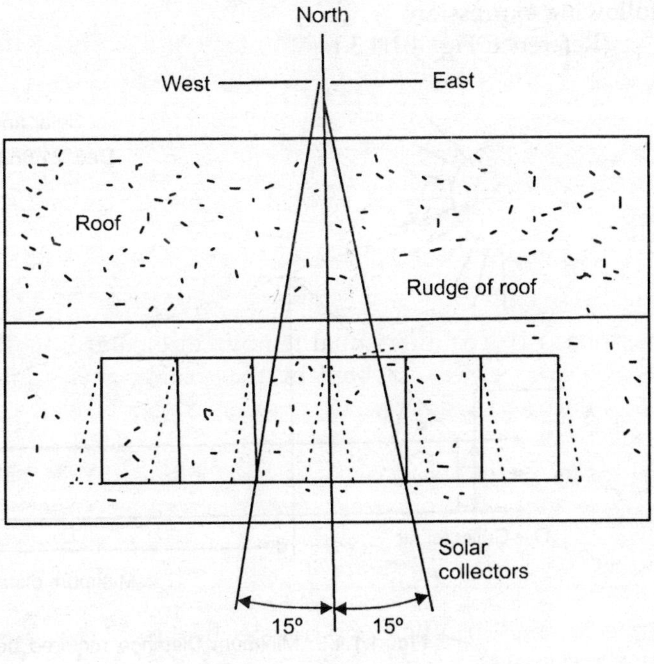

Fig. 4.1.4.1: Collector Orientation

sunshine per unit area. Since the position of the sun changes throughout the day and the year, the correct angle can be achieved if the tilt is changed according to the position of the sun. This can be done by very costly tracking systems. An angle equal to the local latitude is the closest approximation for the required fixed tilt on year round basis. A small variation of 10 degrees one way or the other will respect to the local latitude angle makes little difference in collection of energy. This is shown in Fig. 4.1.4.2.

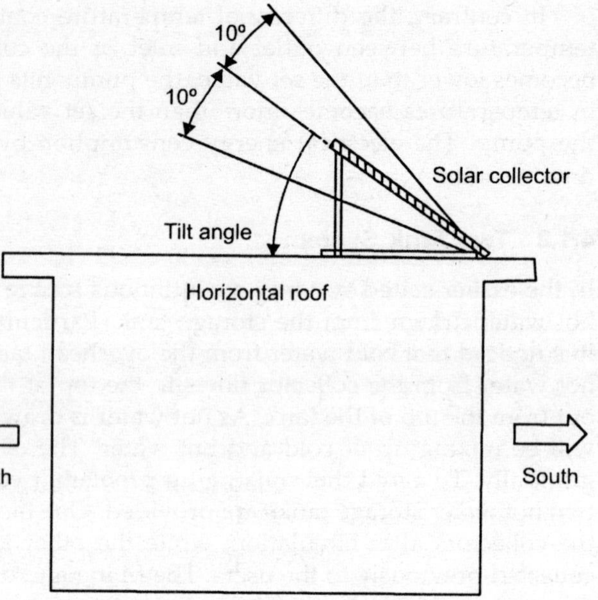

Fig. 4.1.4.2: Collector Tilt

The shading generally occurs due to presence of the house itself such as chimneys, domes and overhangs and also the surrounding buildings and trees located around the collectors. The site for the system should be free from above stated shades throughout the year. The small shading before 9 am or after 4 pm does not affect the performance of the system much.

During placement of large array of collectors in several rows, a minimum distance between rows should be provided to ensure that one row does not cast a shadow on another at any time and to facilitate ease of erection and maintenance. The minimum distance D is given by the following expression:

(Reference Fig. 4.1.4.3.)

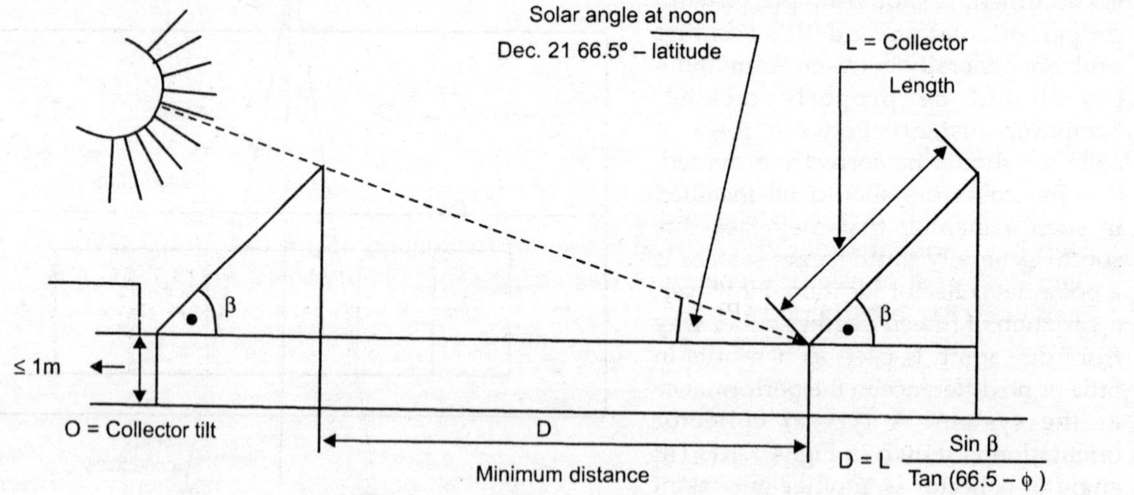

Fig. 4.1.4.3: Minimum Distance required between Rows of Collectors

$$D = \frac{L \sin\beta}{\tan(66.5 - \varphi)} \quad ...(4.1.4.1)$$

where, L = collector length, β = tilt angle, φ = latitude.

4.1.5 Economic Analysis of Solar Water Heating Systems

The use of solar water heating system depends on its cost effectiveness. In general, the investment in solar water heating systems are high when compared to other systems like electrical geysers and furnace oil fired boilers. But the running costs are minimum as the solar energy is freely available. Therefore, the period of return on investment is considered as an important parameter in the economic analysis of the system. In financial terms, an investment is made towards the cost of a solar system to reduce the recurring expenditure on energy bills. The returns should be, therefore, the reduction in monthly energy bills, payback of investment in a reasonable period and free energy for extended period.

In the simplified economic analysis for a domestic solar hot water system, the following parameters should be available:

(i) Capacity of the SWHS in litres/day with available solar insolation (I) in KWhr/sq.m./day.
(ii) Temperature of hot water (t_{fo}) in °C.
(iii) Ambient temperature (t_a) in °C.
(iv) Cost of electricity (C_e) in R_s/KWh.
(v) Efficiency of the solar hot water system η_{th}.
(vi) Efficiency of the conventional system which is supposed to be replaced by SWHS and the calorific value of the fuel if any fuel is used in the conventional system η_c.
(vii) Life of solar hot water system (L), yrs.
(viii) The number of sunny solar days per year (N_d).
(ix) Cost of solar hot water system including collector, storage tank, pipe line and instrumentation (C_H). The followings are the steps used for assessing the required collector area, pay back period and the period for having free energy.

Step A : Total heat energy required for the water (E_1) = quantity of water × specific heat of water × temperature difference, Kcal or KWh.
Step B : Total energy required at the entry of the collector (E_2) = E_1/η_{th}, Kcal or KWh.
Step C : Required area of the solar collector (A_c) = E_2/I in proper units.
Step D : Saving of electricity consumption or fuel energy consumption per day (S_e) in KWh/day, S_e is assessed as $S_e = E_1/\eta_c$, KWh/day.
Step E : Annual saving of electrical energy or fuel energy (S_a) = (S_e) · (N_d), KWh.
Step F : Cost of electrical energy or fuel energy saved in a year (C_s) = (S_a) · (C_e), Rs.
Step G : Pay back period (P) = C_H/C_s yrs.
Step H : Extended period for free energy, (P_e) = $L - P$, yrs.

Example: A domestic solar water heating system of 100 litres per day capacity is used to generate hot water at 60°C replacing a conventional electrical geyser. If the solar insolation is 5.0 KWh/m²/day at ambient temperature of 25°C, estimate the required collector area and the pay back period. Assume that solar collector is working at 50% efficiency. Efficiency of electrical heater is 80%. Cost of electricity per unit is Rs. 3/- . Cost of solar hot water system Rs. 30,000/-. Life of SWHS is 15 yrs. Sunny solar days per year is 300.

Solution: The total heat energy required for the water (E_1) = 100 × (60 – 25) = 3500 kcal/day = 3500/860 KWh/day = 4.07 KWh/day.

Total energy required at the entry of collector (E_2) = 4.07/0.5 = 8.139 KWh/day.

The required collector area (A_c) = 8.139/5.0 = 1.63 ≅ 2m^2.

Saving in electricity consumption (S_e) = [100 × (60 – 25)]/(860 × 0.8) = 5 KWH/day.

Annual saving of electrical energy (S_a) = 5 × 300 = 1500 KWh.

Cost of electrical energy saved in a year (C_s) = 1500 × 3 = Rs. 4500/-

Payback period = 30,000/4,500 = 6.7 yrs.

If the cost of electricity increases to Rs. 4/- per unit, then the pay back period will reduce when all the other parameters remain same.

For this case, the pay back period = 30,000/(1500 × 4) = 5 yrs.

4.2 SOLAR AIR HEATERS

It is conventionally a flat plate collector with an absorber plate, transparent cover at the top and insulation at the bottom and sides. The whole assembly is encapsulated in a sheet metal container. Its principal advantages over SWHS are (i) Need to transfer heat from one fluid to another is eliminated, (ii) Complete elimination of the problem of corrosion, (iii) Freezing of working fluid is non existent. Conventional air heaters are found to have poor thermal efficiency because of low conductive heat transfer coefficient between absorber plate and air flowing through the duct. The high resulting temperature of the plate causes thermal losses and lowers the energy collection efficiency.

Fig. 4.2.1 represents a porous type air heater where the absorber may include slit, expanded metal, over lapped glass and transpired honey comb. In a non-porous absorber, use of fins instead of corrugated absorber is permitted. But the difficulty lies in creation of additional pressure drop. A cross-sectional view is given in Fig. 4.2.2. However, the factors which are critical in both the aforesaid designs to attain acceptable values of heat transfer coefficient for the passage of air are

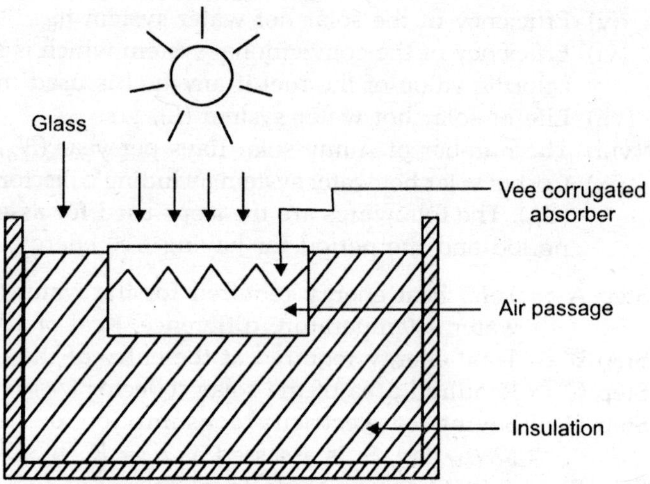

Fig. 4.2.1: Schematic of Porous Type Air Heaters

(a) Porousity of the matrix.
(b) Thickness of matrix.

In fact, the heat transfer coefficient has been found to increase considerably by providing artificial roughness of various geometrics in the form of wires or ribs attached to the underside of absorber plate (Prasad and Saini, 1988), use of corrugated absorber plate (Gupta and Garg, 1967), use of transpiration (Gunnewick, 1996), use of packed bed or porous absorber of different materials and shapes in air flow channel (Sharma et. al, 1991 and Chaudhury and Garg, 1993). The principal reasons of interest in packed bed air heaters are—

(a) High heat transfer area to volume ratio
(b) Increased air turbulence resulting in increase of heat transfer coefficient

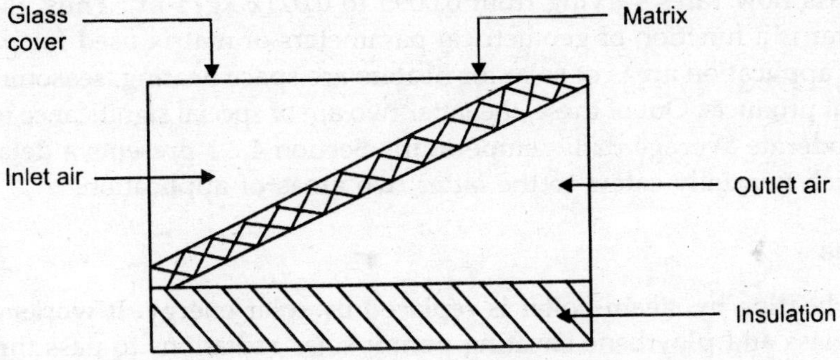

Fig. 4.2.2: Schematic of Non-porous Type Air Heaters

(c) Absorption of energy 'in depth', reducing top losses substantially
(d) Easy to fabricate with little addition of expenses over conventional air heaters.

Recently, Sainey and Varshney of University of Roorkee have carried out detailed experimental studies pertaining to thermal performance of packed bed air heater with wire-mesh screen matrix as packing element. Six matrices designed with different geometrical parameters, i.e., wire diameter, pitch and number of layers have been shown in Table 4.2.1. These systems have been investigated under actual outdoor conditions for typical values of flow rates (G) ranging from 0.0095 to 0.32 kg/s-m^2. Appreciable enhancement in thermal performance of these air heaters over conventional ones can be observed from Table 4.2.2 where values of efficiencies (n_m and n_s) at solar noon for different matrices tested and corresponding smooth collectors have been presented.

Table 4.2.1: Geometrical Parameters of Wire-mesh Screen Matrices

Matrix type	*Wire dia (mm)*	*Pitch (mm)*	*No. of layers*	*Bed depth (mm)*	*Porousity*	*Pitch to diameter ratio*
M1	0.360	2.72	14	25	0.958	7.55
M2	0.450	2.08	10	25	0.939	4.62
M3	0.590	2.23	10	25	0.902	3.77
M4	0.795	3.19	09	25	0.887	4.01
M4a	0.795	3.19	07	25	0.905	4.01
M4b	0.795	3.19	05	25	0.937	4.01

Table 4.2.2: Comparison of Collector Efficiencies with Different Matrices as Packing Element (at Solar Noon)

Matrix designation	*G = 0.0095*		*G = 0.0212*	
	n_m%	n_s%	n_m%	n_s%
M1	42.47	23.58	58.26	34.72
M2	38.40	22.36	55.47	34.45
M3	36.51	23.14	52.45	34.28
M4	40.83	22.97	56.08	34.46
M4a	44.62	24.68	58.69	34.76
M4b	47.07	24.30	59.60	34.74

It can be easily observed from Table 4.2.2 that solar air heaters with matrix M4b exhibits the best performance, which is found to be 71.5 to 93.7% higher compared to conventional smooth air heaters for mass flow rates varying from 0.0095 to 0.0212 kg/s-m^2. Thus, thermal efficiency of a solar air heater is a function of geometrical parameters of matrix used for packing the duct.

The principal application areas of solar air heaters are space heating, seasoning of timber and curing of industrial products. Out of these, the latter two are of special significance to the developing countries with moderate average daily temperature. Section 4.2.1 presents a detailed description of solar kilns which normally caters to the latter two areas of application.

4.2.1 Solar Kilns

In these systems, heating by steam-boiler is replaced by solar energy. It works on the principle that transparent glass and ploythene sheating permit solar radiations to pass through them into the Kiln but obstructs long radiation wavelength emitted by the timber back into the atmosphere. A Solar Kiln consists of three principal parts—(A) The wood seasoning chamber, (B) Flat plate collector and (C) The chimney.

The seasoning chamber is provided on a raised masonary platform. Usually the solar collector is a rectangular wooden box and the chimney is provided with a cover at the top end and a wire mesh to prevent entry of rain water, birds and other drops into it. All these sections have been combined into an isometric view in Fig. 4.2.1.1. The chimney creates a natural draft in the seasoning chamber, which enables proper circulation of hot air around stacked wood. Circulating air transfers heat from solar absorbing surface to the timber stack for evaporation of moisture. Heat emitted by timber in the form of infrared radiation is prevented from escaping by double glass walls.

There exists two principal types of Solar Kilns. In the greenhouse type, timber stack and heat absorbing surface are provided under one structure. Rate of drying is the key parameter of interest depending on the ratio of absorber surface to the timber volume. Value of this parameter for greenhouse Kilns vary from 2 to 4m^2 of absorber surface per cubic metre of timber volume. In Kilns with separate collectors, timber is kept in an insulated box which can attain higher temperature than a greenhouse Kiln. Rate of drying for this second type of Kilns vary around 8 to 9m^2 of absorber surface to timber volume.

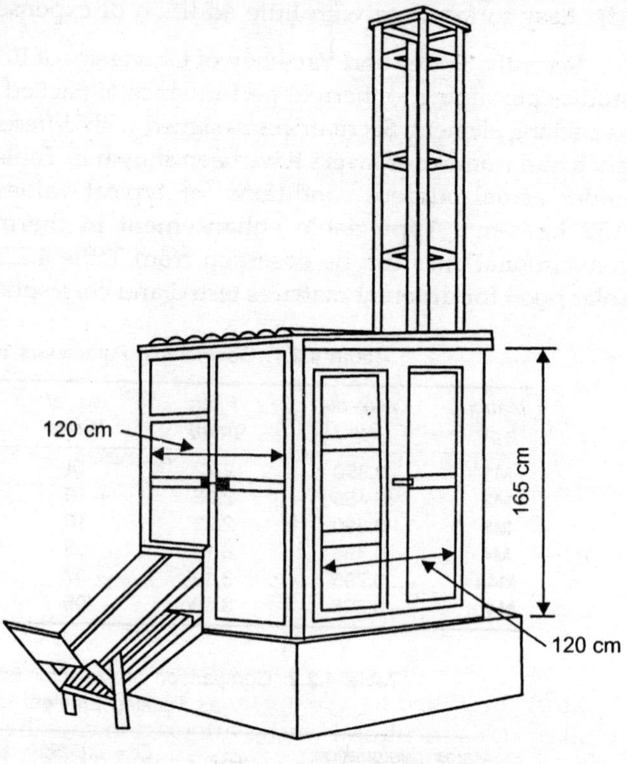

Fig. 4.2.1.1: An Isometric View of a Solar Timber Seasoning Kiln. (After Mital, Non-Conventional Energy Systems, 1998)

Evaluation of the Kilns' performance is based on efficiency of the collector and the capacity of the collector and the capacity to season wood to a required moisture level as compared to open drying. The collector efficiency (E) is given by the simple relation:

$$E = (t_K - t_A)/A_c \cdot I \qquad ...(4.2.1)$$

Where the symbols like t_K and t_A represent Kiln and ambient temperature, while A_c and I stands for surface area of collector (m^2) and radiation intensity (W/m^2) falling on the collector. Based on the recordings by a pyranometer and wet and dry bulb thermometers timber drying rate was found to be higher in the Solar Kiln compared to the open shed. The Solar Kilns attained a maximum temperature of 24°C above the ambient temperature and average efficiency of the heat collector was around 38%. In green timbers high proportion of moisture content has to be reduced and a reduced value around 12% is inequilibrium with atmospheric conditions for long term satisfactory usage. Saw woods (25 × 300 × 360 mm^3) of Mansonia Altissima and Termenalia Superb were dried from 46.16% to 15.02% and from 52 to 15.4% moisture content respectively in 12 days. A side view of timber Solar Kiln is presented in Fig. 4.2.1.2.

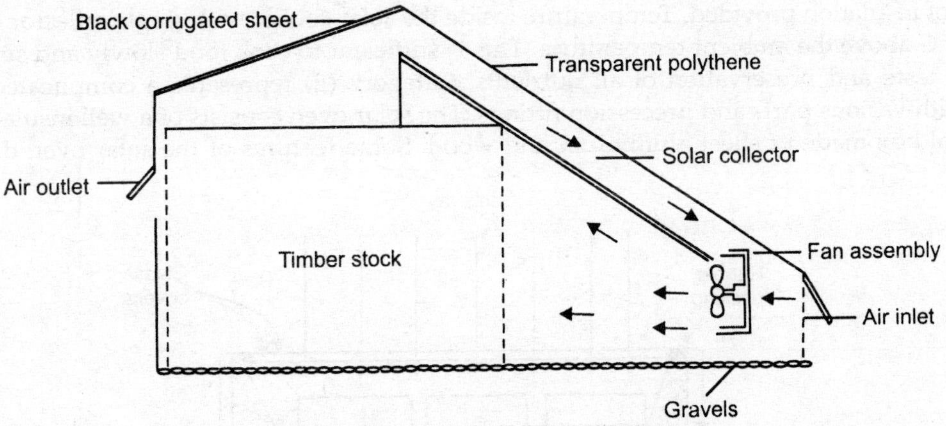

Fig. 4.2.1.2: Side View of Timber Solar Kiln

Conventional air drying is likely to introduce defects such as shrinking, warping, bending circles and infestation by insects leading to considerable loss of timber as well. A Solar Kiln is easy to set-up, maintain and operate free of pollution. Drying process is faster by using Solar Kilns where timbers of higher quality are produced. Principal drawback of these kilns lies in its poor performance during a cluster of repeated cloudy days.

4.3 SOLAR COOKERS

In most of the developing countries, at least 80 to 90% of the total energy consumption in a village can be attributed to cooking only. Variety of fuels like coal, kerosene, cooking gas, firewood, dung cakes and agricultural wastes are used in cooking. Out of these, firewood and cowdung cakes are the two important sources for cooking food. It is well known that the present, pattern of consumption of fuels is largely oriented around the preservation of our environment. This is why harnessing and utilization of solar energy has been naturally promoted for cooking purposes.

Simplest solar cookers have been designed more than fifty years back. But these have not been accepted due to the cheap availability of cooking fuels and little concern for the surrounding environment. Different design of solar cookers developed so far includes

(i) Flat plate box type solar cookers
(ii) Multi reflector type solar cookers
(iii) Parabolic disc concentrator solar cookers

In category (i), a rectangular box enclosure insulated on the bottom and sides contains one or two glass covers on the top. As shown in fig. 4.3.1, solar radiation enters through the top glass cover and heats up the enclosure in which the food to be cooked is placed in separate shallow vessels. The solar radiation entering the box are of short wave-length. Use of two glass covers is to minimize heat losses through reflected path from the absorber black metal tray. Thus, direct, reflected and re-reflected solar radiation striking the cooking containers simultaneously are responsible for the rise in temperature of the containers. As shown in the Fig. 4.3.2, the loss due to convection is minimized by providing a rubber strip all round the periphery between the upper lid and the box. Insulating materials like paddy husk, glasswool or saw dust are employed in between the blackened tray and the outer cover of the box to minimize heat loss due to convection. The temperature attained by the cooking pots depends on (i) intensity of solar radiation and (ii) material of insulation provided. Temperature inside the solar cooker with single reflector is usually 80 to 120°C above the ambient temperature. This is sufficient to cook food slowly and surely with delicious taste and preservation of all nutrients. Category (ii) represents a complicated type of cooker with various parts and precession fittings. The solar oven consists of a well insulated semi-cylindrical box made of sheet aluminium and wood. Some features of the solar oven developed

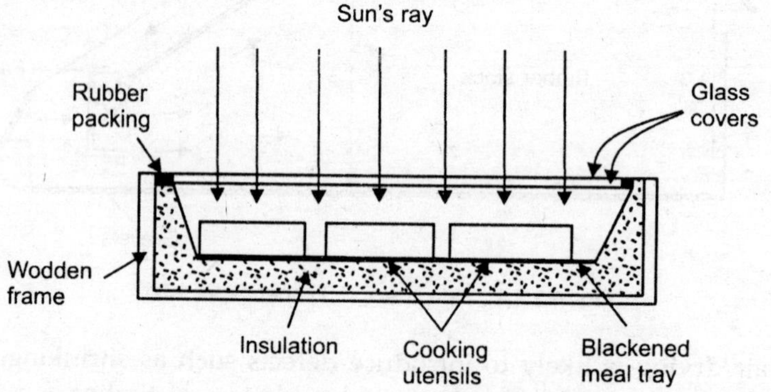

Fig. 4.3.1: Operation of Box Type Solar Cooker

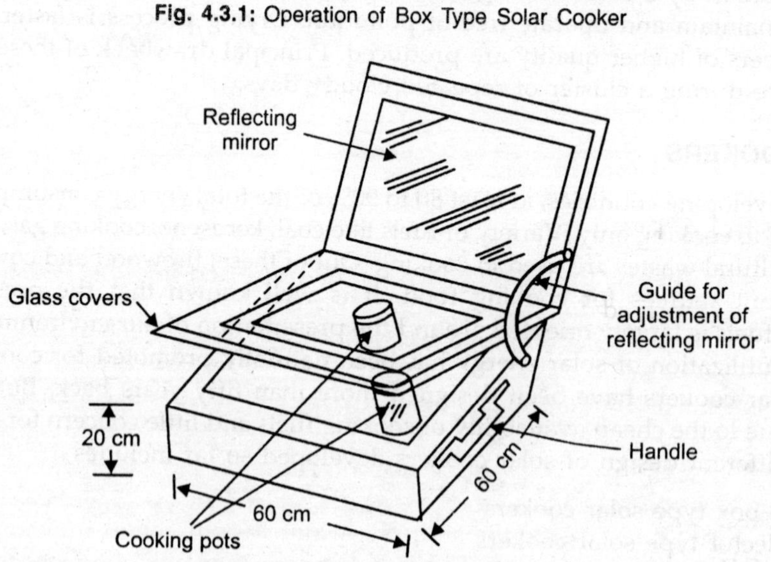

Fig. 4.3.2: Details of a Box Type Cooker

by the Central Arid Zone Research Institute (CAZRI), Jodhpur have been installed in a large number of Indian villages. In this system, eight reflectors made of silvered glass mirrors, four of square shape and four of triangular shape have been used. The system can be manually tilted and oriented towards the sun. A cradle type cooking platform is made in the oven which helps in keeping the vessel containing food horizontal irrespective of the sun. On very clear sunny days, operator expects the inside air temperature to be around 330°C in summer and 240°C in winter.

From various field trials and test reports, the simple box type solar cookers with a single glass reflector have turned out to be the cheapest and widely accepted version of solar cookers. Typical dimensions of the box is about 50 cm square and 12 cm deep and time of cooking a particular item is dependent on the pattern of daily solar radiation. A single glass reflector whose inclination can be varied is often an integral part of box type cookers.

Wide acceptance of these cookers lies in their simple operation and nominal maintenance requirements (an experienced operator only places and takes out the cooking vessels). However, the principal demerits of the system lies in its dependence on the daily available insolation and the inability to prepare some typical foodstuffs like chapatti, puri, fried items, etc. Typical cooking times and temperatures needed for different types of foodstuffs in different seasons are highlighted in Table 4.3.1.

Table 4.3.1: Cooking Times and Temperatures Needed for Different Types of Food in Different Seasons
(After Metal Non-conventional Energy Systems, 1996)

Sl. No.	*Food category*	*Optimum temp. needed to start cooking (°C)*	*Time for cooking after reaching optimum temp. hour*
1.	Very easily cooked food such as cabbage, carrots and leafy vegetables.	70	1.0
2.	Easily cooked food such as root vegetables, mug and rice.	75	1.5
3.	Harder to cook food like some vegetables and special dishes, bread and cake.	80	2.0
4.	Harder to cook like hard pulses, special dishes and meat.	85	2.5

4.4 SOLAR DISTILLATION

With the help of solar energy, the saline water can be converted into distilled water. This is the pure water which is required for battery, automobiles, various industries. The device, which produces pure water by using solar energy, is called as solar water still. The schematic diagram of the solar still is shown in Fig. 4.4.1. It is a basin with blackened surface i.e., basin liner. The basin contains saline water at a shallow depth. The transparent air tight cover encloses completely the space above the basin. Through the filter, the impure water is poured into the basin. The cover is sloped towards a collecting trough. After passing through the transparent cover, the solar energy is absorbed and converted into heat in the basin liner. The saline water is then heated and the water vapour is produced inside the still. The water vapour comes into contact with cooler interior surface of the transparent cover. The water vapour then gets condensed. The condensate is collected through the distillate trough. It can take a wide range of saline water upto sea water,

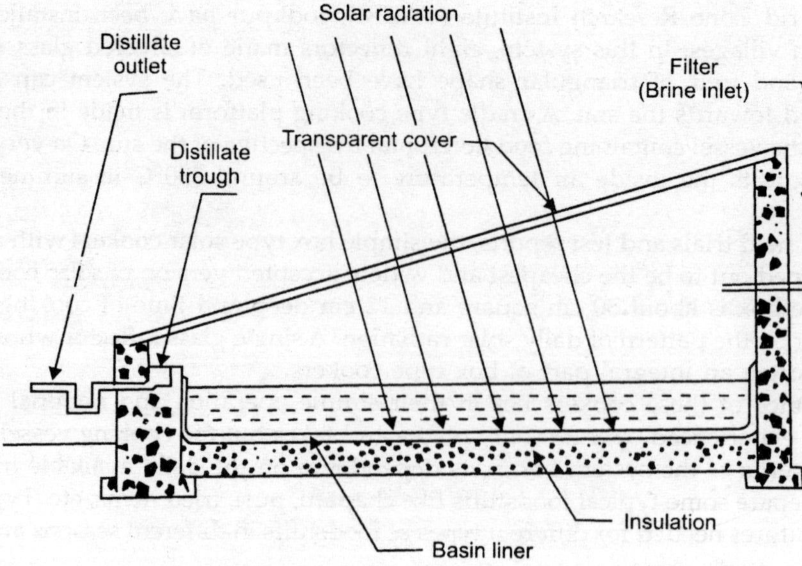

Fig. 4.4.1: Schematic of Solar Still

i.e., about 30,000 mg/lit. It also removes toxic ions and bacteriological contamination. The performance of a solar still is generally expressed as the quantity of water produced by each unit of basin area in a day. The production rate depends on mainly the available intensity of the solar radiation. Apart from that, wind speed, ambient air temperature, atmospheric humidity, sky conditions, etc. also affect the performance of a solar still. Central Salt and Marine Chemicals Research Institute (CSMCRI), Bhavnagar, has done pioneering work in promoting solar stills. The cost of water generally comes to about 50 paise per litre. But the initial cost of the solar still is high. In India, solar stills installed by CSMCRI are given in Table 4.4.1.

Table 4.4.1: Solar Stills Installed by CSMRI

No.	*Location*	*Capacity (Cu.m/day)*	*Evaporating area (sq.m)*	*Year of installation*	*Quality of saline water*	*Present status and other remarks*
1.	CSMRI salt works, Bhavnagar.	1000	350	1965	Sea water	Pilot plant, damaged during salt works 1968. Drinking water for workers in salt works.
2.	Navinar light house near Mundra, Gulf of Kutch	130	49.9	1968	Sea water	Need repairs but still working.
3.	Awania village near Bhavnagar	5000	1866.6	1977-78	Saline Well water TDS : 3000 to 4500 ppm, fluoride; 10 ppm	Supplied drinking water to village. Extensively damaged due to cyclone in November, 1982. Wind mill and solar cell pump used for water pumping.

(Contd.)

Table 4.4.1 *(Contd.)*

4.	Chhachi light house near Mandvi, Gulf of Kutch	250	108.0	1978-79	Sea water	Need heavy repairs. Not properly maintained. Damages by stray dogs, etc.
5.	Narayana Sarovar Dist. Kutch, Gujarat	3000	1244.4		Saline well water TDS: 15,000 ppm	Work started in 1977 but due to lack of saline water discontinued. Half of the plant commissioned by 1983 summer using saline water from new source.
6.	Bhaleri Dist. Churu, Rajasthan	8000	3110.0	1979	Saline well water TDS: 3800 ppm Nitrates: 340 ppm Fluoride 4-5 ppm	Constructed by PHED, Rajasthan faces problems of supply of saline water. Wind mill for pumping saline water. PHED is considering its repairs.
7.	Bitra island, UT of Lakshadweep	2000	750.0	1983	Sea water	Constructed by PWD. Provides drinking water to Islanders.

The desalination of water can also be done using solar pond and different solar thermal collectors for providing process heat for desalination process. The collectors may be flat plate collectors, evacuated tubular collectors, parabolic concentrators. Solar photovoltaics and Wind energy can also be used to produce fresh water from brackish water. Table 4.4.2 states the renewable energy based desalination plants across the world.

Table 4.4.2: Some of the Renewable Powered Desalination Plants Across the World

Location	*Type of Energy Source*	*Energy Source Characteristics*	*Type of Desalination Plant*	*Capacity (m^3 per day)*
Ranking, Texas, USA	Central receiver, heliostat	20,448 m^2	Single stage RO	–
Yabu, Saudi Arabia	Wind, Fresnel line focusing collectors	12,800 m^2	Two stage RO in series	–
Yabu, Saudi Arabia	Line focus concentrators + Photovoltaics	5,000 m^2	One stage RO series with ED	250
Yabu, Saudi Arabia	Point focus thermal collectors	43,800 m^2	Indirect freezing	210
Yabu, Saudi Arabia	Central receiver, heliostats	22,800 m^2	Two-stage RO in parallel with 24-effect distillation	3,478 (RO) 2,520 (Distillation)
Sulaibiya, Kuwait	Point focus parabolic collector	1,000 m^2	RO, MSF	20 (RO) 25(MSF)
Abu-Dhabi	Evacuated tube collectors	1,862 m^2	Multiple-effect (18-effect)	120

(Contd.)

Table 4.4.2 *(Contd.)*

Location	*Type of Energy Source*	*Energy Source Characteristics*	*Type of Desalination Plant*	*Capacity (m^3 per day)*
Al-Ain, UAE	Flat plate collectors	5,180 m^2	20-effect MSF 55-effect MED	600 (MSF) 500 (MED)
Arabian Gulf	Parabolic concentrators	53,960 m^2	MED	6100
Takami Island, Japan	Flat plate collectors	–	16-effect MED	–
Ohsima Island, Japan	Photovoltaics	25 kWp	Electrodialysis	10
Innoshima Island, Japan	Photovoltaics	25 kWp	Reverse Osmosis	10
La-Paz, Mexico	Flat plate collector, parabolic concentrator	194 m^2 + 160 m^2 (flat plate) 160 m^2 (concentrator)	10-stage MSF	10
Conception del ore, Mexico	Photovoltaics	2.5 kWp	RO	1.5
La Desirade, French Antil.	Evacuated tube collecotors	670 m^2	MSF	40
El-Paso, USA	Solar Pond	90°C	Falling film ME distillation	19
Thar Desart, India	Photovoltaics	450W	Electrodialysis	1
Tripoli West 2, Libya	Parabolic concentrators	–	HTME distill	500

4.5 SOLAR PUMPING SYSTEMS

The solar pumping system utilizes the power generated by solar energy for pumping of water in assisting irrigation and drinking water facility. Solar thermal energy conversion system or photovoltaic system can be used for solar pumps.

Solar thermal energy conversion system can be used for water pumping with cost effectiveness for large scale application. In this system, the water or thermic fluid is heated by solar energy and the heated fluid is used to drive a pump or turbine. Fig. 4.5.1. shows the simplified view of a flat

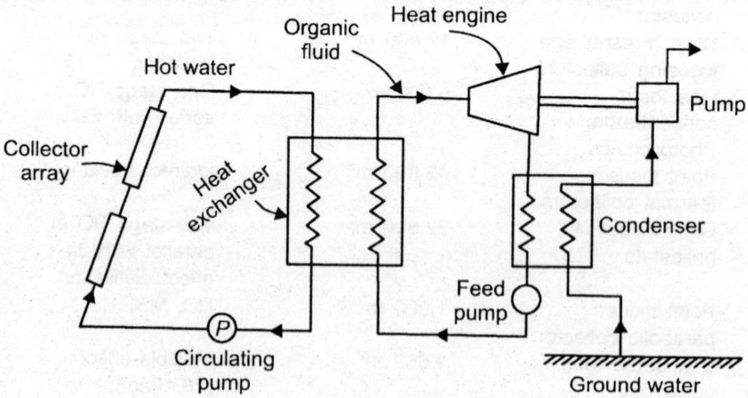

Fig. 4.5.1: Schematic of Solar Pump

plate collector based pump. Water/thermic (heat transport) fluid is heated by a flat plate collector or the concentrating collector depending on the required temperature to be achieved for the working fluid. By flat plate collector, the temperature of the fluid can be attained upto about 93°C. By concentrating collector, the fluid is heated upto about 215°C. The hot fluid is allowed to flow to the mixing tank/storage tank and then to the boiler or heat exchanger of a heat engine to convert the state of the working fluid of the heat engine from liquid to vapour. The vapour then drives a turbine which in turn operates a pump set. The working fluids of the heat engine may be refrigerent – 113, refrigerent – 11, toluene (CP – 25), more – chlorobenzene (MCB), trifluoro ethanol (TPE – 100 and 85), hexafluoro benzene (HFB), pyridine (CP – 32) and thiopene. Among them, R – 115 is considered to be working fluid and it yields high cycle efficiency and it is non-toxic in nature having low cost. The water or heat transport fluid is again fed to the collector by a circulating pump. In the heat engine cycle, the vapour coming out from the turbine is passed through the condenser where the vapour gets condensed. The liquid working fluid of the cycle is then sent to boiler or heat exchanger with the help of a feed pump to complete the cycle. The water which is pumped from the ground or irrigation well is used as the coolant for the condenser.

The overall efficiency of the system, η_o, is assessed as $\eta_o = \eta_e \times \eta_c$

Where, η_e is the efficiency of the engine.

η_c is the efficiency of the collector.

A higher temperature in the boiler, offeres higher engine efficiency. But for this, higher collector temperature is needed. This decreases the collector efficiency. Therefore, there should be an optimum range of operating temperatures for the solar pumping system, using solar thermal energy conversion technology, to achieve the maximum efficiency of the whole system. The energy efficiency, i.e., percentage of energy collected that is converted into useful work, is about 13 to 14 percent.

In a solar *PV*-based water pumping system, *PV* modules convert sunlight directly into electricity and this energy is used to run an electric motor pump set for pumping water. In this system, the solar photovoltaic modules feed electricity into storage batteries which operates a pump. The pumps may be of positive displacement reciprocating type or centrifugal type. Any pump requires a minimum power input to start it. For a positive displacement reciprocating pump, a higher input power is required in order to overcome peak starting torque of the pump. In case of a centrifugal pump, water will not be lifted until a threshold power level inspite of it's a much lower level as input power requirement. The typical value of starting threshold power is 300 W/m^2. For water depths upto eight metres, started centriful pumps are used. In case of higher depths, positive displacement pumps are preferred. Centrifugal pumps achieve a close natural match with photovoltaic array over a broad range of operating conditions.

A power conditioning equipment is needed to protect the storage batteries from overcharging or discharging during the non-sunshine hours and for converting *DC* into *AC* if the system is to be connected with an *AC* motor pump set. This equipment normally contains a voltage regulator, a blocking diode and a *DC* to *AC* inverter wherever necessary, storage batteries are required to supply power during non-sunshine hours. Lead acid batteries of 12V and 60 ampere-hour or 12V and 120 ampere-hour capacity in suitable numbers connected together are usually used. For pumping system, the use of batteries is not essential but depends on system design and needs. The batteries can be charged and discharged many number of times. In this system, there may be a need of using a water storage tank when water is not utilized simultaneously with the pump operation. For meeting drinking water supply, the tanks are needed. For irrigation purpose, use of water tanks may not be necessary.

An SPV water pumping system consists of a photovoltaic panel mounted on a stand including a motor pump set compatible with the panel. The motor pump sets may be d.c. surface mounted centrifugal pump set, d.c. submersible pump set, d.c. floating pump set and a.c. submersible pump set. The *PV* panel converts available solar energy into electricity. It runs the motor pump set. The pumping system draws water from the open well, bore well, stream, pond, canal or a rivulet. The range of ratings of pumps are of ¼ hp to 3 hp. Selection of *PV* array capacity, specification of motor pump set and power conditioner, capacity of storage batteries and overhead tanks can be made after considering few important parameters. Those are daily water pumping rate, depth of water table, average solar insolations, system efficiency and water needs during non-sunshine hours. The followings are the major advantages of a solar *PV* based water pumping system

(i) *PV* arrays can be installed easily at site without laying long distance transmission and distribution lines,
(ii) Capacity of the pump can be built up or reduced conveniently following the modular nature of *PV* arrays,
(iii) It does not require any recurring expenditure on fuel and its transportation,
(iv) It requires minimum maintenance,
(v) No expert technical manpower is required,
(vi) *PV* arrays are generally trouble-free and pollution free long life of 20-25 years with excellent reliability and
(vii) Solar cells assembly can be done at the village level.

Central Electronics Ltd., Ghaziabad is considered to be a pioneer organization in developing *PV* based pumping systems in India. *PV* based water pumping systems developed by CEL, Ghaziabad is stated in Table 4.5.1.

Table 4.5.1: Photovoltaic Water Pumping Systems Developed by the CEL, Ghaziabad

System	*PV modules (pk watts)*	*Motor-pump set*	*Elec. Storage capacity*	*Water depth (m)*	*Pumping rate (lit/day)*	*Applications*
A	165	24V (DC) 125 watts	–	4	12,000	At CEL, Ghaziabad (1977),
B	112	24V (DC) 125 watts	24V → 120 A-hr	2.5	7,000	At Awania, Bhavnagar (1979), drinking water
C	225	30V (DC) 180 watts	30V → 120 A-hr	4	25,000	At Sci. & Tech. Bhavan, New Delhi (1980), drinking water
D	350	350 watts	–	5	1,65,000	For irrigation of 1-1.5 hectares

Source: CEL, Solar Photovoltaic Project, Annual Progress Report no. 4, 1979-80, sponsored by the DST, CEL, Ghaziabad, 1980.

4.6 SOLAR COOLING

One of the thermal applications of solar energy is for the purpose of cooling. The cooling with the help of solar energy can be utilized for air conditioning or refrigeration systems. The objective of the systems is to provide comfortable living conditions or to preserve the food products. The common techniques used for refrigeration systems are vapour compression and vapour absorption. In both the systems, the solar energy is used to obtain heated water or any liquid which finally vaporizes the refrigerant. Among the two systems, the absorption cooling systems are considered to be more practical because cooling is needed mostly in summer and the availability of heat from sun is easy.

There are two types of solar operated vapour absorption coolers. The first is continuous cooler and the second one is intermittent cooler.

4.6.1 Continuous Vapour Absorption Cooling System

Continuous vapour cooling system is similar to a conventional vapour absorption refrigeration system.

Fig. 4.6.1.1 shows a diagram of a simple solar operated vapour absorption refrigeration system. In this system, the water is heated with the help of array of solar collectors (in the fig. flat plate collector array has been shown). The hot water is passed through a heat exchanger which is called as generator. In the generator, the hot water transfers heat to a solution mixture of the absorbent and refrigerant. The solution is rich in the refrigerant. The refrigerant vapour at high pressure is generated in the generator and the vapour is fed to the condenser. The vapour is condensed here to high pressure liquid. With the help of expansion valve, the high pressure liquid is throttled to a low pressure and low temperature vapour. This vapour is passed through the evaporator coil, where the vapour absorbs heat from the space surrounding the coil to maintain the cooling effect in the space. The refrigerant vapour is then absorbed into a solution mixture in the absorber. The solution mixture is taken from the generator and it is weak in refrigerant concentration. After absorption a rich in refregerant concentration solution is produced and it is then pumped back to the generator with the help of a liquid pump. The cycle completes here. Usually, the rich

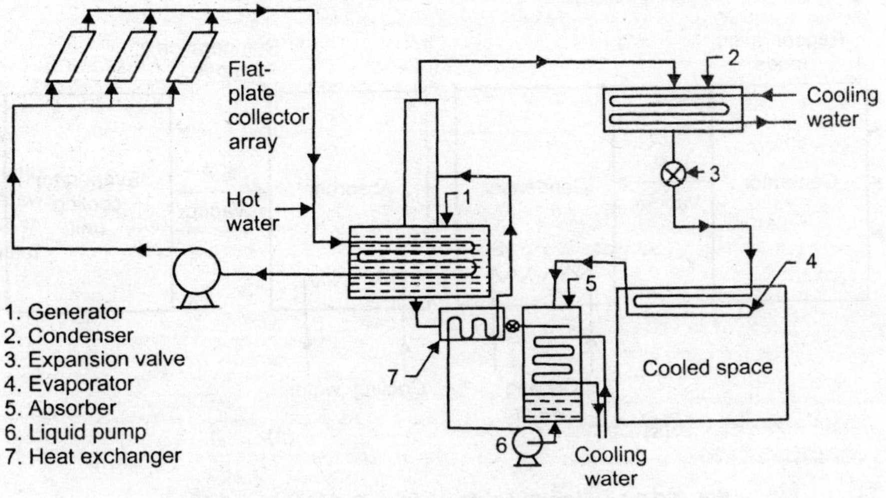

Fig. 4.6.1.1: Solar Absorption Refrigeration System

solution flowing from the absorber to the generator is heated in a heat exchanger by the weak solution withdrawn from the generator to achieve the better performance of the cycle.

The common types of vapour absorption coolers are the lithium-bromide water (LiBr-H_2O) system and the ammonia-water (NH_3-H_2O) system. Among these two systems, the LiBr-H_2O system is widely used. The LiBr-H_2O system requires the lower generator temperature of the order of 85° to 95°C. This can be achieved by using flat plate collectors. It also offers higher coefficient of performance (COP). In this system, water vapour is a refregerant. But it is known that the water can not exist in liquid form below the triple point temperature. It means that practically, it cannot operate much below 4°C. So for air conditioning purposes, the LiBr-H_2O system is equally chosen. The LiBr-H_2O system's COP is often in the range of 0.6 to 0.8. The generator temperatures usually remain in the ranges of 75 to 95°C.

In the NH_3-H_2O refrigeration system, a rectifying section is added at the top of the generator to reduce the water vapour content in the vapour stream going to the condenser. The generator temperatures for this system is in the ranges of 120 to 150°C. This temperature range is higher than the temperatures which can be achieved by flat plate collectors. So, in this system the costlier concentrating collectors are used. Compound parabolic concentrators (CPC), evacuated tubular collectors, other focussing collectors and well designed flat plate collectors can be used to maintain the required higher temperature for the generator. For food preservation purpose, NH_3-H_2O system is chosen.

4.6.2 Intermittent Absorption Cooling System

The basic approach of the intermittent absorption cooling system is that it is allowed to operate intermittently and during the periods of availability of solar energy. Fig. 4.6.2.1. shows the simplified block diagram of the intermittent absorption refrigeration system which operates in two alternative modes. One is regeneration mode and the other one is refrigeration mode. In this mode, two vessels are used. One is generator and the other one is condenser (Fig. 4.6.2.1(a)). In this phase, heat is supplied to the generator by gas, oil, steam or solar energy. In the refrigeration mode, the first vessel is the absorber and the second vessel is the evaporator (Fig. 4.6.2.1(b)). Whole system operates in the regeneration mode for a few hours and is then switched over to refrigeration mode either manually or automatically.

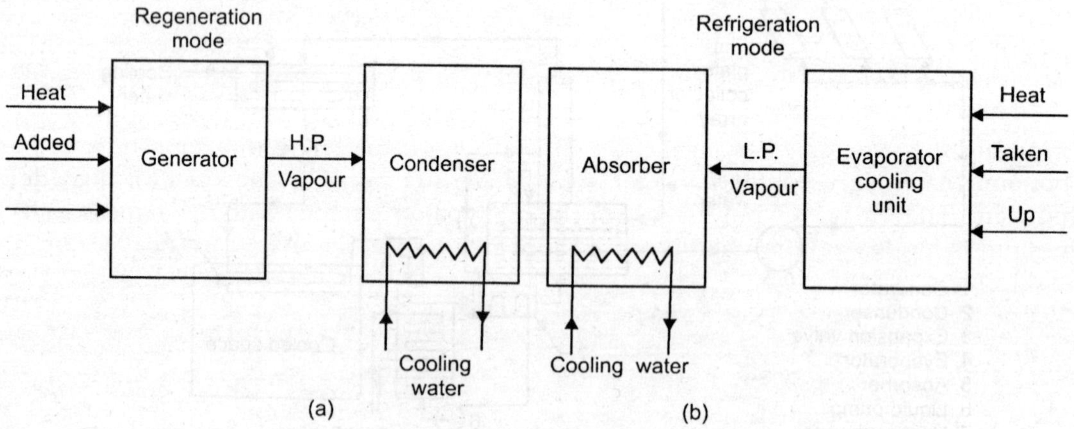

Fig. 4.6.2.1: Intermittent Absorption Refrigeration System

In this system, the used refrigerant absorbent combinations commonly are lithium bromide water, ammonia-water and ammonia-sodium thiocynarate. This technique is usually used for the food preservation purpose in rural areas where electricity is not available.

4.7 SOLAR PONDS

In order to reduce the capital cost of solar thermal installations, the area of the collecting surface has to be optimized and also economical ways of collecting and storing the energy during the same period have to be proposed. When large stretches of water of small depth have been used for simultaneous collection and storage, natural convection currents prevent the rise in temperature of such a pond more than only a few degrees. Thus, an artificially designed pond in which significant temperature rises in the lower regions are possible by preventing the phenomenon of convection is called a 'Solar Pond'. The usual technique followed for preventing convection mechanism is to dissolve a particular salt in water and maintain a definite concentration gradient. Such a concept of 'salt-gradient solar ponds' are very common while concepts like gel solar pond, honey comb solar pond and equilibrium solar pond will only be touched briefly.

A schematic diagram of a solar pond is shown in fig. 4.7.1, where three distinct layers can easily be visualized as—

(I) Surface convective zone (SCZ),
(II) Concentration gradient zone (CGZ),
(III) Lower convective zone (LCZ).

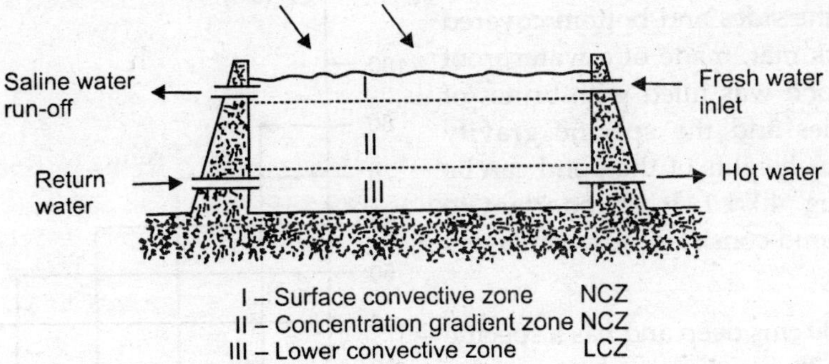

Fig. 4.7.1: Schematic Diagram of a Solar Pond

Typically, the structure is about 1 or 2 metres deep with a thick durable plastic liner laid at the bottom. Materials used for the liner are low density polyethylene (LDPE), high density polyethylene (HDPE), woven polyester yarn (XB-5), hypalon enforced with nylon mesh. Salts like magnesium chloride, sodium chloride or sodium nitrate are dissolved in water to result in a concentration varying from 20 to 30 percent at the bottom compared to almost zero at the top. Due to the upward diffusion of the salt, this concentration gradient will decay over a certain period of time. In order to maintain this gradient, fresh water is replaced through inlet at the top of the pond while saline water is run off through another inlet. At the same time, estimated quantity of concentrated brine is also added to the bottom of the pond.

For extracting the heat energy from the pond, hot water is removed continuously from the bottom and then returned back through a heat exchanger. Alternatively removal of hot water may be done through a submerged heat exchanger coil. As a result of continous movement and mixing of the fluid at the top and bottom, the solar pond is characterized by three zones:

(I) Surface convective zone (SCZ) , having a small thickness around 10 to 20 cms with a low, uniform concentration and a temperature close to ambient air temperature.

(II) The Non-convective zone (NCZ) is much thicker and occupies more than half the depth of the pond. It serves principally as an insulating layer for preventing heat losses in the upward direction. Thus it is also an integral part of the thermal storage.

(III) Lower convective zone (LCZ), having a thickness of the same order as the NCZ, is characterized by constant concentration and temperature throughout its depth. It is thus the principal heat collection as well as thermal storage medium. Typically, the LCZ of a large solar pond operating in West Indian environment can enjoy a maximum temperature between 85° to 95°C in summer and a minimum of 50° to 60°C in winter with an annual collection efficiency ranging from 15 to 25 percent. Although the efficiency values are much low compared to flat plate collectors, the large area ponds (specially 1000 m^2 or more) have an edge over collectors with respect to lower value of cost/sq. metre.

4.7.1 General Features of the Reserves

It is known that a major portion of the southern Tunisia consists of salted lakes called Sebkhats and Scholts. These lakes are aqueous solutions, highly concentrated in Na^+, K^+, Mg^{++}, SO_4^{--}, Cl^-. The concentration in Ca^{--} is low, typically less than one gm/litres. An experimental pond having an area of $4m^2$ and a depth of 1m was fabricated with the sides and bottom covered with a 5mm thick mat, made of a waterproof material. This pond was filled with brines of different densities and the specific gravity profiles at various heights of the pond can be observed from fig. 4.7.1.1. It is important to realize that the pond consisted of seven layers out of which

(a) The LCZ is 30 cms deep and has a specific gravity of 1.28.

(b) The upper zone is 20 cms deep and is filled with fresh water.

(c) The gradient layer is formed of five layers of equal thickness of ten cms each.

(d) It is equipped with seven thermocouples to take temperature maeasurements of aforesaid seven layers.

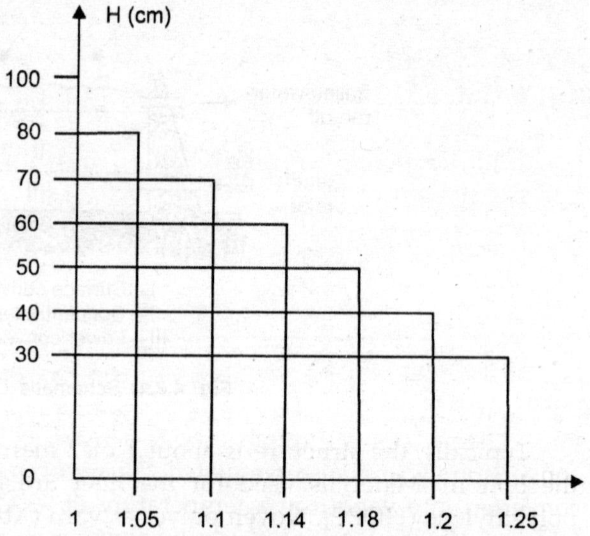

Fig. 4.7.1.1: Specific Gravity Profiles at Various Heights of Pond

Finally, Table 4.7.1 potrays some characteristics of the natural brine reservoiers located in different parts of Tunisia. With an abundance of natural brine and a high potential of mean solar radiation, the arid region of south east Tunisia is a favourable candidate for a thermal power plant site. The heat collected in these solar ponds can be utilized as potential heat sources for heating greenhouses, industrial plants and water desalination.

Table 4.7.1: Characteristics of Natural Brine Reservoir in Southern Part of Tunisia

Location	*Area in Km^2*	*Total reserves in situ 10^6 m^3*	*Reserve recovery 10^6 m^3*	*S in g/l*	*Na^+ in g/l*	*Mg^{++}*	*K^+*	*Ca^{++}*	*SO_4^-*	*Cl^-*
Sabkhat chott jerid	5000	5000	600	330	100	12	9	0.7	12	197
Sabkhat El melah Zazis	150	600	160	335	45	50	7.7	0.2	28	200
Sabkhat oumel kharat	75	50	17	150	43	3.3	0.8	0.4	74	24
Sabkhat El hadibat	125	245	100	260	82	10	3.5	0.6	28	140
Sabkhat El briga	22	78	18	330	84	24	6.3	0.4	32	185

4.8 SOLAR THERMAL POWER PLANT

In case of solar thermal power production, energy is first collected by using a solar pond, a flat plate collector or a focusing collector. This energy is supplied for increasing the temperature of a fluid directly used in any one of the commonly known cycles such as Rankine, Brayton or Sterling or passes through a heat exchanger to heat a working fluid which is used in the cycle to produce mechanical power. The final step will be conversion of this mechanical power to electrical power. Solar thermal power cycles are classified as:

(A) Low temperature cycles using flat plate collectors/solar ponds.

(B) Medium and high temperature cycles usually employing the line focusing parabolic collector technology.

(C) High temperature systems using paraboloidal dish collectors or central receiver system.

(D) Distributed collector systems.

4.8.1 Low Temperature Systems

The flat plate collector and solar ponds are classified as low temperature collectors because temperature achieved using these systems is in the range of 60° to 100°C with collection efficiency of 30 to 50%. Technology and principle of operation of flat plate collectors have been detailed out in chapter 3. The hot water at temperature close to 100°C is stored in a well insulated storage tank. Then it flows through a heat exchanger through which the working fluid of the energy conversion cycle is also circulated. This working fluid (e.g. R-11) having a low boiling point is converted into vapour at a temperature of 90°C and a pressure of a few bar. The vapour, then, forms a regular Rankine Cycle by flowing through a prime mover, a condenser and a liquid pump. However, the overall efficiency of such systems usually ranges from 2 to 3% and these suffer also from high cost of collectors resulting in an installation cost of the order of Rs. 300000 per kW for only six to eight hours of daily operation.

To reduce aforesaid cost of large solar thermal installations, a low temperature power production scheme is shown schematically in Fig. 4.8.1. Here, a solar pond is responsible for both collection and storage of heat energy. This energy drives a Rankine Cycle heat engine. Hot water from the bottom level of the pond is pumped to the evaporator where the organic working fluid is vaperised. As the vapour flows under high pressure to the turbine it expands through the

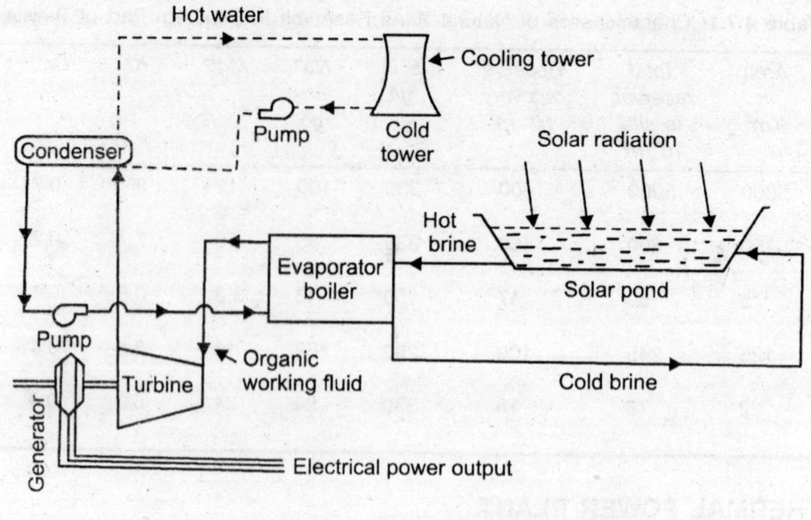

Fig. 4.8.1: Flow Diagram of Solar Pond Electric Power Plant

turbine wheel. It then travels to the condeser where cold water from the cooling tower condenses the vapour back to liquid. Finally, this liquid is pumped back to the evaporator for the whole cycle to be repeated.

Another low temperature engine using heated water from flat plate collector but Butane as the working fluid is shown in Fig. 4.8.2. Here, the high pressure butane vapour, formed by water at 70°C from heat exchanges, runs a butane turbine which operates a hydraulic pump for irrigation.

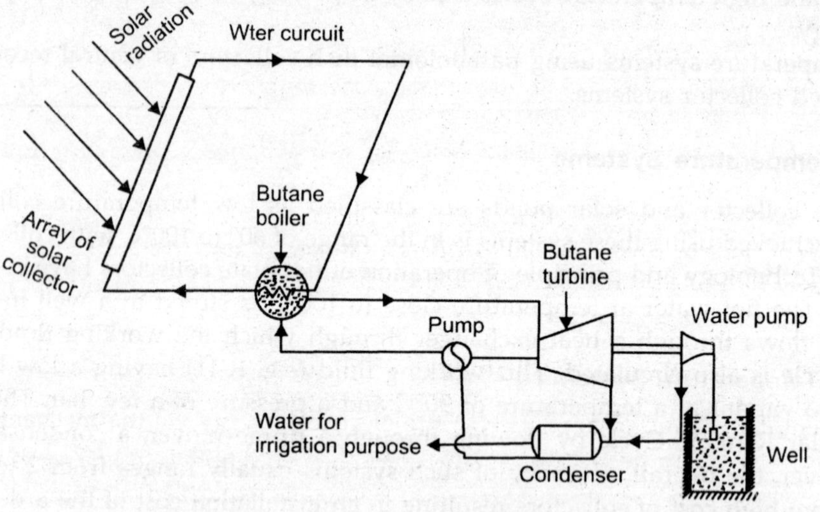

Fig. 4.8.2: Schematic of a Low Temperature Solar Power Plant

4.8.1.1 Solar Chimney Power Plants

This consists of a central chimney structure surrounded at its base by a circular greenhouse that has transparent plastic materials supported a few metres above the ground by a metal frame.

Sunlight passing through the plastic causes the air trapped in the greenhouse to heat up. A convection system is set up in which the hot air is drawn up through the central chimney to turn a turbine located near the base of the chimney. The hot air, in the process, is continuously replenished by fresh air drawn in at the periphery of the greenhouse. The very first solar chimney power plant fabricated was a 100 kW pilot plant in Spain, details are provided elsewhere. The energy conversion efficiency of such power plant is low (of the order of 1 to 2%).

4.8.2 Medium Temperature Systems

These systems generally employ an array of parabolic trough concentrating collectors, which gives temperatures above 100°C and a general range of temperature within 250 to 400°C for the absorber fluid. Solar thermal power plants operating on medium temperature cycles, therefore, inevitably uses the line focusing parabolic collector technology as the most cost effective and successful technology. Here, a parabolic cylindrical reflector with polished inner surface reflects the incident sunlight selectively on to a focal plane where a cylindrical absorber is situated within a pyrex on glass envelope. A selective coating of suitable material to the absorber piping is used for minimizing infra-red emission. Also, proper suntracking arrangement is made so that maximum amount of solar energy can be focussed on the absorber pipe line during the day. Typical features of some 80 MW plants commissioned in California is given in Table 4.8.1.

Table 4.8.1: Technical Specifications of a 80 MW Power Plant

1.	Collector array-area	–	464340 m^2.
2.	Aperture	–	6m.
3.	Axis orientation	–	North-south.
4.	Absorber tube	–	Steel body encapsulated by a glass cover in vacuum.
5.	Concentration ratio	–	80.
6.	Absorber fluid	–	Synthetic oil at 400°C.
7.	Efficiency of collector	–	70% (Beam radiation).
8.	Efficiency of Rankine Cycle	–	38%.
9.	Current generating cost	–	Rs. 2.00/KWh.

It is interesting to note from Table 4.8.1 that the installed cost of this plant has reduced significantly over the last decade to the current value of about Rs. 2.00/KWh because of the increased installed capacity.

The Indian experience with the line focusing parabolic collector technology have been restricted so far to about 30 MW in Rajasthan under the auspicies of the Department of Non-conventional energy sources.

4.8.3 High Temperature Systems

For the efficient conversion of heat energy into electrical energy via the mechanical domain, the working fluid has to be delivered to the turbine at high temperature. Use of focusing or concentrating collectors are, thus essential for concentrating direct solar radiation (Beam Radiation) for attaining temperatures around and above 175 to 180°C. Two such common arrangements have been out lined briefly in the following sub-sections.

4.8.3.1 Paraboidal Dish Collectors

In these systems, the rays from the beam radiation are brought to a common focusing point. Any working fluid moving through this point is heated and this heat utilized in driving a prime mover. Typically sterling engines have been favoured as the prime movers with systems having efficiencies around 30 to 35% and power generation in the range of 10 to 100KW have been possible. On the Indian scenario a small 30KW power station has been installed near Hyderabad. Here, four paraboloidal dish collector modules were used to generate steam for running a steam engine. Due to obvious limitations in size, these collectors are capable of generating at best power in kilowatts and thus achieving economy in case of large scale generation is not feasible. However in USA, a number of 5MW power plants utilizing the steam generated by seven hundred dishes have been erected. Out of these dishes, six hundred are used to obtain saturated steam at 275°C, while the remaining one hundred dishes are used to superheat the steam to 400°C. This steam, then runs two turbine generator sets—one main set of 3.6MW and another peaking set of 1.24MW.

4.8.3.2 Central Receiver Collectors

In these power plants, solar radiation reflected from arrays of large mirrors (called HELIOSTATS) is concentrated on a receiver situated at the top of a supporting tower. A fluid flowing through the receiver absorbes the concentrated radiation and transports it to the ground where it is used to operate a Rankine Power Cycle. A schematic diagram showing the main components of a Central Receiver Power Plant in which water is converted into steam in the receiver itself is shown in Fig. 4.8.3.2.1. Alternatively, the receiver may be used to heat a liquid metal or a molten salt and this fluid is passed through a heat exchanger in which steam for the power cycle is generated. A number of power plants commissioned in the eighties had outputs in the range of 0.5 to 10MW.

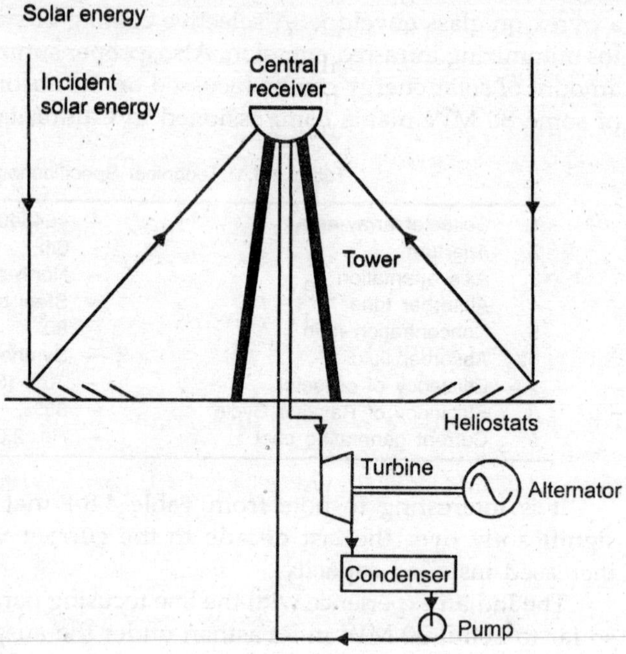

Fig. 4.8.3.2.1: Central Receiver Power Plant

Although these power plants have been demonstrated successfully, they are characterized by high capital investment. For example, one of the largest of these power plants, SOLAR ONE at Barstow, California costed approximately $14,000/KWh, the salient technical features of which are presented in Table 4.8.2.

Table 4.8.2.: Technical Information on Solar One

Output	:	10 Mwe.
Location	:	Barstow, California, USA.
Mirror field	:	1818 heliostats each having 39.3 m^2 area comprising a total reflective area 71447 m^2.
Central tower	:	80 m high.
Receiver	:	External type (7 m diameter, 13.5 m high, produces super heated Steam at 102 bar, 510°C).

On the basis of aforesaid classified description, it is easy to find that both the flat plate collector system as well as the paraboidal dish collector are not suitable for large scale power generation. They are essentially suitable as stand-alone systems. On the other hand, the line focussing parabolic collector technology, the solar ponds and central receiver systems have shown adequate potential for large scale power generation. One of the low temperature systems involving the solar chimney concept has so far been tried out only on small scale demonstration levels and are also far from being cost effective. Considering the present status, the line focussing parabolic collector technology stands out to be reasonably promising.

4.9 AN INTEGRATED SOLAR DEVICE

Solar devices like solar water heater, cooker, drier and stills have been discussed in details in the previous sections. Most of these have been developed by several R&D institutes and entrepreneurs and are commercially available in various forms. Significantly, these appliances have not been adopted to the extent it should have been for supplementing the energy requirements of rural and urban areas. Well known reasons for this situation are rooted principally in a few of their operational difficulties, non availability of technical personnel for maintenance and sometimes the higher system cost. It is also imperative that any of these devices cannot be used throughout all the seasons at a particular Indian site. For example, the SWHS are useful during the winter months at major locations while solar driers are applicable only when fruits and vegetables are available. A solar cooker also sits idle during monsoon months and other intermittent days. Moreover, three different devices for three separate domestic purposes consume a much bigger area for installation and use. With these constraints in view clearly, some dual purpose solar appliances viz, solar drier cum water heater [P.C. Pande et al., energy conversion and management, 31(5), 419-424, 1991], solar cooker cum drier [P.C. Pande et al, Int. J. of energy 12, 539-545, 1988] and solar cooker cum still [P.C. Pande et al., NSEC-1997, Chennai] have been designed, fabricated and rigorously tested. The experience gained by the R&D group at Central Arid Zone Reseach Institute was accumulated to develop an integrated Three-in-one solar device with a view to using the same device for water heating, cooking and drying of fruits and vegetables. A schematic of the integrated device is shown in Fig. 4.9.1. and the performance in different modes is summarized in Table 4.9.1.

Table 4.9.1: Performance of Integrated Device in Different Modes

Mode of operation	*Ambient temp (°C)*	*Max temp (°C)*	*Solar radiation (KWh/m²/day)*
As water heater	21	54 (water)	5.6
As cooker	25	120 (cooking tray)	5.8
As drier (loaded with 2 kg spinach)	27	65 (air)	6.1

As a solar cooker the stagnation temperature without load was found to be above 120°C. On loading with 1 kg of water at 11 am the water temperature reached 98°C within two hours during winter month while it took only one hour during summer. Food (250 gm rice + 250 gm lentils) could be cooked within two hours. Familiar results of use as solar drier are the following:

(a) 2 kg of spinach could be dehydrated from 90% moisture content to 3% within 28 hours.
(b) 3 kg ber could be dehydrated from 80% moisture content to 8% in some 90 hours during February.

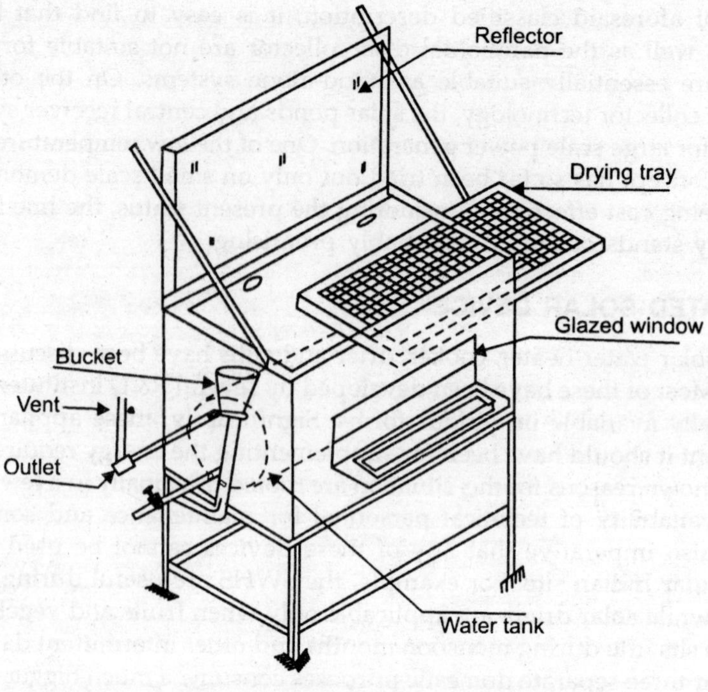

Fig. 4.9.1: Schematic Diagram of Integrated Solar Device (After Pandey, Proceedings of NREC-2001)

Interestingly, when the device was used as a solar drier, the dehydration process continued even during night through the heated water. The cost of the system has been estimated around Rs. 6,000. It can save some 200 KWh equivalent electricity in a year as a water heater, about 25-30% fuel for cooking and also 80 to 100 kg of fruits/vegetables could be dried.

EXERCISE

1. With a neat sketch, mention the different basic components of a typical solar water heating system. Classify it. Discuss in detail the working of forced circulation water heating systems.
2. How the temperature of the water in the load side of a solar water heating system can be controlled? What are the important parameters considered during the installation of a solar collector? Discuss in brief how those parameters affect the performance of the collector.
3. In a solar water heating system of 50 litres per day capacity, the water is generated at 55°C. The ambient temperature is 28°C. The solar insolation is 5.2 KWh/m^2/day. The system is supposed to replace an electrical geyser. If the efficiency of the solar collector is 45%, Calculate (i) the required collector area and (ii) the pay back period. Consider the cost of solar hot water system as Rs.35,000/-. Life of the system is 12 years, sunny solar days per year is 250. Efficiency of electrical heater is 75%. Cost of electricity per unit is Rs. 3/-.
4. With the neat sketches, describe in brief how the water can be pumped by using a solar thermal energy conversion system and a solar photovoltaic system.
5. How the solar energy can be utilized for the purpose of cooling? Describe in detail the working of continuous vapour absorption cooling system and intermittent absorption cooling system. Why the absorption cooling system is considered to be more practical for the purpose of cooling by utilizing solar energy?

6. Discuss briefly with the aid of block diagrams the scheme for employing solar energy for pumping of water.
7. With a neat sketch, explain the operation of a solar still and list briefly the factors affecting its performance.
8. What are Solar Air Heaters? How do these differ from solar water heating systems? What are the principal reasons of interest in packed bed air heaters?
9. (i) Identify clearly the constituents of a Solar Kilns and mention the names of two different types of Kilns.
   (ii) Estimate the collector efficiency of a Solar Kiln where Kiln temperature attained is 60°C, ambient temperature is 36°C, surface area of collector is $25 \times 300 \times 360$ mm^3 and radiation intensity is 840 w/m^2.
10. (a) Give only the names of
    (i) Different designs of solar cookers developed so far
    (ii) Insulating materials employed between the blackened tray and outer cover of the box.
    (iii) Large arrays of mirror reflecting solar radiation in high temperature power cycles.
11. What are solar thermal power plants? How are these classified? Use a neat sketch to describe briefly the operation of a low temperature power production scheme.
12. Write short notes on:
    (a) Solar chimney power plant.
    (b) Salt gradient solar ponds.
    (c) Medium temperature solar thermal power cycles.

SUMMARY

This chapter has dealt with the detailed discussions on the different types of solar water heating systems in relating to use in practice. The discussion has also been carried out in the area of effective installation of the collector. The economic analysis of the solar water heating systems has been presented to facilitate to expose the readers in computing the different components e.g. saving of electricity consumption or fuel energy consumption per day, pay back period, extended period for free energy etc. Solar air heaters, solar kilns, solar cookers, solar stills, solar pumps and solar cooling systems have been discussed in detail. The description of the respective systems have been presented and their application have also been mentioned. During the discussion, their use as rural energy systems have also been stated in the appropriate places. The working principle of solar ponds has been explained illustratively along with the relevant typical data. Different types of power plants using the solar energy through the thermal route have been explained. The recent approach in using the integrated solar devices in urban, semi-urban and rural areas has been highlighted along with the typical data regarding the performance of an integrated Three-in-one solar device in different modes, used for heating of water, cooking and drying of fruits and vegetables.

REFERENCES

1. G.N. Tiwari and S. Suneja, Solar Thermal Engineering Systems, Narosa Publishing House, New Delhi, 1997.
2. S.P. Sukhatme, Solar Energy-Principles of Thermal Collection and Storage, (2/e) Tata Mc-Graw Hill, 1996.
3. C. Palaniappan, Fish Drying Through Solar Heating, pp. 19-23, IREDA, NEWS, vol. II, no. 4, Oct-Dec 2000.

4. S. Rajagopal, Renewable Energy for Rural Development, pp. 37-43, IREDA NEWS, vol. 10, no. 4, Oct-Dec 1999.
5. B.K. Saxena, Energy Efficient Buildings and Renewable Energy in the Building Sector, pp. 81-86, IREDA NEWS, vol. 10, no. 4, Oct-Dec 1999.
6. K.M. Mital, Non-conventional Energy Systems-Principles, Progress and Prospects, Wheeler Publishing, New Delhi, 1997.
7. J.S. Hsieh, Solar Energy Engineering, Prentice Hall, New Jersy, 1986.
8. J. Nagaraju, et al., 1 MWth Industrial Hot Water System for Egg Powder Plant at Hyderabad, Renewable Energy Applications to Industries, Narosa Publishing House, pp. 73-86, 1998.
9. A. Swarup, et al., Performance Evaluation of one of the largest Solar Water Heating Systems in the World, SESI Jl., 7(2), pp. 69-73, 1997.
10. Manual for Solar Thermal Systems, IREDA Ltd., New Delhi, 1997.
11. T.K. Dhamodaran, et al., Solar Kiln for Timber Drying, pp. 57-60, IREDA NEWS, vol. II, no. 1, Jan-Mar 2000.
12. I.A. Fewape and J.A. Fewape, Construction and Evaluation of A Timber Drying Solar Kiln, Bio-resource Tech., 52, pp. 283-285, 1995.
13. A.E. Akachuku, Solar Kiln Dryers for Timber and Agriculteral Crops, Int. Jl. of Ambient Energy, 7(2), pp. 95-101, 1986.
14. C. Chowdhuri and H.P. Garg, Evaluation of A Jet Plate Solar Air Heater, Solar Energy, 46, p. 199, 1991.
15. C. Choudhuri, et al., A Solar Air Heater for Low Temperature Applications, Solar Energy, vol. 40, pp. 335-343, 1988.
16. C. Palaniappan and S.V. Subramanian, Performance of 500 m^2 Area Solar Air Heater for Spices Drying, Renewable Energy Application to Industries, Narosa Publishing House, pp. 38-47, 1998.
17. J.A. Duffie and W.A. Beckman, Solar Engineering of Thermal Processes, Wiliey-Interscience, NY, 1991.
18. G.N. Tiwari, et al., Design Parameters of a shallow Bed Solar Crop Dryer with reflector, Energy Conversion and Management, 35 (6), pp. 535-542, 1994.
19. C. Tiris, et al., Experimental Testing of A New Solar Dryer, Int. Jl., of Energy Research, 18 (4), pp. 483-490, 1994.
20. J. Nagaraju, A Solar Tobacco Drier, SESI Journal, 7 (2), pp. 81-88, 1997.
21. L. Varshney and J.S. Saini, Heat Transfer and Friction factor correlation for Rectangular Solar Air Heater Duct Packed with wire mesh screen Matrices, Solar Energy, 62 (4), pp. 255-262, 1998.
22. L.H. Gunnewick, et al., Flow distribution in unglazed transpired plate Solar Air Heaters of large area, Sol. Energy, vol. 58, no. 4-6, pp. 227-237, 1996.
23. C. Choudhuri and H.P. Garg, Performance of Air Heating Collectors with Packed Air Flow Passage, Sol. Energy, 50 (3), pp. 205-221, 1993.
24. K.B. Muluwork, et al., Studies on discrete Rib Roughened Solar Air Heaters, Proc. of NSEC-98, pp. 75-84, University of Roorkee, 1998.
25. G.N. Tiwari and R.K. Goyal, Green House Technology, Narosa Publishing House, New Delhi, 1998.
26. Amita Gupta and G.N. Tiwari, Performance Evaluation of Greenhouse for Different Climatic Zones of India, SESI Journal, 12 (1), pp. 45-57, 2002.
27. TERI, Solar Cookers in India: A Profile, TIDE, 4 (1), pp. 18-20, 1994.
28. G. Naik and G. Sharan, Assessing Consumer Preference for Product Feature Selection, Solar Cookers, Vikalpa, 22 (4), pp. 49-54, 1997.
29. G. Sharan and T.K. Choudhuri, Estimating Strength of Electrical Back up for Box Solar Cooker Using Simulation, SESI Journal, 8 (2), pp. 69-78, 1998.
30. S.K. Philip, et al., Testing of Solar Box Cookers, 3rd Int. Conference on Solar Cookers—Use and Technology, Coimbatore, 1997.
31. T.K. Chowdhuri, Estimation of Electrical Back up for Solar Box Cooker, Renewable Energy, 17, p. 569, 1999.
32. A.K. Binark and N. Turkmen, Modelling of A Hot Box Solar Cooker, Energy Conversion and Management, 37, p. 303, 1996.
33. R.S. Patel, et al., Theoretical-cum-Experimental Studies on the Performance Parameters of A Box Solar Cooker, Proc. of 26th NREC-2002, Coimbatore, India, pp. 176-181, Jan 2003.

34. M.A.S. Malik, et al., Solar Distillation, London, Progamon Press, 1982.
35. S.D. Gomkale, Operational Experience with Solar Stills in An Indian village and their contribution to the Drinking Water Supply, Desalination, 69, pp. 177-182, 1988.
36. G. Sharan and S. Kumar, Feasibility of Integrating Solar Desalination with Polyhouse in Kach, SESi Jl., 8(1), pp. 57-68, 1998.
37. M. Fard and F. Hamad, Performance of A Solar Still, Proc. of Renewable Energy, vol. 3, no. 1, pp. 75-83, 1993.
38. N. Nijegorodob, Thermal-Electrical, High Efficiency Solar Stills, Renewable Energy, vol. 4, no. 1, pp. 123-127, 1994.
39. T.V. Arjunarn, Design and Performance Evaluation of A Solar Still, Proc. of 26th NREC-2002, Coimbatore, India, pp. 148-155, Jan 2003.
40. C.Wu, et al., Optimization of Steady Flow Heat Pumps, Energy Conversion and Management, 39 (5/6), pp. 445-453, 1998.
41. S.C. Kaushik, et al., A Parametric Study Based on Finite Time Thermodynamic Analysis of A Solar Engine Driver Air Conditioning System, SESI Journal, 10 (2), pp. 63-78, 2000.
42. K. Kant, et al., Thermal Comfort in A Multistoried Building in Delhi using Evaporative cooling: Effect of Air Change Rate and Bypass Factor, SESI Journal, 10 (1), pp. 27-39, 2000.
43. D.K. Dixit, et al., Analysis and Simulation of an Aqua-Ammonia Vapour Absorption Refrigeration System, SESI Journal, 2 (2), pp. 81-89, 1992.
44. A. Venkatesh, Solar Cooling Devices, Proc. of Three day Seminar on Solar Energy-Latest Technologies from Germany and India, MaxMueller Bhavan, Madras, pp. 35-51, Dec. 15-17, 1994.
45. A. Venkatesh, Solar Refrigeration and its Applications, Renewable Energy Applications to Industries, pp. 87-94, Narosa Publishing House, 1998.
46. TERI, Solar Pond Technology: A Profile, TIDE, 3 (4), p. 348, Dec 1993.
47. TERI, Asia's First Solar Pond Commissioned in India, TERINIS, 2 (18), May 1994.
48. T.A. Newell, et al., Characteristics of a Solar Pond Brine Reconcentration Systems, Jl. of Solar Energy Engg., 116, pp. 69-73, 1994.
49. MNES, Annual Report, 1995-1996.
50. J. Srinivasan and A. Guha, The Effect of Bottom Reflectivity on the Performance of a Solar Pond, Solar Energy, 39, p. 361, 1987.
51. E. Wilkins, Operation of a Commercial Solar Gel Pond, Solar Energy, 46, p. 383, 1991.
52. Z. Harel, et al., The Equilibrium Solar Pond; A Laboratory Model for the Gradient Layer, Jl. of Solar Energy Engg, Trans. ASME, 115, p. 32, 1993.
53. F.A. Banat, Carnalite Salt Gradient Solar Ponds: An Experimental Study, Jl. of Renewable Energy, 4, pp. 265-269, 1994.
54. M. Lodhi, Solar Ponds in Alkaline Lake and Oil Well Regions, Jl. of Energy conversion and Management, 37 (12), pp. 1677-1694, 1996.
55. M. Hassairi, et al., Natural Brine Solar Pond: An Experimental Study, Solar Energy, 70 (1), pp. 45-50, 2001.
56. R. Jayaprakash, et al., GEL Shallow Solar Pond: Simulation Model, Proc. of 26th NREC-2002, Coimbatore, India, pp. 182-188, Jan 2003.
57. P.C. Pandey and K.P. Thanvi, Design and Development of Solar Dryer cum Water Heater, Energy Conversion and Management, 31 (5), pp. 419-424, 1991.
58. P.C. Pandey and K.P. Thanvi, Design and Development of Solar Cooker cum Dryer, Int. J. Energy, 12, pp. 539-545, 1988.
59. P.C. Pandey, Design and Development of Solar Cooker cum Still in towards commercialization of Clean Energy, Proc. of 21st National Solar Energy Convention, Chennai, pp. 322-326, Dec 1997.
60. P.C. Pandey, Design and Development of An Integrated Solar Device, Proc. of 25th National Renewable Energy Convention, Osmania University, Hyderabad, pp. 88-93, Dec 2001.
61. D. Sakthi, et al., Development of Hybrid Solar Collector-Cooker cum Dryer, Proc. of 26th NREC-2002, Coimbatore, India, pp. 189-194, Jan 2003.
62. R.Z. Wang, M. Li, Y.X. Xu and J.Y. Wu, An Energy Efficient Hybrid System of Solar Powered Water Heater and Adsorption Ice Maker, Solar Energy, vol. 68, no. 2, pp. 189-195, 2000.

5

Solar Photovoltaic Devices and Systems

INTRODUCTION

When photon energy initiating from the sun (Solar Energy) falls on a solar cell, a fraction of the incident energy with appropriate wavelengths is converted into electricity. The technology of conversion is known as photovoltaics (*PV*) and this necessitates the use of one or other variety of homojunctions, hetrojunctions and metal-semiconductors junctions in bulk and thin film forms. Several solar cells can be interconnected to form *PV* modules, which are the basic building blocks for a Solar Photovoltaic System (SPS).

These *PV* systems appear in stand-alone, grid connected and hybrid forms to solve medium and large scale non-conventional energy needs in the present day terrestrial scenario. The era of space-age requirements giving berths to the early solar cells, is now history. It now remains to be seen how far the *PV* system prices becomes cost effective in comparison to conventional electricity by using *PV* generators comprising of appropriate thin film (tandem) modules.

5.1 PHOTOVOLTAIC CONVERSION

Becquerel in 1839 first observed that a photovoltage developed when photons were directed on to one of the electrodes in an electrolyte solution. Similar effect was later observed in selenium in cuprous oxide photovoltaic cells, which eventually led to the development of photo electric exposure meter that is a vital component in photography. It was not until 1954 that a group of scientists namely Chapin, Fuller and Pearson from the RCA group demonstrated the photovoltaic effect for converting solar radiation into electrical energy.

The Photovoltaic (*PV*) effect can be defined as the generation of a potential when light (photon) radiation strikes the region in on near the built in barrier of a semiconductor *p-n* junction. It is characterized by a self-generated e.m.f. and the ability to deliver power to a load considering that primary power comes from the striking photons. Two common and simple techniques of realizing a potential barrier are (1) The metal semiconductor or the schotky barrier which can be formed by vacuum evaporation, spraying or sputtering onto a semiconductor, (2) *p-n* junction formed by introduction of a suitable type impurity into the opposite impurity type semiconductor.

The mechanism of creation of photovoltaic potential when photons strike a semiconductor *p-n* junction can be seen from Fig. 5.1.1. and 5.1.2. Fig. 5.1.1. represents the energy band diagram of a *p-n* junction in dark conditions where the amount of band bending for both conduction and valence band is dictated by the amount of contact potential (V_o) and the fermi-level occupies a constant position (E_f) on both *p* and *n* sides. This fermi-level is split up into two quasi fermi levels E_{fn} and E_{fp} on the *n* and *p* sides in Fig. 5.1.2. It is interesting to note here that the creation of a photovoltage (V_{oc}) is given by

$$qV_{oc} = E_{fn} - E_{fp} \quad ...(5.1.1)$$

Now, it is well-known that when a *p* and *n*-type material are brought in contact, the average energy of the carriers (Fermi-level) be maintained at the same level in the p and n sides. Thus, eqn. 5.1.1 indicates how the quasi fermi-level E_{fn} and E_{fp} are responsible for the generation of the open-circuit voltage (V_{oc}). A mathematical result based on equivalent circuit will be presented in 5.1.1.

5.1.1 Equivalent Circuit

Electrical characteristics of a photovoltaic cell can be easily understood by considering a simple equivalent circuit of Fig. 5.1.1.1, which consists of a constant current generator, a non-linear junction impedance

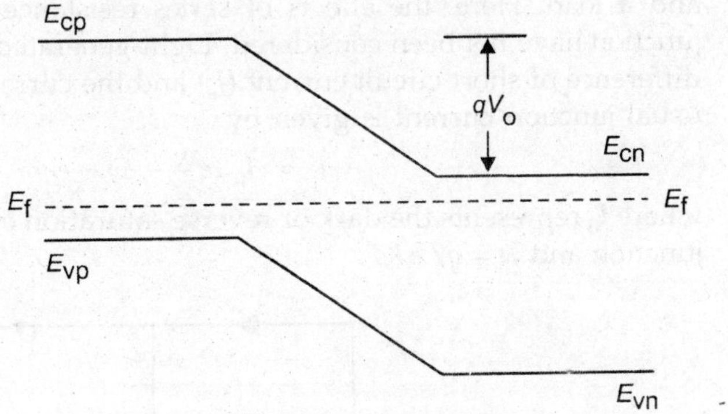

Fig. 5.1.1: Energy-band Diagram of *p-n* Junction at Equilibrium

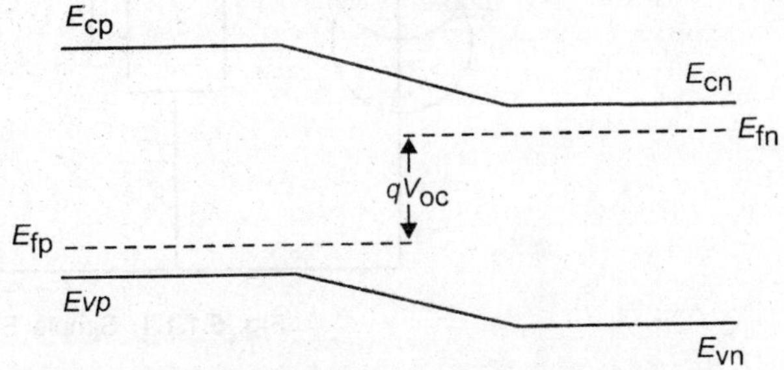

Fig. 5.1.2: Energy-band Diagram of an Illuminated *p-n* Junction

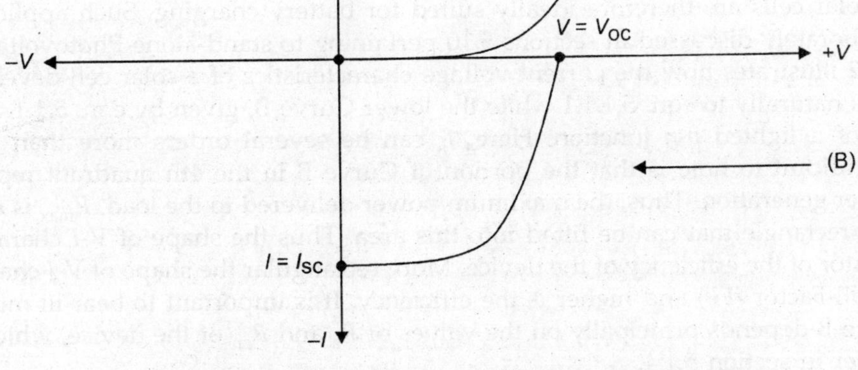

Fig. 5.1.1.2: Effect of Light on I-V Characteristics of *p-n* Junction

and a load. Here, the effects of series resistance (R_s) and shunt resistance (R_{sh}) with the *p-n* junction have not been considered. Light-generated current (I) flowing in the load is naturally the difference of short circuit current (I_s) and the current flowing in the non-linear junction (I_j). I_j, the usual junction current is given by

$$I_j = I_0 (e^{\lambda V} - 1) \qquad ...(5.1.1.1)$$

where I_0 represents the dark or reverse saturation current, V is the photo-voltage generated by the junction and $\lambda = q/AKT$.

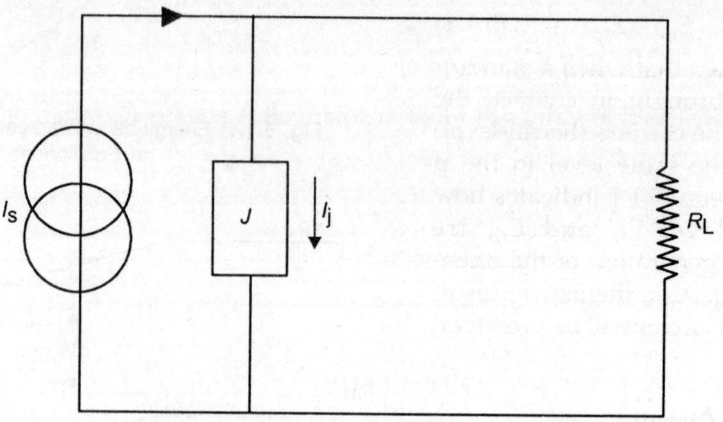

Fig. 5.1.1.1: Simple Equivalent Circuit

Here all the symbols have their usual significance except that A stands for the diode quality factor.

$$\because \quad I_j + I = I_s \qquad ...(5.1.1.2)$$

$$\therefore \quad I = I_s - I_0 (e^{\lambda V} - 1) \qquad ...(5.1.1.3)$$

Eqn. 5.1.1.3 shows that I_s is the maximum light generated current at zero voltage. This is directly proportional to the (a) intensity of available sunlight and (b) the area of the device exposed to light.

Now $V = V_{oc}$ (the maximum open circuit voltage) when $I = 0$. Therefore in eqn. 5.1.1.3

$$V_{oc} = (AKT/q) \ln [(I_s/I_0) + 1)] \qquad ...(5.1.1.4)$$

Eqn. 5.1.1.4 shows that V_{oc} increases logarithmically with short circuit current and intensity of sunlight. Solar cells are therefore ideally suited for battery charging. Such application areas have been elaborately discussed in sections 5.10 pertaining to stand-alone Photovoltaic Systems.

Fig. 5.1.1.2 illustrates how the current voltage characteristics of a solar cell develops. Curve A corresponds naturally to eqn. 5.1.1.1 while the lower Curve B, given by eqn. 5.1.1.4 shows the performance of a lighted *p-n* junction. Here, I_s can be several orders more than I_0. What is extremely significant to note is that the portion of Curve B in the 4th quadrant represents the region of power generation. Thus, the maximum power delivered to the load, P_{max} is represented by the largest rectangle that can be fitted into this area. Thus the shape of *V-I* characteristics is the true indicator of the efficiency of the device. More rectangular the shape of *V-I* characteristics, better is the Fill-Factor (*FF*) and higher is the efficiency. It is important to bear in mind that the shape of Curve-B depends principally on the values of R_s and R_{sh} of the device, which has been introduced later in section 5.1.4.

5.1.2 Fill Factor

The fill factor of the device is best understood from Fig. 5.1.2.1 which indicates V_m as the maximum power point voltage and I_m as the maximum power point current corresponding to Curve I. Here, Curve I represents *V-I* characteristics at lower light intensity while Curve II represents *V-I* characteristics of the same device at higher intensity. V_{oc_1} and V_{oc_2} denotes the open circuit voltages while I_{sc_1} and I_{sc_2} denotes the short circuit currents at the two insolation levels. Fill Factor for Curve I is given by

$$FF = (V_m I_m)/V_{oc_1} \cdot I_{sc_1} \quad ...(5.1.2.1)$$

Eqn. (5.1.2.1) indicates that values of *FF* of a solar cell is always less than one. (It varies from 0.3 to 0.8 for all practical purposes). It is easy to see that for an ideal solar cell, $V_{oc_1} = V_m$ and $I_{sc_1} = I_m$. Here, Fill Factor has a value of unity and the shape of *V-I* characteristics is a rectangle as shown in Fig. 5.1.2.2. For each point on any one of *V-I* curves, the product of current and voltage represents the power output for that Operating condition. The maximum power Output of a cell is graphically the largest rectangle that can be fitted under the *I-V* curve i.e. $d/dV(IV) = 0$ giving

$$V_m = V_{oc} - (AKT/q) \ln [\{V_m/(AKT/q)\} + 1] \quad ...(5.1.2.2)$$

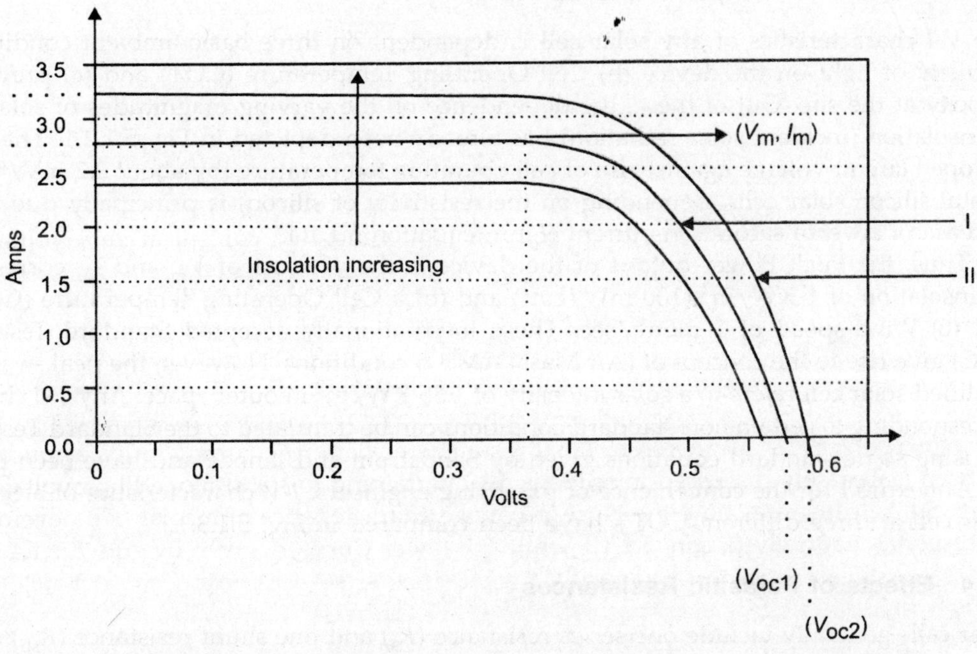

Fig. 5.1.2.1: I-V Characteristics of a Typical Silicon Solar Cell at Different Levels of Insolation

From a knowledge of typical values of V_{oc} for a particular type of solar cell and its diode quality factor (A), Eqn. (5.1.2.2) can be used to estimate the difference between V_m and V_{oc}.

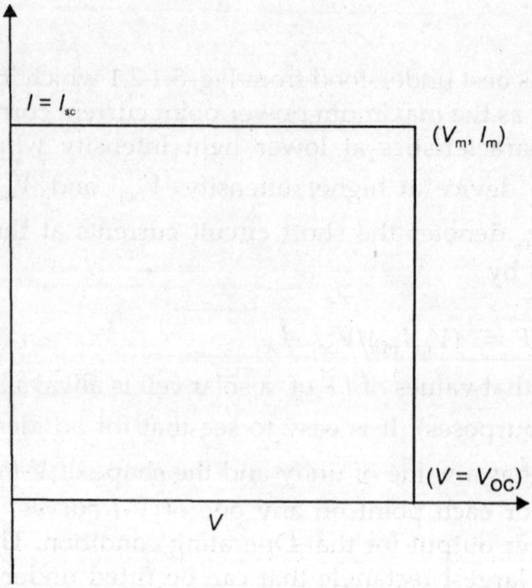

Fig. 5.1.2.2: Voltage Current Characteristics of an Ideal Solar Cell

5.1.3 Peak Power Output of the Cell (pW)

The *V-I* characteristics of any solar cell is dependent on three basic ambient conditions (a) the intensity of light on the device (b) Cell Operating Temperature (COT) and (c) prevailing wind velocity at the site. Out of these, the dependence on the varying magnitudes of solar irradiance or Insolation (incident solar radiation) has already been depicted in Fig. 5.1.2.1. The decrease in the open circuit voltage against rise of cell operating temperature (by about 2.2 mV/°C for single crystal silicon solar cells, depending on the resistivity of silicon) is principally due to the rapid increase of reverse saturation current (I_o) in equation 5.1.1.4.

Thus, the Peak Power output of the device is the product of V_m and I_m corresponding to (a) Insolation of 1 KW/m^2 (100 mW/cm^2) and (b) a Cell Operating Temperature (COT) of 25°C and (c) Wind-speed of 1 metre/sec. These internationally accepted Standard Test Conditions (STC) give rise to the concept of (Air Mass) AM 1.5 conditions. However, the peak-watt of a space qualified solar cell refers to a solar intensity of 1.35 KW/m^2 in outer space. Any *V-I* characteristics corresponding to certain non-standard conditions can be translated to the Standard Test Conditions by using some standard equations given by Sandstrom and Janoch and have been presented in the Annexure I for the convenience of practicing engineers. *I-V* characteristics of a typical silicon solar cell at three different COT's have been compared in Fig. 5.1.3.1.

5.1.4 Effects of Parasitic Resistances

Solar cells generally include one series resistance (R_s) and one shunt resistance (R_{sh}) as shown in the extended equivalent circuit of Fig 5.1.4.1. These are the parasitic resistances which are responsible for a behaviour different from that of an ordinary *P-N* junction. Significantly, both these resistances control the value of the fill-factor of the device to a very large extent. This can immediately be understood from the following two equations used for analytical estimate of R_s and R_{sh}

$$R_s = (dV/dI) \text{ at } V = V_{oc} \qquad \text{...(5.1.4.1)}$$

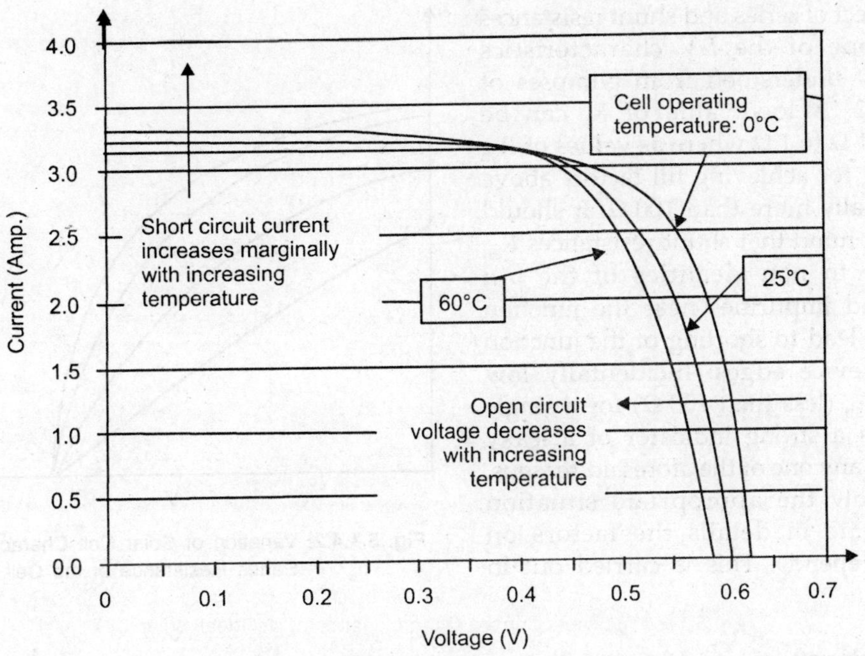

Fig. 5.1.3.1: I-V Characteristics of a Typical Silicon Solar Cell at three Different Cell-operating Temperatures of 0°C, 25°C and 60°C

and $$R_{sh} = (dV/dI) \text{ at } I = I_{sc} \quad ...(5.1.4.2)$$

Eqn. 5.1.4.1 can easily be interpreted as the slope of any *V-I* characteristics at the point of operation $I = 0$; $V = V_{oc}$ while eqn. 5.1.4.2 refers to the slope of the same *V-I* characteristics having the other extreme point of operation i.e. $V = 0, I = I_{sc}$.

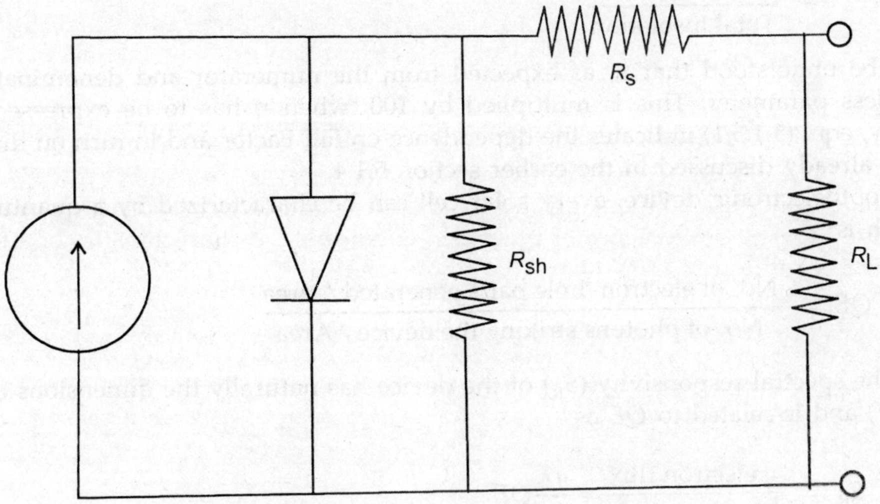

Fig. 5.1.4.1: Role of R_s and R_{sh} in the Equivalent Circuit

The effect of series and shunt resistances on the slope of the *I-V* characteristics can be best understood from glimpses of Fig. 5.1.4.2. A low value of R_s can be between 0.1 Ω to 1 Ω where as values of R_{sh} satisfactory for achieving fill factors above 0.6 are usually more than 100 Ω. It should be borne in mind that shunt resistances R_{sh} arises due to non-idealities of the *p-n* junction and impurities near the junction which may lead to shorting of the junction near the device edges. Incidentally low values of R_{sh} (less than 100 Ω) for a single solar cell is a strong indicator of a leaky junction for any one of the aforesaid reasons.

It is now the appropriate situation to investigate in details the factors on which R_s depends. This is carried out in section 5.2.

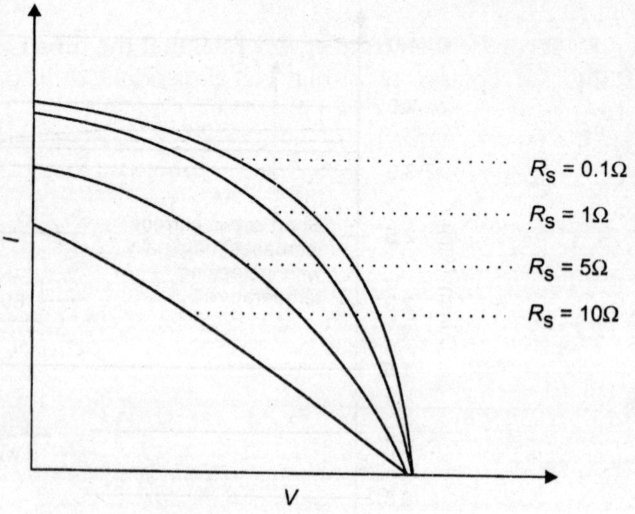

Fig. 5.1.4.2: Variation of Solar Cell Characteristics with Series Resistance of the Cell

5.1.5 Efficiency

There can be two simple ways of having an insight into this important parameter of the device. From the point of view of energy conversion, a solar cell is a device that converts optical energy into electrical energy. Thus, the efficiency (η) should be

$$\eta = \frac{\text{Maximum Power Output}}{\text{Incident Intensity} \times \text{Area of the device exposed}} = \frac{V_m \cdot I_m}{\text{Insolation} \times \text{Area}}$$

$$= \frac{FF \times V_{oc} \cdot I_{sc}}{\text{Total Insolation}} \quad \text{...(5.1.5.1)}$$

It can be understood that η, as expected from the numerator and denominator will be a dimensionless parameter. This is multiplied by 100, when η has to be expressed in percent. Incidentally, eqn. (5.1.5.1) indicates the dependence on Fill Factor and in turn on the parameters R_s and R_{sh} already discussed in the earlier section 5.1.4.

As an optoelectronic device, every solar cell can be characterized by a quantum efficiency (*QE*), which is

$$QE = \frac{\text{No. of electron-hole pairs generated / Area}}{\text{No. of photons striking the device / Area}}$$

Now, the spectral responsivity (S_R) of the device has naturally the dimensions of Amperes/Watt (*A/W*) and is related to *QE* as

$$S_R = \frac{q \text{ electron flux}}{\frac{hc}{\lambda} \text{ photon flux}} = \frac{q\lambda}{hc} QE \quad \text{...(5.1.5.2)}$$

It can be seen from eqn. 5.1.5.2 that the direct wave-length dependence of responsivity makes the performance of a solar cell dependent duly on the spectral content of sunlight.

5.2 GRID DESIGN

To appreciate the factors on which R_s depends, it is important to visualize the effective cross-section of a photovoltaic device or solar cell as presented separately in Fig. 5.2.1.

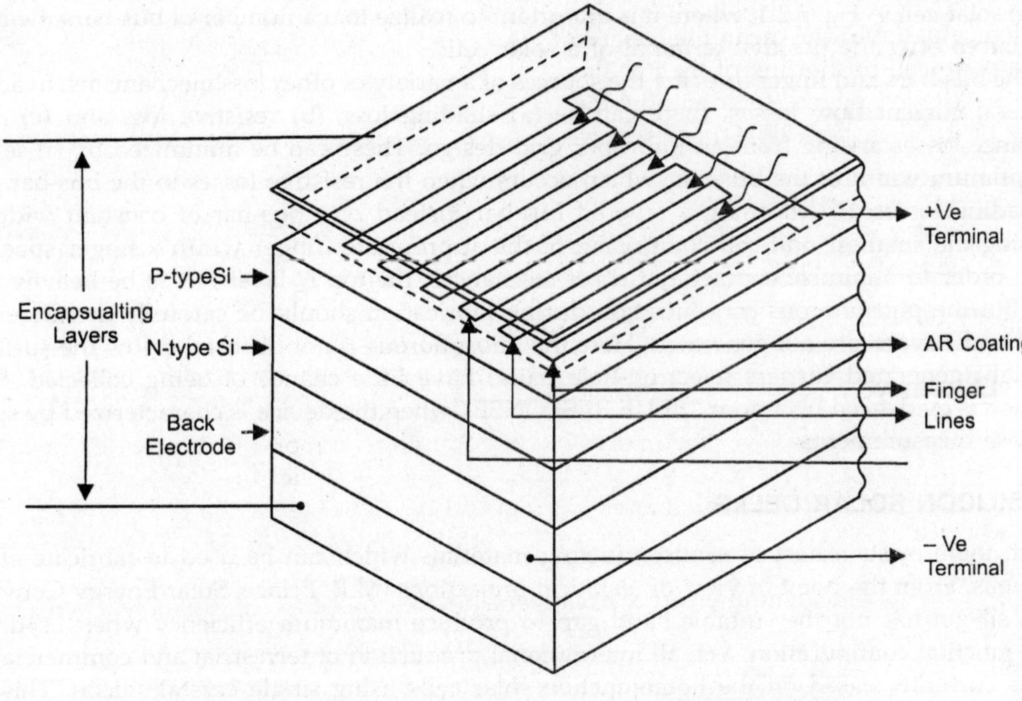

Fig. 5.2.1: Essential Components of an Encapsulated Solar Cell

Factors controlling R_s are—(i) Bulk resistance of the *N* and *P* type semiconductor layers, (ii) Ohmic nature of the front grid and back electrode, (iii) Grid Design or Top Contact Design, which in turn is largely dictated by the sheet resistance of the top semiconductor layer.

The material chosen for the front and back electrodes are particularly important to ascertain whether a metal/semiconductor contact is ohmic or non-ohmic. As familiar examples, aluminium is a very good ohmic contact to P-type silicon where as it is fairly non-ohmic to semiconductors such as CdS, CdTe or Copper Indium Diselenide (CIS) used in hetrojunction solar cells.

Non-ohmicity will lead to higher values of R_s which in turn deteriorates the fill factor and efficiency of the device. Suitable ohmic electrodes to CdS can be the use of Nickel or Indium. A combination like Ti/Pd/Ag is suitabley ohmic for *N*-type silicon. The Grid design of a solar cell (Fig. 5.2.1.) has been a subject of several Post Graduate and Doctoral studies. It is essential that metallic finger lines of appropriate—(a) thickness, (b) width and (c) interline spacing should be fixed on to the top semiconductor layer (on which photons are incident) by techniques like photolithography or laser scribing. A simple and low-cost technique have been practiced by Ghosh (B. Ghosh, Ph. D. thesis, Engg. Faculty of J.U., 1987) on thin film CdS/Cu_2S solar cells in course of his Ph.D. work at Jadavpur University. The number of finger lines used per centimeters on the top semiconductor along with their geometrical considerations is largely dependent on the sheet resistance of the top semiconductor, which can be experimentally determined by the Four Probe measurements. In four probe set-up, sheet resistance can be estimated from voltage and current readings as

$$\text{Sheet Resistance} = \rho = \pi/\ln 2.\ V/I = 4.53\ V/I \qquad ...(5.2.1)$$

Typical values of sheet resistance for silicon solar cells is in the range of 30 to 100 $\Omega/^2$. The combination of a bus-bar and the finger lines forming the Top Contact design can be seen on a silicon solar cell in Fig. 5.2.1, where it is important to realize that a number of bus-bars themselves combine to offer the positive terminal of a solar cell.

The bus-bars and finger lines are the sources of a variety of other loss mechanisms. In addition to lateral current flow losses, there can be (a) shading loss, (b) resistive loss and (c) contact resistance losses arising from an improper grid design. These can be minimized by (i) selecting the optimum width of the bus-bar, which occurs when the resistive losses in the bus-bar equals its shading losses, (ii) choosing a tapered bus-bar instead of a bus-bar of constant width, (iii) choosing the smallest unit cell comprising of the appropriate (finger width × finger spacings).

In order to minimize contact and sheet resistances, the top N layer has to be heavily doped (N^+). But the phosphorous concentration during fabrication should be carefully monitered such that 'Dead Layers' are not generated. Here the phosphorous is not diffused below the surface, so that light generated carriers (electron-hole pairs) have little chance of being collected. Such a situation is evidenced by a poor "BLUE RESPONSE" when the device is characterized by spectral response measurements.

5.3 SILICON SOLAR CELLS

In fact, there exists a host of semiconducting materials which can be used to fabricate efficient solar cells. From the point of view of technical projections (M.B. Prince, Solar Energy Converters, 1955), silicon has not the suitable band-gap to produce maximum efficiency when used in the homo-junction configuration. Yet, all international production of terrestrial and commercial solar cells is currently based on the homojunction solar cells using single crystal silicon. This is so, because—(i) Almost 26% of the earth's crust contains silicon mostly in the form of silica (SiO_2) and silicates. Other semiconductor materials like Ga, In, As, Cd, Tc, etc. are not available in such abundance, (ii) It is a safe material to handle and is environmentally benign, (iii) Silicon device technology has attained the highest degree of technological maturity, (iv) With space and terrestrial applications over three decades, these cells have successfully stood the test of 'Degradation' compared to its hetrojunction and Schotky-barrier counterparts.

Under laboratory conditions using the current state of the art technology, it is possible to produce single crystal silicon solar cells with efficiency values in excess of 30%. However, commercially mass produced cell efficiencies vary in the range of 13 to 20%. The difference can be reconciled by simply considering that, in general, laboratory techniques are unsuited to industry. It is expected that in the recent future, the *PV* industry will be immensely benefited by using modules having efficiencies of 20% or more. This will mean a terrific boost to the cost of *PV* electricity as fewer modules will have to be incorporated corresponding to power output requirements. A typical cross-section of a silicon solar cell has been projected before in Fig. 5.2.1. In spite of several merits, three distinct disadvantages of bulk silicon solar cell are (i) Efficiency value is limited, (ii) The quantity and cost of material consumption is sizeable, (iii) The energy pay back period (EPP) is quite high. Some of these drawbacks have been removed through the photovoltaic R&D Project in progress at National Renewable Energy Laboratory (NREL) and Sandia National Laboratory (SNL).

Researches at the Georgia Institute of Technology and Sandia National Laboratories (SNL) are of the opinion that crystalline silicon material is capable of yielding higher photovoltaic performance than that obtained with commercial fabrication processes. Thus, there is ample opportunity to improve the performance of crystalline Si modules also. Two novel concepts that has been identified for realizing high efficiency and low-cost cell structures are (i) Emitter Wrap Through (EWT) cell and (ii) Self aligned selective emitter cell. The EWT cell has both contacts on the back surface which is achieved by wrapping emitter through laser drilled holes from the front surface to the back surface (Fig. 5.3.1). The back contact geometry has potentially higher performance due to no grid obscuration. Gee et al. (25th IEEE photovoltaic specialist conference) have projected efficiencies more than 20% for 100 cm^2 EWT cells using *PV* grade silicon. Self aligned selective emitter cell uses plasma processing to achieve a low recombination passivated emitter with commercial metalisations. Reasons for achieving lower cost of fabrication with this structure are (a) Use of screen printed grid, (b) Emitter etch is self aligned and (c) Plasma etch and deposition is normally executed in the same chamber. The most significant advantage of the aforesaid configurations is essentially one-step simplification and cost reduction in the *PV* module assembly. The present geometry with contacts on front and back surfaces is difficult to automate. As a result, module fabrication (inclusive of labour and material charges) now accounts for nearly 50% of the finished module cost.

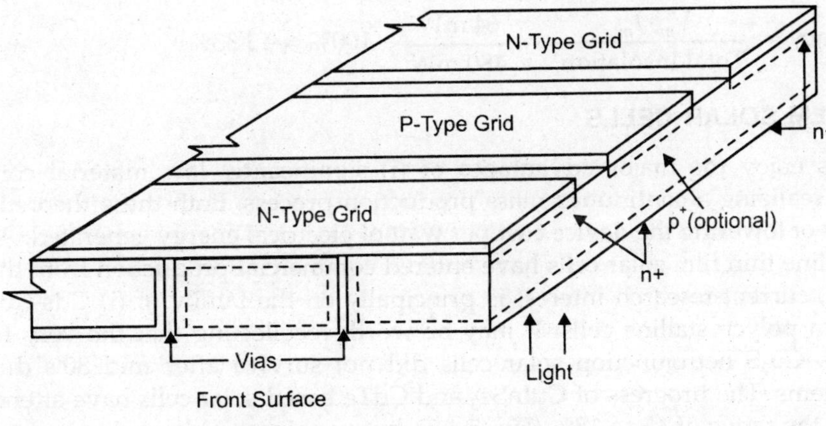

Fig. 5.3.1: Illustration of an EWT cell

5.3.1 Ribbon Silicon Solar Cells

The wafer slicing process with single and polycrystalline bulk silicon accounts for a material loss not less than 40 to 45%. A significant saving in this direction was achieved by eliminating this step and growing Si Ribbons directly from the polycrystalline melt. Optimized growth processes may lead to lower energy consumption, thus reducing the EPP. These advantages have not worked out as the production yields of these cells are extremely low which results in higher cost of production compared to theoretical estimates. Although a technological marvel, the desired commercial status of ribbon silicon solar cells could not be achieved and commercial production and sale of PV modules were discontinued before 1990.

Illustrative Examples (1): A silicon solar cell (band gap 1.2 ev) is uniformly illuminated with monochromatic light of wave-length 800nm and incident intensity 20 mW/cm^2. Given that its quantum efficiency at this wavelength is 0.80, calculate the short circuit current of the cell, if its area is 4 cm^2.

Solution:

Responsivity (S_R) of a solar cell is $S_R = \frac{q\lambda}{h.c} \cdot QE = \frac{1.6 \times 10^{-19} \times 800 \times 10^{-9} \times 0.80}{6.6 \times 10^{-34} \times 3 \times 10^{8}}$

$\because$ S_R has the dimensions of Amperes/Watt,

$$\therefore \text{Short Circuit Current} = \frac{1.6 \times 10^{-19} \times 800 \times 10^{-9} \times 0.80}{6.6 \times 10^{-34} \times 3 \times 10^{8}} \times 80 \times 10^{-3}$$

($\because$ Effective Insolation $= 20 \text{ mW/cm}^2 \times 4 \text{ cm}^2 = 80 \text{ mW} = 80 \times 10^{-3}$ W)

$= 41.37$ mA.

Illustrative Example (2): A hetrojunction solar cell of active area 6 cm^2 gave the following results: V_{oc} = 400 mV, short circuit current = 200 mA under an insolation equal to 0.8 Sun. What is the energy conversion efficiency of the device? (Assume Fill Factor of 80%).

Solution:

Maximum power output of the device $= V_{oc} \cdot I_{sc} \cdot FF = V_m \cdot I_m$

$= 400 \text{ mV} \times 0.2 \text{ Amps} \times 0.80 = 64 \text{ mW}.$

Total Incident Insolation = 0.8 SUN = 80 mW/cm^2 × 6 cm^2 = 480 mW.

$$\therefore \text{Efficiency} = \frac{V_m \cdot I_m}{\text{Total Insolation}} = \frac{64 \text{ mW}}{480 \text{ mW}} \cdot 100\% = 13.33\%.$$

5.4 THIN FILM SOLAR CELLS

These devices enjoy the major advantages of (i) significantly low material consumption, (ii) possibility of realizing a continuous mass production process. Both these theoretically points to the possibility of lowering the device cost per watt of electrical energy generated. A small amount of polycrystalline thin film solar cells have entered commercial production from the beginning of 1990's, where, current-research interest is principally on the family of (i) Cds-$CuInSe_2$ (ii) Cds-CdTe thin film polycrystalline cells. It may be worth recollecting that the very first alternative family of Cds–Cu_2S hetrojunction solar cells did not survive after mid 80's due to unsolved stability problems. The progress of $CuInSe_2$ and CdTe based solar cells have attended laboratory efficiencies in the range of 15 to 18% (Fig. 5.4.1), but nagging problems do persist with regard to their commercial exploitation. A current status of *PV* parameters for different families of solar cells is presented in Table 5.4.1. A schematic diagram of a typical $CuInSe_2$/CdZnS thin film hetrojunction cell can be seen in Fig. 5.4.2. This device has yielded higher efficiency compared to the CdS/Cu_2S counterpart as the (i) stoichiometry of $CuInSe_2$ is responsible for a higher short circuit current and (ii) the lattice mismatch between CdS and ZnS is also very low so as to yield a higher open circuit voltage. CdTe, when chosen to replace Cu_2S yielded higher efficiency in view of its near optimum direct band-gap (1.5 eV) and higher value of absorption coefficient.

As can be seen from the common examples of thin film solar cells, these are essentially hetrojunctions, which are poised with degradation problems right from their birth even if they are encapsulated with great care. The term 'Degradation' indicates a deterioration of some or all the essential *PV* parameters like I_{sc}, V_{oc}, FF etc. with the passage of time. This effect, unfortunately, is also persistent in amorphous thin film silicon solar cells to be discussed in the next section.

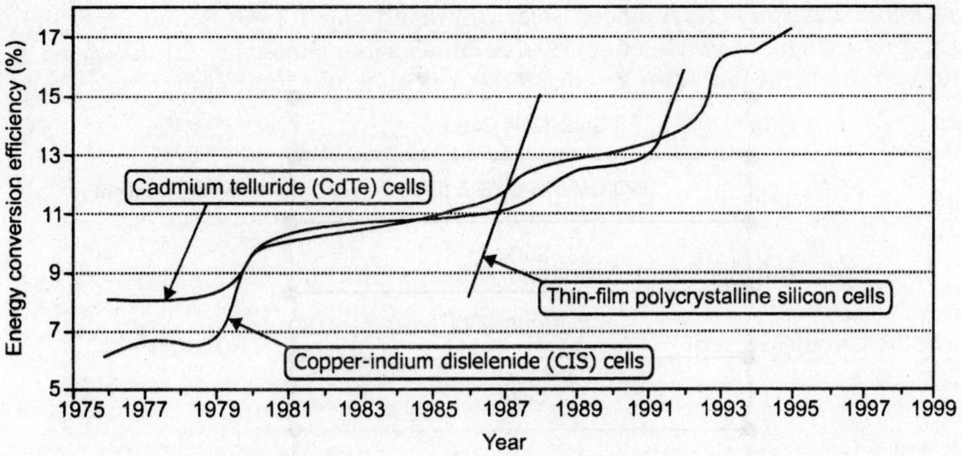

Fig. 5.4.1: Achieved Laboratory Conversion Efficiencies of Various Thin-film Solar Cells

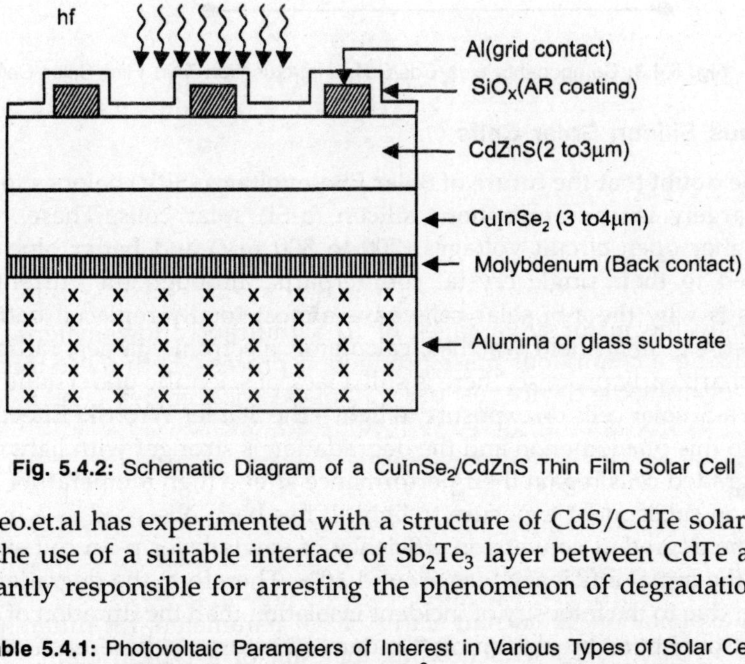

Fig. 5.4.2: Schematic Diagram of a $CuInSe_2/CdZnS$ Thin Film Solar Cell

Recently N. Romeo.et.al has experimented with a structure of CdS/CdTe solar cell presented in Fig. 5.4.3. Here, the use of a suitable interface of Sb_2Te_3 layer between CdTe and Molybdenum has been significantly responsible for arresting the phenomenon of degradation.

Table 5.4.1: Photovoltaic Parameters of Interest in Various Types of Solar Cells [Global AM 1.5 spectrum—1000W/m^2 at a temperature of 25°C]

Classification (MCZ)	J_{SC} *(mA/cm^2)*	V_{OC} *(volts)*	*FF*	E_{ff}	*Test Centres*
Si (Crystalline)	41.6	0.704	83.5	~25	Sandia
Si (Thin film transfer)	30.6	0.634	80.3	~16	University of Stuttgart
CIGS(Thin film)	35.7	0.669	77.0	~19	NREL
PEC Cells (Nanocrystalline dye)	19.4	0.795	71.0	~11.5	EPFL
CdTe	25.9	0.848	74.5	16.9	NREL
a-Si	19.4	0.887	74.1	13.0	NREL Sanyo
GaInP/GaAs	14.22	2.488	85.6	30.3	Japan Energy (Monolithic)

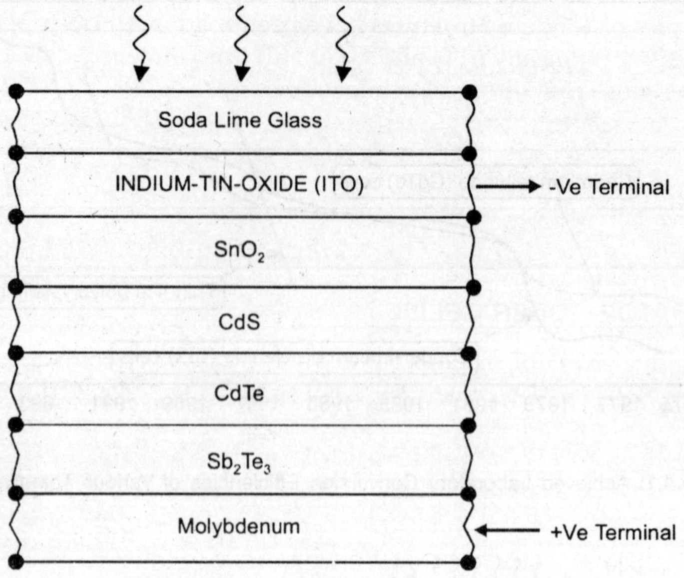

Fig. 5.4.3: Components of a Cds/CdTe Hetrojunction Thin Film Solar Cell

5.4.1 Amorphous Silicon Solar Cells

There is very little doubt that the future of Solar Photovoltaics (SPV) belongs to the thin film solar cells—perhaps largely in the amorphous silicon (a-Si) solar cells. These cells have initially demonstrated higher open circuit voltages (700 to 800 mV) and better blue response (5000 to 6000°A) compared to their single crystal counterparts, although the current density was not appreciable. This is why the a-Si solar cells have almost totally replaced batteries in consumer electronics market, e.g. hand-held watches, calculator, electronic games, radio receivers.

In spite of a sharp early promise, there is still a lack of scientific understanding of the problem of degradation in a-Si solar cells on exposure to light – the Stabler Wronski Effect. All hydrogenated a-Si is subjected to this phenomenon and the degradation is stronger with light of higher intensity. Although the degraded cells regain their performance after a high temperature annealing, it again starts degrading on subsequent exposure to light. It has been observed as a matter of laboratory experience that the degraded conversion efficiency of single layer p-i-n cell stabilized at about 4 or 5%, although the initial efficiency was around 10%. Therefore, the degradation effect is much more pronounced due to the intensity of incident insolation than the duration of exposure. Solution to this has been sought for by designing Tandem structures, where each of the two or more separate a-Si cells tackles only a fraction of the incident sunlight.

5.5 TANDEM STRUCTURES

The word 'tandem' means an intelligent combination. Accordingly, Tandem Solar Cells involves a number of structures, each of whose band-gap has been tailored to match photons of a certain wave-length range only. The idea is to obtain a uniform Spectral Response characteristics compared to non-uniform one from a single solar cell. This means higher short circuit current leading to higher efficiency. Another method of increasing the efficiency of *PV* devices is the use of concentrator cells which will be discussed in the next section 5.6.

Several examples of Tandem Structures have been tried out terrestrially with improvement of cell efficiency. These are typically (i) Si and GaAs, (ii) upto three p-i-n structures of a-Si solar cells, (iii) other combinations from Gr. II-VI semiconductors.

Out of these, the second variety demonstrated an efficiency of 13% initially with an active area of 1 cm^2 with the very important promise of minimizing the degradation problem in outdoor applications. Unfortunately, the difficulty creeping in with basic p-i-n structures is a higher series resistance contributed to the equivalent circuit of the device. Consequently, this sets a limit to the production of a-Si solar cells of large surface area and consistent quality.

5.6 CONCENTRATOR SOLAR CELLS

As understood from a study of sections 5.3 to 5.4.1, the *PV* devices can be categorized as silicon, hetrojunction/thin film and a-Si solar cells. Till now, it has not been mentioned whether these devices have been tested under normal available sunlight or concentrated sunlight. Terrestrial solar cells designed for use in normal outdoor sunlight are called Flat-Plate Solar Cells while those meant for use in concentrated sunlight (principally the space solar cells) are called concentrator cells.

Commercially available large-area terrestrial solar cells are quite unsuitable as concentrator cells. The features which has to be incorporated into these cells are:

(i) To operate efficiently under an input energy density around 100 × 1KW/m^2, the internal series resistance should be designed much lower than terrestrial cells so as to handle a maximum cell current density of 3 to 4 amps/cm^2.

(ii) Condition (i) for above cells needs for special attention to be paid in grid designing. Here, the grid design results in a higher grid coverage, often as high as 30% against only 5 to 10% for flat plate solar cells.

Concentrator cell operation is accompanied by the nagging problem of rise of cell operating temperature leading to a reduction of energy conversion efficiency. Thus, efficient integral heat sinking arrangements are essential for operation. Such arrangements along with the need for continuous tracking and control mechanism makes concentrator *PV* modules very complex and unsuitable for use with small capacity simple stand alone *PV* systems. Such situation will be clarified in sections dealing with *PV* applications.

Although concentrator cells have been successfully fabricated with Silicon and GaAs semiconductors, GaAs have an edge over silicon as it exhibits lower dark current at higher operating temperatures. Concentration ratios with silicon cells are generally restricted around 100 while that for GaAs under laboratory conditions has been as high as 1000.

5.7 PHOTOVOLTAIC (PV) MODULES

In practice a single solar cell cannot be used for outdoor energy generation. Irrespective of its size and shape, it can generate a maximum open circuit voltage of 0.6 volts. This voltage would be further reduced when power is delivered across a load. For example, a 4 inch (100 mm) diameter single crystal silicon solar cell can produce a peak power of 1 pW under STC. It is essential to have a fixed number of cells interconnected in series. A suitably interconnected series string of solar cells when hermetically sealed with a transparent cover and a durable back cover constitutes a solar *PV* module. It has been the accepted international practice to connect 34 to 36 silicon solar cells in series so that the resulting module is capable of charging a 12 volt storage battery. (The value of V_m lying around 15v to 17v at STC without considering temperature effects). It is easy to understand that such a module must have the following features:

(a) It must withstand all possible outdoor conditions such as rain, moisture, dust, pollutants and undesirable gases.
(b) It has to be environmentally benign and asthetically pleasing.
(c) Interconnected solar cells must be well protected to withstand external and accidental shocks leading to micro-cracks (not visible to bare eye) and open circuit failures;
(d) Air-gap/Air-pockets inside the module must be avoided so that cell operating temperature (COT) does not exceed the ambient temperature due to green house effect.
(e) Lamination Technology should be cost-effective and easily adaptable to mass production.
(f) It must be lightweight for ease of transporting. A schematic representation of the LAY-UP details of a *PV* module before lamination is presented in Fig. 5.7.1. A white glasssheet is first kept upside down on the Table Top, so that its inner surface faces upwards. A pre-sized sheet of Ethylene Vinyl Acetate (EVA) is then placed on the top of inner surface of this glass. The solar cell string is next placed upside down on top with the active surface facing down. This is followed by placing a sheet of pre-sized crane glass. Another pre sized sheet of EVA is placed on the back surface of the crane glass. Finally a pre-sized sheet of Tefzel (Tedlar Polyster-Tedlar) is placed on top. The positive and negative electrical terminals of the interconnected string of cells are then manually brought out through a pair of perfectly aligned holes in the last three layers.

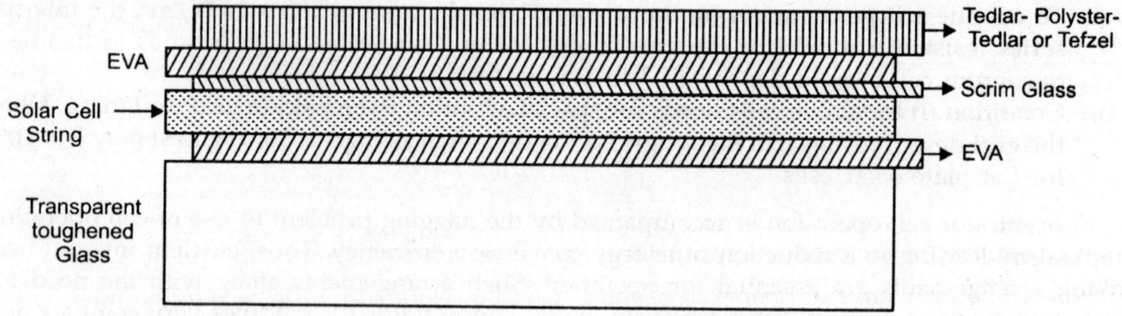

Fig. 5.7.1: Schematic Representation of the Lay-up Details of a PV Module before Lamination

5.7.1 Field Degradation

A solar *PV* module is normally required to satisfactorily perform outdoors for at least two decades. A number of failure mechanisms arise when modules are combined either in series or parallel to form a panel and will be elaborately discussed under the cover of 'Mismatch losses' in section 5.8.1. Apart from these, a module on a singular basis may be subjected to the following 'Degradation' mechanisms:

(A) *Short Circuited Solar Cells:* Inside a module, it is a standard procedure to connect each cell to the bottom of next cell as a part of series connection. This is followed for all cells except the two end cells of the series string. Module degradation occurs when such an interconnection gets shorted with the next cell.

(B) *Open Circuited Solar Cells* may be caused by—(i) Micro-cracks in cell surfaces, (ii) opening or rupturing of cell surfaces.

(C) *Soiling of Module Front Cover* will result in reduced transmission and degraded value of I_{sc} and V_{oc}, if not cleaned periodically.

(D) Failure of associated components.

5.7.2 Packing Factor

PV modules differ from one another in respect of the type of Solar Cells used in them and the number of them interconnected in series. Packing factor is the ratio of the total area of all the interconnected solar cells to the total area of the front glass cover of the module. This is obviously less than unity and the power output (*PW*) of a module is directly dependent on this factor. The type of solar cells in use with a module are—

(i) Nominal 100 mm diameter single crystal silicon.

(ii) Off-square or pseudo-square single crystal silicon with maximum edge to edge length of 100 mm (shaped from 125 mm diameter round cells).

(iii) Numerical 100 mm × 100 mm square polycrystalline silicon.

(iv) Numerical 125 mm × 125 mm (or larger) square polycrystalline silicon.

(v) Edge defined film fed (EFG) poly-silicon ribbon cells where commercial production and marketing has been discontinued after 1988.

5.7.3 Temperature Effects

Module temperature affects naturally output current, voltage and power. Generally, for modules rated at standard test conditions (STC), the reference temperature to be borne in mind by a designer is 25°C. The open circuit voltage of each cell decreases by about 2.3 mV per degree centigrade rise of ambient temperature. Thus when a standard *PV* module manufactured by BHEL or CEL is to be used in a *PV* panel and subsequently in arrays during summer months in India, the operating module voltage at a typical ambient temperature of 45°C = 17 volts – 36 × 2.3 × (45 – 25) mV = 15.344 volts.

Illustrative Example: Assuming that each of the single crystal silicon solar cell delivers an open circuit voltage of 600 mV under STC, estimate the actual open circuit voltage of a non-standard module containing 18 identical interconnected cells at an ambient temperature of 40°C.

Solution: The difference of temperature that has to be tolerated by the cells in the module = 40 – 25 = 15°C. The open circuit voltage of the module without temperature correction = 18 × 0.6 volts. Actual open circuit voltage at an ambient temperature of 40°C = 18 × 0.6 – 18 × (2.3 mV/°C × 15°C) = 18(0.6 – .034) volts = 18 × 0.566 volts = 10.188 volts.

5.7.4 Module Efficiency (η_M)

The energy conversion efficiency of a module can in principle, be estimated in the same way as the efficiency of a single solar cell shown in section 5.1.5. However, the actual value to be adopted in the design of photovoltaic systems depends on (i) Peak-Power output at the reference operating cell temperature (P_R), (ii) Change in output power/area/°C (T_p), (iii) Cell operating temperature (C_T), (iv) Reference temperature (T_R) and (*v*) Module area (A_M), in the following way—

$$\eta_M = \frac{P_R \cdot T_P (C_T - T_R)}{A_M \cdot I_{Ref}} \qquad \text{...(5.7.3.1)}$$

In eqn. (5.7.3.1) a designer assumes the following (i) T_P = 0.039mW/°C/Cm^2 for silicon single crystal modules and (ii) I_{Ref} = 1 KW/m^2 = 100mW/cm^2. With these assumptions, it is possible for a designer to estimate monthly values for η_M. This parameter is essential for designing *PV* systems and will be considered once again in connection with overall efficiency of a *PV* system.

5.7.5 New Module Assembly

This assembly concept encapsulates and electrically connects all the cells in the module in a single step. The key features of this new process includes the following—(a) Back contact cells, (b) A module back plane that contains both the electrical circuit and encapsulation material in a single piece, (c) A single step process for assembly of these components into a module (Fig. 5.7.2). There is a significant cost reduction by eliminating low throughput (e.g. individual cell tabbing, cell stringing, layout, etc.) steps and by using complete planar process which are easy to automate. This process has been refered to as 'Monolithic Module Assembly' since it translates several advantages of monolithic module construction of thin film *PV* to wafered c-Si *PV*. Recently, the c-Si photovoltaic group at SNL and NREL have demonstrated a two-step assembly process where (a) The back contact cells are soldered to the back-sheet, (b) The cells/backsheets are then encapsulated in the module. Here, Kapton^{TM} was used for the backsheet and the circuit on the backsheet consisted of copper strips that was applied by a proprietory selective plating technique, without including an encapsulation layer.

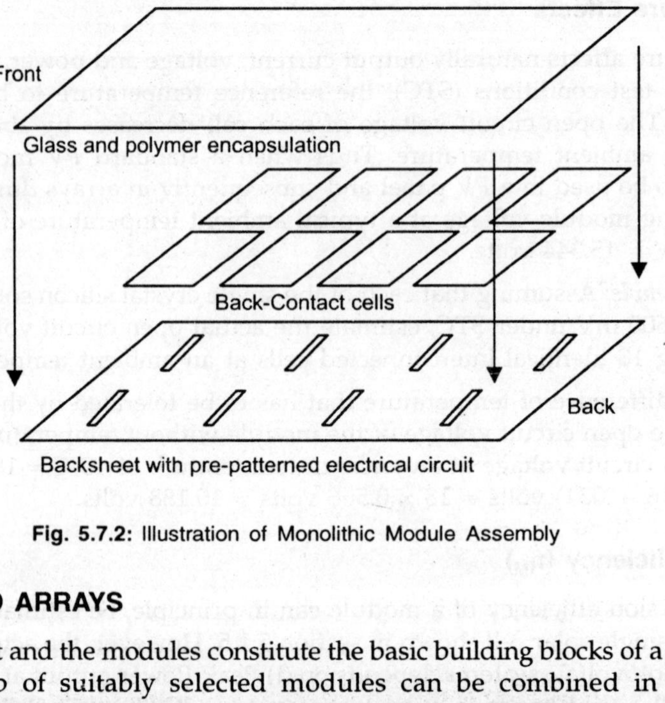

Fig. 5.7.2: Illustration of Monolithic Module Assembly

5.8 PV PANELS AND ARRAYS

Solar cells form the heart and the modules constitute the basic building blocks of a solar photovoltaic system (SPS). A group of suitably selected modules can be combined in a series/parallel combination to form a *PV* panel, while an adequate number of panels are grouped in series/ parallel combination to generate a sub-array. In fact, a sub-array is a configuration whose voltage should be equal to the desired system load voltage. Thus, to generate an array, a few sub-arrays are grouped in parallel only to meet up the required load current. This constitution from solar cells to modules, modules to panel, panels to sub-array and sub-array to a *PV* array exemplifies the total Modularity in a *PV* generating system.

Now, to understand why the modules constitute the basic building blocks of an SPS, it is time to visualize the relation between a *PV* array and a solar *PV* system (SPS). This has been attempted in the section 5.9.

5.8.1 Mismatch Losses

The interconnection of modules to form a panel is governed by similar loss mechanisms as those valid for cell to cell interconnections inside a single module. The analytical study in the case of a panel is more complicated as a *PV* module cannot be represented by an equivalent circuit of a single *PV* cell presented in Fig. 5.1.1.1. However, interconnection losses practically occurring are of two different origins:

(A) *Relative Power Loss (RPL),* which occurs whenever the peak-power point voltage (V_m) or the peak-power point current (I_m) or both these parameters of a number of individual modules (connected to form Panel or Array) are different.

(B) *Hot Spot Losses,* which occurs whenever part or full of one or more modules in a panel connection is shaded for some natural reasons.

5.8.1.1 Relative Power Loss

PV modules are known to be connected in series parallel combination to generate a desired current, voltage and power levels. But these modules, for all practical purposes are non-identical with respect to the basic photovoltaic parameters like I_{sc}, V_{oc}, I_m and V_m. As a result, the maximum power to be drawn theoretically from a series or parallel string of modules is always higher than the actual estimate of maximum power point from the *V-I* characteristics of the string. The theoretical estimate of the maximum power output of the string is done by finding

$$\sum_1^N P_m = \sum_1^N V_m . I_m = V_{m1} . I_{m1} + V_{m2} . I_{m2} + V_{m3} . I_{m3} + \dots \text{ and } P_{mp} = V_{mp} . I_{mp} \qquad \text{...(5.8.1)}$$

where V_{mp} and I_{mp} are the peak power point voltage and current respectively selected from the V-I characteristics of the series string/parallel string of modules. Each term in the right hand side of eqn. (5.8.1) denotes the maximum power's estimated from the individual V-I characteristics of the modules. It is natural that, for a combination of N modules in series or parallel, $\sum_1^N P_m > P_{mp}$ when the modules are non-identical and $\sum_1^N P_m = P_{mp}$ when the modules are perfectly identical.

To generate the best quality or ideal panel, where there is no mismatch with regard to the basic parameters I_{sc}, V_{oc}, I_m and V_m,

$$\text{Relative Power Loss (RPL) in percent} = \frac{\sum_1^N P_m - P_{mp}}{\sum_1^N P_m} \times 100 \qquad \text{...(5.8.2)}$$

Eqn. (5.8.2) can be used to estimate *RPL* values of any *PV* Panel or Array, where the *PV* module is the reference building block. It is important to note that for the best quality Panel or Array, *RPL* is nil. Thus, it is apparent that in order to minimize the *RPL* value for a series/parallel combination of modules/panels, a designer should concentrate on selecting modules with identical values of V_m and I_m.

Illustrative Example (4): Configurations (1) to (3) in the Table below specifies I_{sc}, V_{oc}, I_m and V_m of three individual modules while configuration no. (4) represent the same parameters drawn from their series combination. Estimate the value of *RPL* for the series configuration.

Configuration	I_{sc}(Amp)	V_{oc}(volts)	I_m(Amps)	V_m(volts)
1	1.97	21.6	1.66	16.0
2	1.88	27.6	1.615	16.9
3	1.87	21.6	1.49	17.16
4	1.76	65.0	1.44	50.0

Solution: Here, clearly $N = 3$

$$\therefore \sum_{1}^{3} P_m = 16.0 \times 1.66 + 16.9 \times 1.615 + 17.16 \times 1.49 = 79.421 \text{ and } P_{mp} = 1.44 \times 50.0 = 72.$$

$$RPL = \frac{\sum_{1}^{N} P_m - P_{mp}}{\sum_{1}^{N} P_m} \times 100 = 13.08\%$$

5.8.1.2 Hot Spot Losses

This effect occurs whenever part or a full module is shadowed owing to some natural reasons. The effect is best understood by casting a glimpse at Fig. 5.8.1 which depicts one fully shadowed module in a series string containing 5 modules. It is important to realize first that the series string would not be equal to an open circuited one. It is clear from Fig. 5.8.1 that the shadowed module in this case acts as a reverse biased rectifier diode. Thus, the actual effect of shadowing would be to deliver a reverse voltage (V_R) across the shadowed module, where $V_R = \sum_{1}^{4} V_{oc}$ ...(5.8.3)

Typically the value of V_{oc} at STC for a module containing 36 single crystal silicon solar cells is 18 volts. Thus, for four nos. of identical modules, $V_R = 72$ volts. This is a situation which will send a high reverse current (of the order of short circuit current of the module) through the shadowed module. As a result, due to prolonged shadowing, the temperature will rise. When the temperature reaches 100°C, there is a high possibility of burning of the module. This is precisely the *Hot-Spot Effect.* Such a phenomenon may be prevailing for one or more cells in a series connected module, where each shadowed cell is subjected to a Peak Inverse Voltage $\sum_{1}^{N} V_{oc}$.

Under steady state condition, the temperature of a cell can be determined by energy flow to and from the cell. If P_s, P_E and P_u denotes the input solar power, the electrical power output from the cell and the residual power in the cell, then

$$P_u = P_s - P_E \qquad \text{...(5.8.4)}$$

As soon as a particular cell is reverse biased, P_E becomes negative in eqn. (5.8.4), so that

$$P_u = P_s + P_E \qquad \text{...(5.8.5)}$$

If J_R denotes the incident radiation energy/area/time on the unshadowed cell of area A_c, then,

$$P_s = J_R \cdot A_c \qquad \text{...(5.8.6)}$$

Now, the power input on a shadowed cell receiving J_{shad} amount of radiation percent area per unit time is

$$P_s = J_R \cdot A_c \,(J_{shad}/J_R) \qquad \text{...(5.8.7)}$$

Where the fraction $J_{shad}/J_R = I_{sc}'/I_{sc}$ with I_{sc}' and I_{sc} representing the shadowed and illuminated short circuit currents respectively. Thus,

$$P_s = J_R \cdot A_c \,(I_{sc}'/I_{sc}) \qquad \text{...(5.8.8)}$$

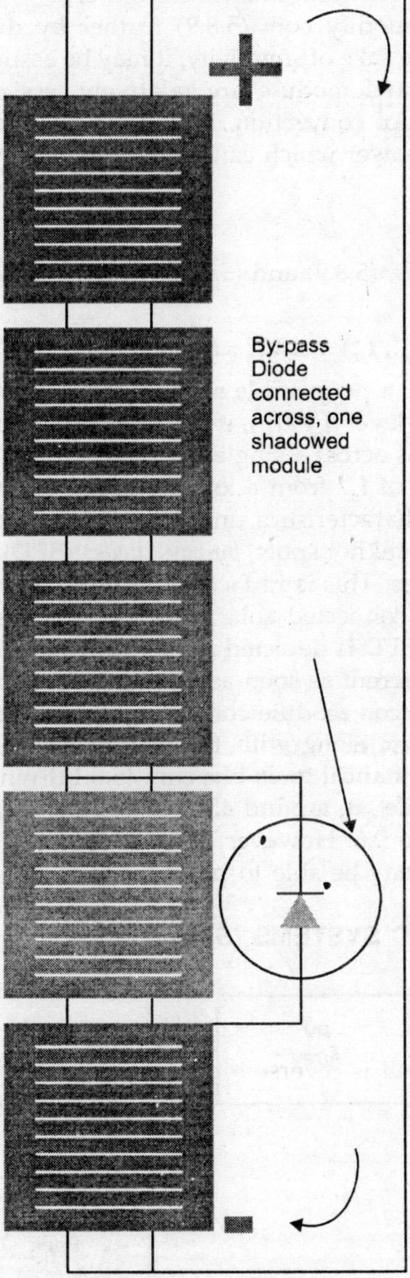

Fig. 5.8.1: A Series String of 5 PV Modules Depicting one of them as Fully Shaded Module.

The maximum power dissipation under worst case condition occurs when $P_E = I_m \cdot V_R$ i.e., when $P_E = n\, V_m \cdot I_m$ so that eqn. (5.8.5) becomes

$$P_u = J_R \cdot A_c\,(I_{sc}'/I_{sc}) + n\, V_m \cdot I_m \qquad ...(5.8.9)$$

Eqn. (5.8.9) is generally valid for estimating power dissipation in a *PV* module containing n no. of uniformly illuminated cells and one shadowed cell.

It is now attempted to modify eqn. (5.8.9) further by due consideration of heat transfer characteristics of them. For the sake of simplicity, it may be assumed that conductive and radiative heat losses in the encapsulated module are relatively less compared to the heat lost to its surroundings by the process of convection. If α is the heat transfer co-efficient of the module (temperature co-efficient of power which can be determined by experiments) for the convection process, then

$$P_u = \alpha\,(\theta_c - \theta_a) \qquad ...(5.8.10)$$

Thus, combination of eqns. (5.8.9) and (5.8.10) yields a steady state power balance condition as—

$$\alpha\,(\theta_c - \theta_a) = A_c \cdot J_{rad}\,(I_{sc}'/I_{sc}) + n\, V_m \cdot I_m \qquad ...(5.8.11)$$

eqn. (5.8.11) indicates that for a permissible safe limit of cell temperature (θ_c), specified ambient temperature (θ_a) and shadow level (I_{sc}'/I_{sc}), it is possible to predict the maximum number of cells that can be connected in series across a single By-pass Diode. There is an approximate geometric method estimating the value of I_{sc}' from a knowledge of the value of constant shunt resistance (R_{sh}) from a set of reverse characteristics under fixed levels of shadowing. In order to ensure minimum damage by accidental hot spots, *by-pass diodes* (BPD) are usually employed as integral part of manufactured modules. This is in fact an appropriate rectifier that is suitably connected parallel to a group of series connected solar cells (or modules) in a module (or a panel). This principle of operation of the BPD is depicted in Fig. 5.8.1, where it automatically acts as the load to the path of high reverse current as soon as a module or part of it is shadowed. For example, the standard single crystal silicon module containing 36 solar cells normally contains two BPD's placed inside the terminal box along with the back cover of the module. This guideline has emerged as a result of mathematical tools like eqn. (5.8.11) which predicts a value of n = 18 for Indian prevailing conditions i.e., θ_a around 45°C and wind velocity not exceeding 5 metres/sec. and α having values of 0.4 to 0.6. However, for lower values of θ_a and higher values of wind velocity, one by-pass diode may be able to protect one complete module also.

5.9 SOLAR PHOTOVOLTAIC SYSTEMS (SPS)

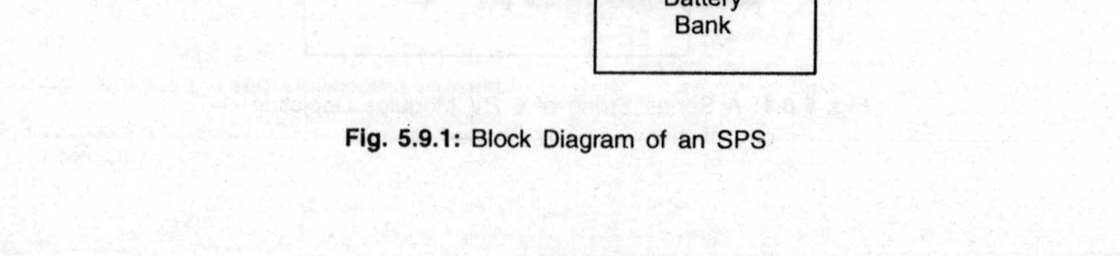

Fig. 5.9.1: Block Diagram of an SPS

A *PV* array, a battery-bank, a Power Conditioner and the electrical load constitutes an SPS, as represented in Fig. 5.9.1. During the daytime, the d.c. power generated by the *PV* array may cater to the daytime electrical load or charge the battery-bank, which is supposed to store charge for use after sunset (evening). All the devices, components and sub-systems that are necessary in a *SPS* other than the *PV* array and battery-bank constitutes the Balance of Systems (BOS). Even the mechanical supporting structure for all the modules of the Array is a BOS component. However the Power Conditioner (PC) is the most important BOS component as it essentially acts as an intelligent switch for sequentially controlling the flow of d.c. charge from the *PV* array. A *PC* normally consists of the following sub-systems and components : (a) Inverter circuits for converting d.c. voltages to a.c. voltages of desired frequency (b) Sensors (c) Circuit breakers or timers for automatically disconnecting the battery-bank, electrical load or back-up generator or grid line from the *PV* array (d) Maximum Power Point Trackers (MPPT) for maintaining the operation of the entire *PV* array at the maximum power point, irrespective of day time. For this a microprocessor and its allied control may also be an integral part of the power conditioner.

It is thus apparent that special attention should be paid to design and selection of various BOS components and sub-systems. This is so as field experience with different *PV* systems in various parts of the universe has clearly shown that most frequent failure mechanism is primarily due to the malfunctioning or outright failure of its electronic BOS. Finally, it is worthwhile to categorize commonly used electronic BOS in a *PV* system into two broad categories.

(A) *Energy Conversion Sub-system,* which comprises of – (i) D.C. to A.C. Inverter or (ii) D.C. to D.C. Inverter.

(B) *Energy Control and Management Sub-system,* which may consist of (i) Power Conditioner and Synchroniser (ii) Charge-controller to protect battery overcharging and discharging (iii) Load Management electronics including load interfacing and switching (iv) Timer and energy management circuitry (v) Switch gears, contractors, fuses, etc. (vi) Indicators, meters, counters and associated monitoring electronics.

5.9.1 Grouping of SPS

Solar *PV* electricity has the potential to be used both in a conventional centralized mode as also in a decentralized and distributed fashion. In a very general way table 5.9.1.1 shows the possible way of grouping various SPS.

Table 5.9.1.1

	Group	*Indicative Capacity*	*Brief Comments and Description*
I.	Central Power Station	10 to 500 MWp	This involves conventional principle of centralized power generation to feed utility grids. This is technically feasible and tried out upto capacities of few tens of MWp. Likely to meet day peak loads principally.
II.	Roof-top Grid Interactive Systems	1 to 10 KWp	PV array in this category owned by house owner and placed at his rooftop. It is interconnected with Utility grid through a two way metering system. It is a fully feasible proposition meeting daytime loads without storage.

(Contd.)

Table 5.9.1.1 *(Contd.)*

III.	Stand-alone Systems	10Wp to 100 KWp	Distributed but dedicated energy systems to meet all electrical loads of a village/community or a specific set of loads. This has to be operated at a fixed center with local management. Without any back-up source, energy storage is essential here.
IV.	Consumer Application Systems	Microwatts to about 10Wp	PV energy catering to consumer devices such as electronic calculators watches, games etc. This is designed for indoor applications of low-wattage compared to outdoor applications.

Now looking from an entirely different and practical view point, PV systems may be categorized according to four distinct types:

(i) Stand-alone systems, (ii) PV Hybrid systems, (iii) Grid Interactive systems, (iv) Consumers support systems.

There exists some fundamental differences in the character of conventional grid electricity and solar *PV* electricity. This has been presented in a summarized form in Table 5.9.1.2.

Table 5.9.1.2: Major difference between Solar *PV* and Conventional Grid Electricity

Solar PV Power	*Conventional Grid Power*
1. Energy generation influenced by local site parameters	1. Energy generation influenced by fuel input and not local conditions
2. Load and site specific design	2. Capacity specific design
3. Additional storage facility needed for a reliable system	3. No other storage needed than the grid itself
4. D.C. current and voltage is produced from PV array	4. International standards of single phase 110v, 60 c/s or 220v, 50 c/s
5. Variable d.c. output is produced depending on time of day and seasons unless charge controlling units and battery-bank (BB) are used	5. Steady output irrespective of season and time of day
6. Very low maintenance requirement other than (a) Distilled water supply to BB. (b) Cleaning of the module surface at intervals	6. Continuous attendance needed
7. Very high initial capital cost	7. Lower capital cost
8. Ecologically benign and not associated with environmental pollution	8. Associated with environmental pollution, may cause ecological imbalance

5.10 STAND-ALONE SYSTEM

As the name suggests, a stand-alone system works entirely on its own. It is supposed to meet all the load requirements without help from any other back-up energy source. These systems may or may not have an electrical energy storage sub-system like a battery-bank. Since these systems are small units deployed in a decentralized fashion in remote areas, the reliability envisaged is very high taking into consideration very little maintenance or attendance at remote sites.

Stand-alone systems *without storage battery* covers the applications where energy is needed only during the sunshine hours, such as—

(a) Water-pumping for drinking and irrigation purposes;
(b) Battery charging stations located in a central community center for supplying charges to Solar *PV* lights, torches and Solar *PV* Lanterns;
(c) Day only radio receiver, insecticide sprayer and other such equipments.

Stand-alone systems *with storage battery* covers well-known applications like Lighting, Refrigeration, TV and Radio, Telecommunication, Village Power Supply, Water desalination, Electric fencing, Cathodic protection for Meterological equipment, Railway and Traffic control etc.

As stated categorically, a day time radio receiver or an agricultural equipment like an insecticide sprayer constitutes low power battery less stand alone applications. In the late 1970's, organization like UNICEF carried out programme on education and entertainment by providing a large number of solar powered community radio receivers in remote villages of many developing countries in Asia and Africa. Small and portable insecticide sprayers have used small and light weight PV module resting on the operator's head. This acts as the power source and also an obvious sunshade protector. One major application area of solar PV for reducing environmental hazards lies in its ability to charge batteries used in fuel driven vehicles. This is specially significant in view of the growing concerns about emission of harmful gases like CO_2, SO_2, NO_2 and other hydrocarbons. Considering that a car is driven no more than an average of 8 to 10 km a day, a typically small (two person) electric vehicle may need an energy intake of about 11 KWh per 100 km of its run. Solar PV electricity seems ideally matched to this level of electricity consumption. Charging efficiencies of small PV arrays in the range of 500 Wp to 1KWp have been under routine investigation by scientists in various industrialized countries to make the dream of Solar *PV* vehicles a reality.

5.10.1 Lighting

Next to water pumping, lighting is the next most important and extensive application of stand-alone solar *PV* systems. Requirement of lighting continues to be the crying need in many unelectrified villages of the developing world. Current requirements in remote sectors are met either by kerosene oil or burning of wood. Problems posed by these conventional sources are (i) Creation of unwanted pollution and ecological imbalance (ii) Energy inefficient methods which is unable to provide a comfortable level of light flux. Solar Photovoltaic lighting systems have also been currently on the look out for need-based mass-scale applications that are acceptable socio-economically. A block diagram of a Solar *PV* lighting system is shown in Fig. 5.10.1.1. All the lamps shown in the diagram are energy efficient ones. The term 'energy-efficient' can be examined in some technical details. Luminous flux of a light source is the total amount of light emitted by the light source. The unit of luminous flux is Lumens. Thus, Luminous efficiency of a light source is normally expressed in lumens/watt. Since *PV* electricity at a glance is more expensive than conventional electricity, it is all the more necessary that *PV* lighting applications should compulsorily consist of energy efficient or high efficiency lamps. The glaring examples in the market are (i) three-phosphor compact fluorescent lamps (CFL), (ii) Low-pressure sodium vapour lamps (LPSV). For driving these loads, the power conditioner should contain suitable dc to ac invertors. The two minimum technical requirements to be met by invertors are: (i) High frequency (10 to 40 KHz), (ii) High efficiency (more than 90%). Problem exists with the associated electronics, which is prone to failure.

In the developing world, the national railways have benefited immensely from the use of solar *PV* electricity. Applications recently promoted by Indian Railways include semaphore

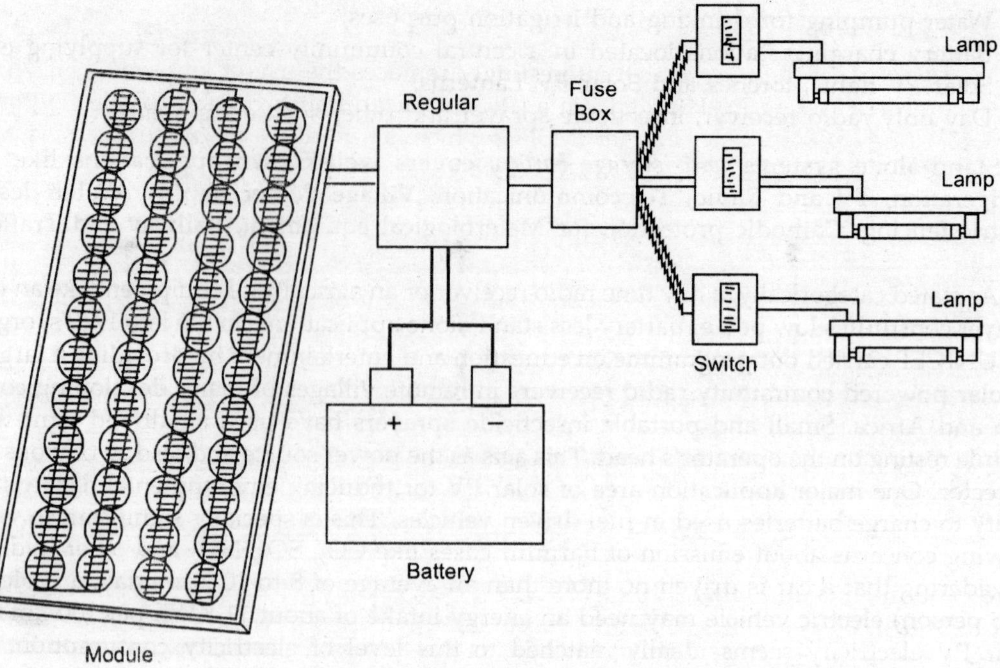

Fig. 5.10.1.1: Block Diagram of a Solar PV Lighting System

signaling, automatic radio controlled level crossing, audio-visual alarms, wayside small station electrification. In several situation diesel generator sets have been replaced and the semaphore signaling have proved to be highly cost effective as compared to kerosene lights used earlier. Another notable application in the Indian sector have been the use of very high frequency (VHF) remote radio telephone links. Nearly 15 lakhs of remote *PV* operated radio telephone links have been installed in India by 1999 where nearly 10 MWp of *PV* modules have been deployed accordingly.

Next most widely acclaimed application of solar photovoltaics is in the various telecommunication (long distance communication) sectors like the provision of electrical energy to remote telephone exchanges, Radio and TV broadcasting and receiving units and microwave communication links. It is well-known that telecommunication in the true sense is poised with problems primarily due to dispersed population settling in hilly and mountainous terrains. It is natural that signals may get blocked or reflected back by the undulating terrain. A practical solution in these cases is in the use of low power or very low power television transmitters (LPT or VLPT) at or near the site. The LPT or VLPT receives signals from the Central Broadcasting Stations via a space satellite and will re-transmit it for the local population. At present, there are more than 200 *PV* powered VLPTS (T.K. Bhattacharya, Terrestrial Solar *PV* (Narosa), 1998) commissioned in India in places that are beyond the reach of standard TV transmitters due to natural obstruction and shadowing. Each VLPT is designed to operate in a completely unattended mode and its typical feature includes:

(a) One operating and one stand-by 10 watt transmitter.
(b) A TV Receive Only (TVRO) dish antenna.
(c) The mean daily energy consumption is less than 4 KWh, which matches the limited energy availability from *PV* systems.
(d) A large storage autonomy of six days is assumed.

5.10.2 PV Lanterns

This is a form of *PV* lighting which interestingly replaces the use of kerosene. Instead, the user is accustomed to buying *PV* charge from a (i) Centralized battery bank during evening hours or (ii) Mobile Solar Photovoltaic Charger (MSPC) which is basically a mobile unit which can charge individual batteries at respective doorstep during daytime.

The conventional scheme for rural electrification through the extension of transmission grid lines have not succeeded in providing electricity to the rural population. Latest schemes for providing electricity to the individual households from a centralized *PV* power plant or building up of small residential *PV* power packs have remained unsuccessful owing to high initial investment and subsequent maintenance problems. In order to popularize large scale application of SPS and also to create some self-employment, a novel idea of using Solar *PV* Lanterns through a centralized charging system has been proposed (Project on Dev. & Evaluation of SPS for Rural Applications, DNES No. 31/228 189-PV, J.U., Calcutta). A prototype system based on these concepts have been designed, installed and tested in the Kaikhali Island of Sunderbans, Gangetic Delta, West Bengal, India (K. Mukhopadhyay et al. Solar Energy Mat. & Solar Cells, 31(1993), pp. (437-446).

The Socio-economic concept in the above proposition is very interesting. A village entrepreneur sets up a small centrally located charging shop which consists of modules, charge distribution boards and maintenance kits. A fixed number of villagers in this scheme have a solar *PV* Lantern with built-in batteries, but not *PV* modules. These villagers are the fixed customers who buy charge for their lanterns from the centralized charging shop. The shop-keeper will sell charge to these customers at a rate, which is cheaper compared to kerosene used in a conventional lantern. Maintenance of the *PV* lantern from time to time will also be done by the shop-keeper. A tentative list of essential components to be housed in a PV lantern, schematically presented in Fig. 5.10.2.1 is given in Table 5.10.2.1.

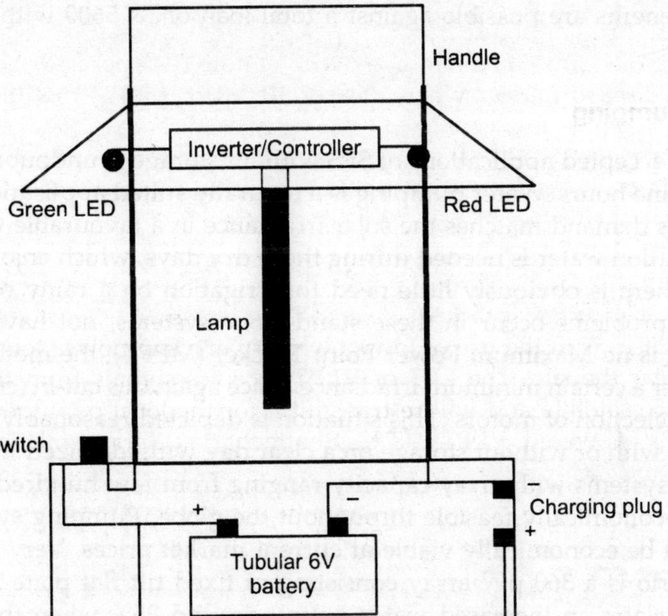

Fig. 5.10.2.1: Schematic Diagram of a PV Lantern

Table 5.10.2.1: Components of a Solar PV Lantern

	Component	*Specification*
1.	Lamp	7W giving an illumination of 400 lumens equivalent to light output by a 40W tungsten lamp.
2.	Battery	6 volts, 12 Amp-hours with—(a) Tubular positive grid (b) Lead Acid Electrolyte (c) Spill proof feature
3.	Inverter	(a) Efficiency above 80% (b) High frequency
4.	Overcharge and deep discharge protections for battery bank and under voltage firing protection for the lamp.	

5.10.2.1 Self-employment Potential

Interestingly, the use of *PV* lanterns has immense potential for generation of self-employmer. within a village cluster of around 60 families and such self-sustainable schemes are in operation at the Sunderban Island of West Bengal. The benefits of the scheme has been worked out in details by K. Mukhopadhyay et al [(Solar Cells & Solar Energy Materials (1993)] and can serve as a tutorial to the rural electrification society/group of any developing nation. Besides offering self-employment, this scheme saves considerable amount of kerosene burning, which in turn keeps the environmental pollution under control and ensures a better quality of electric lighting in everyday life. The attractive features of the scheme can be summarized as:

(A) Entrepreneurs of co-operative society can generate income of nearly Rs. 30,000/- per year.
(B) Pay back period not to exceed 10 years.
(C) Saving of 30,000 litres of kerosene in 10 years.
(D) Expenses for charging plus maintenance of lanterns = Rs. 16 per week
(E) Initial Capital investment = Rs. 2,000/-

The above benefits are possible against a total loan of Rs 5600 with a compound interest of 6% bank interest.

5.10.3 Water Pumping

The most widely accepted applications of SPS without storage is undoubtedly the water pumping during the sunshine hours. Water pumping is a naturally suited application of Solar Photovoltaics as the time of this demand matches the solar irradiance in a favourable way. For example, in the case of crop irrigation water is needed during those dry days, which enjoy plenty of sunshine and low humidity. There is obviously little need for irrigation on a rainy or cloudy day. However, certain inherent problems occur in these stand-alone systems, not having provision of storage batteries. If there is no Maximum Power Point Tracker (MPPT), the motor starts operating in the morning only after a certain minimum irradiance. Once again this cut-in (critical) value of irradiance depends on the selection of motors. This situation is depicted reasonably in Fig. 5.10.3.1, showing the water output with or without storage on a clear day with idealized input irradiance. Solar *PV* Water pumping systems with array capacity ranging from few hundred pW to KWp have been technically and economically feasible throughout the globe. Pumping systems with higher array capacity may not be economically viable at current market prices. Very common specification in the Indian scenario is a 360 pW array consisting of fixed tilt flat plate modules. Manufacturers of such systems claim an increased water output around 25% when three times a day manual array tracking mechanism is deployed. Both active and passive tracking arrays continue to receive sufficient R&D attention. Typical specifications of a solar *PV* water pumping system is presented through an illustrative example.

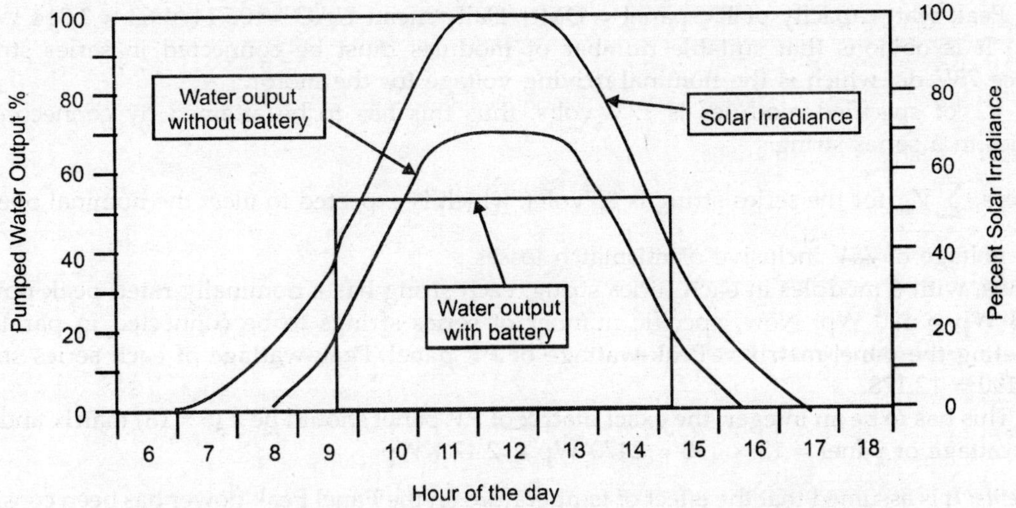

Fig. 5.10.3.1: Solar Irradiance and Water Output with or without Battery Storage

Water-pumping constitutes an important application of Solar Photovoltaics as it is estimated that more than 13,000 Solar *PV* Water pumping systems have been installed till end of 1994 in various parts of the world. Interestingly, the largest number of *PV* Water pumping systems are in India, where typically 2500 systems suitable for deep and shallow water tubewells have been installed in between 1980 to 1994.

Illustrative Example: A stand-alone *PV* Water pumping system installed at Tripura uses an AC induction motor coupled to a multi-stage centrifugal pump with an intermediate DC to AC inverter. It is provided with the following technical specifications:

(A) Nominal motor operating voltage = 75V D.C.
(B) Daily average efficiency of the motor pump set = 35% (without inverter).
(C) DC to AC conversion efficiency of the inverter = 80%
(D) Dynamic head presented to pumping set = 55 metres
(E) Average unit array output is 1690 Wh/Wp at optimum array tilt for the location.

Estimate (i) the peak-watt capacity of *PV* panel to run the system, (ii) the exact panel matrix, consisting of modules (38 Wp ± 5%) having peak-power point voltage V_m = 17.0 ± 0.2 volts.

Solution: (i) The energy contained in the daily pumped water = $E = V_{phg}$ joules.

Where p = Density of water = 1000 Kg/m^3
V = Volume of daily water to be pumped = 20 m^3
h = Dynamic water pumping head = 55 m
g = Accn. due to gravity = 9.81 m/sec^2

$$\therefore E = \frac{20 \times 1000 \times 55 \times 9.81}{3.6 \times 10^6} = 3.0 \text{ KWh/day} \quad [\because 1 \text{ KWh} = 3.6 \times 10^6 \text{ Joules}]$$

Balance of system efficiency = $\eta_{BOS} = \eta_{MP} \times \eta_{INV}$ = 28%
Where, η_{MP} = 35%, η_{INV} = 80%

$\therefore$ The daily mean load (DML) seen by the *PV* array is DML = $E/(28/100) = (100 \times 3)/28$ = 10.714 KWh/day.

Thus, the *PV* array must produce an average daily electrical energy to the tune of 10.714 KWh/day in order to meet the specified system performance.

$\because$ Annual UAO for Agartala = 1690 Wh/Wp/day.

∴ Peak watt capacity of PV panel = DML/Daily mean UAO =10.714/4.63 = 2.314 Wp

(ii) It is obvious that suitable number of modules must be connected in series string to produce 75V dc, which is the nominal driving voltage for the motor.

∵ V_m of specified modules is 17.0 volts, thus this has to be attained by connecting five modules in a series string.

Here, $\sum_{1}^{5} V_m$ for the series string is 85 volts, which is expected to meet the nominal operating motor voltage of 75V inclusive of mismatch losses.

Thus, with 6 modules in each series string, each string has a nominally rated peak power of 5 × 38 Wp = 190 Wp. Now, specific number of series strings to be connected in parallel for completing the panel-matrix = Peak-wattage of *PV* panel/Peak-wattage of each series string = 2314/190 = 12.178.

∵ This has to be an integer, the exact matrix of *PV* panel should be a (5 × 13) matrix and Exact Peak-wattage of panel = 13 × 190 = 2470 Wp = 2.47 KWp.

Comments: It is assumed that the effect of temperature on the Panel Peak power has been considered in the specified UAO. Thus, no temperature correction of Panel/Array Wattage is needed.

5.11 GRID INTERACTIVE APPLICATIONS

The conventional method of providing electrical energy in all industrialized countries has been to generate electricity in a central location and then distribute it to load centers using a complicated interconnected national network of utility grid lines. In almost all industrialized countries all consumers are generally connected with these utility grids. Thus, the electrical energy requirements of the consumers should be met through these grid lines. Thus a *PV* power plant or a centralized stand-alone system is connected to the utility.

Alternatively, the *PV* array of a grid interactive system may be located on the rooftop of the owner's house to cater to his day time requirements. Here, the *PV* system supplies fuel or a part of the day time energy requirements with the remaining requirement being met from the utility grid. Such configurations often uses a two-way energy meter between the *PV* array and the grid lines to take care of the amount of charge flowing to the grid when the *PV* array power is not required by the owner.

Similar cases may also arise for centralized *PV* charging stations, whose power conditioner indicates a fully-charged battery bank during anytime of the day.

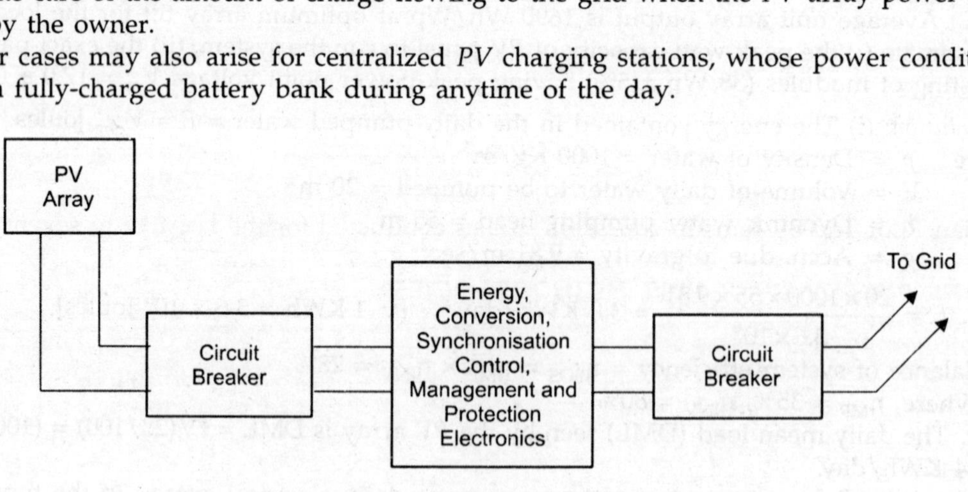

Fig. 5.11.1: Schematic of a Grid Connected PV System

Different types of grid-interactive systems are being tried out in countries where extensive utility grid lines are available. A *PV* array is connected and synchronized to the utility grid using an appropriate power conditioning sub-system (shown in Fig. 5.11.1) which connects the input DC energy to AC energy synchronized to the grid energy. No additional energy storage is therefore necessary in Fig. 5.11.1. The utility grid itself is the storage medium for such a grid interactive system, which delivers energy to the grid as long as enough sunshine is available. Two distinct approaches of grid interactive systems have been under consideration so far.

The category of *PV* hybrid systems has been elaborately discussed in Chapter X, since such a category has been placed under Hybrid Energy Systems.

5.11.1 PV Roof Top Programme

It is well known that the decentralized use of *PV* is debated as an important tool to cope up with current environmental problems related to generation of electricity and its use. Yet, dissemination of *PV* systems is still very limited owning to high capital cost of *PV* systems. Thus, a couple of financial incentives are under active consideration for effective market penetration. Some of these are—

(A) Enhanced Pay back tariffs,
(B) Net Metering,
(C) Solar stock exchange.

In this context, the first comprehensive international dessimination programme was the '1000 Roofs Programme' launched in Germany in 1990, for which plant installations were completed in 1995. Some 2250 German Roofs were equipped with *PV* systems having average size of 2.6 KWp. However, the largest of such program till date has been launched in Japan during 1994 and continued till the end of 1999. Here, more than 10,000 grid connected systems with an average capacity of 2 KWp had been installed (Renewable Energy News). The latest in this trend is the Austrian 200 KWp PV roof top program. (Brettentest'—the Austrian name).

Obviously, any *PV* system is considered technically reliable if they can work with minimum maintenance, which is a sign of maturity of this technology. This is judged best by examining the Performance Ratio (PR) and is a characteristic feature depending on the yearly solar insolation.

$$PR = \frac{\text{Yearly generated electricity (KWh) per KWp installed}}{\text{Amount of Solar insolation in KWh/m}^2\text{ of module area}}$$

Significantly enough, typical *PR* values ranged from 0.6 to 0.8 $\frac{\text{m}^2}{\text{KWp}}$ for the German and Austrian Roof Top Systems in course of studies conducted for the last five to seven years.

5.12 PV SYSTEM DESIGNING METHODOLOGIES

Familiarity with the essential components of a Photovoltaic System and Stand-alone Photovoltaic System (SAPV) in particular leads to the realization that two most crucial steps in designing a SAPV system is the sizing of the *PV* array and the Battery-bank. In order to supply the required energy on a long-term (20 to 25 years) basis, two more features have to be simultaneously born in mind—

(a) The life cycle cost of the system should be minimized,
(b) Acceptable performance figure for Loss of Load Probability (LOLP) should be determined.

A number of array and battery sizing approaches are rooted in the literature. Out of these, the NASA method is the pioneering one where a designer starts from a knowledge of the standard deviation of insolation. In case this data is not available, the best quick-sizing approach has been worked out and illustrated by Saha et al. In both these cases, it is essential for the designer to carry out certain trial and error computations with a well interpreted balancing parameter.

5.12.1 NASA Approach

Array-sizing methods have been elaborately covered by H.L. Macomber et al of NASA, where the Array-area (A) for each month is given by—

$$A = \frac{DL}{[n_{ov}\,(I - M.S.)]} \qquad \text{...(5.12.1)}$$

where *DL* (KWh/day) is the estimated daily load demand for a particular month and n_{ov} is the overall *PV* system efficiency including the module, battery and power conditioner efficiencies. *I* is the average insolation (KWh/m^2/day) in the concerned month and S is its standard deviation.

M is a crucial balancing parameter in eqn. (5.12.1) and can be interpreted as the fractional monthly average insolation difference, where $M = (I - I_D)/S$ ...(5.12.2)

Here I_D is the exact value of the insolation required to meet the load demand. Battery-banks provide the following advantages to a SAPV system.

(A) Ability to provide energy for no sunidays (NSD) and at night time,
(B) Ability to meet momentary peak power demands,
(C) Stabilize system voltage and increase the reliability in a continuous way.

Thus, battery-banks are an integral part of every SAPV system. The amount of nominal battery capacity (KWh) will depend on load demand, parameter characteristics of *PV* system components and weather patterns at the site. Simply, total battery capacity (*Q*) can be expressed as

$$Q = Q_1 + Q_2 \qquad \text{...(5.12.3)}$$

where the long-term battery capacity Q_1 is the energy needed to satisfy the load during the cloudy days of the month. The short term capacity Q_2 corresponds to the daily peak demand that cannot be tackled by the sized array output. Q_2 is actually the difference between the area under the load curve and the area under the array output curve. In order to estimate Q_1, it is needed to define the capacity appreciation factor (CAF) of the battery-bank which is

$$CAF = \frac{1}{(\text{DOD})_{max}.\,\eta_B} = \frac{1}{\eta_B\,(1 - \text{SOC})} \qquad \text{...(5.12.4)}$$

where DOD is the battery depth of discharge representing the percentage of capacity below the manufacturer's rated value or measured total capacity. State of charge (SOC) of the battery is the ratio of energy presently stored to the maximum rated value. The SOC capacities are usually expressed in Ampere-hours for a specified cut-off voltage and discharge temperature. Thus, it is easy to see that the present battery capacity (Q_p) in KWh is

$$Q = \text{Rated Capacity (RC)} \cdot \text{SOC} \qquad \text{...(5.12.5)}$$

Thus, $Q_1 = \text{(CAF) C (DL)}$ ...(5.12.6)

and $Q_2 = \text{(CAF) (DL) (WSD)}$ ...(5.12.7)

where $\text{NSD} = \text{No Sun Days} = \dfrac{\text{Night Load (KWh)}}{\text{Daily Load Demand (KWh/day)}}$

Here C represents the number of consecutive days for which the battery should be able to support the load. This is dependent on the DL and the choice of M. Thus, at a particular site having a specified load demand and a required level of reliability, a quick sizing methodology or algorithm can be formulated by testing eqns. (5.12.1) to (5.12.7) for estimating array-area and storage capacity requirements. A value of M as small as 0.1 will yield a set of values of A and Q. Other higher values of M are chosen leading to increase in value of A and decrease in Q_1 and hence Q. The final task of the designer will be to select the values of A and Q yielding the minimum life cycle cost of a *SAPV*. All the essential components of life cycle testing have been listed up and illustrated in section 5.13.

5.12.2 Quick-sizing Approach

A simple semi-empirical rule for selecting the right size of a *PV* array (A) is

$$A = (1/X)\,[DL + (DL{\cdot}C/C_R{\cdot}C_B)\cdot 100] \qquad ...(5.12.8)$$

Where A is the array size in peak watts and the symbols DL and C have their significance as defined in the NASA approach. C_R is the charge recovery period of the battery while C_B represents its watt-hour efficiency. $\bar{X}$ can normally be accepted as the annual average equivalent peak hours per day. In fact, this is the annual average watt-hours received per peak watt of the flat plate *PV* module per day at the site of installation.

It is evident that the value of $\bar{X}$ varies from one site to another and can be estimated on a monthly average basis from $\bar{X} = 1/12 \cdot \sum_{m=1}^{12} X_m$ ...(5.12.9)

with

$$X_m = (n_{ov}/n_M)\,\bar{I}_m \qquad ...(5.12.10)$$

Here, the parameters n_{ov} and n_M have their usual significance while I_m is the average insolation on the horizontal array surface at the location. The monthly average daily horizontal global insolation for 12 months at any desired location can be obtained from the standard hand book of Mani. A designer following this method can initiate the estimates starting from the above hand book. The average value $\bar{X}$ for use in quick estimates turns out to be 3.5 typically for an Indian site.

It is well-known that storage batteries are an integral part of all SAPV system which has to run reliably during the cloudy daytimes and cater to night time loads. The battery-bank size C_B (ampere hours) is given as,

$$C_B = \frac{DL\cdot C}{V_s\cdot DOD/100} \qquad ...(5.12.11)$$

While V_s is the system voltage (top of the bus bar voltage) and DOD is the maximum permissible depth of discharge.

Variation of parameters like DL, C can be incorporated into eqn. 5.12.11 on a monthly basis. So long as the performance analysis of a battery-bank is concerned, this can be felt like the pulse of a patient by sensing the State of Charge (SOC) on an hourly basis.

5.13 LIFE CYCLE COST ESTIMATES

It is understood from 5.12 that a realistic and rigorous life cycle cost analysis has to be executed based on monthly estimates of *PV*-array and battery-bank sizes in order to arrive at an optimum size of these two essential components of a SAPV system. This cost is the sum of the initial cost plus the total present worth of annual costs. The Initial Cost (*IC*) includes the cost of *PV* array, battery-bank, the power conditioning system, engineering installations and project management costs. Let us address each one of them separately as—

a = *PV* array unit cost ($\$/m^2$)

b = Battery-bank unit cost ($/KWh)

c = Power conditioning unit cost ($\$/m^2$)

d = Ratio of engineering cost to total hardware cost

e = Ratio of installation cost to total hardware cost

f = Ratio of management cost to total hardware cost

It is now easy to think dimensionally and write

total hardware cost $= T_H = (a + c) \cdot A = b \cdot Q$ ...(5.13.1)

With the plausible assumptions that—

(i) Initial cost of *PC* system is proportional to *A*

(ii) *a* and *c* are independent of *A*

(iii) Also, *b* is independent of *Q*

(iv) *d, e, f* are each independent of *T*

Thus, the cost of engineering, installation and project management

$$T_{EIP} = (d + e + f) \cdot T_H \quad ...(5.13.2)$$

Thus, initial cost (*IC*) $= T_H + T_{EIP}$; i.e, $IC = T_H (1 + d + e + f)$ ...(5.13.3)

The present worth of annual costs includes maintenance annual costs and present value of all necessary replacements. These have to be examined separately and added up. If X_{pv} is the present value of operation and maintenance costs and X_o the value in first year only,

Then, $X_{pv} = X_o\,[(1 + g_o)/(k - g_o)]\,[1 - \{(1 + g_o)/(1 + k)\}^n]$ ...(5.13.4)

Where, g_o = Escalation rate for operations; R = Money interest rate

n = System life in years and $X_o = h\,(a + c)\,A + jb \cdot Q$ ...(5.13.5)

Where, h = Ratio of X_o to array cost, j = Ratio of X_o to battery cost

Significantly, the 1st term in eqn. (5.13.5) is due to energy generation while the second term is due to storage requirements. Next, if R_{pv} represent the present value of battery replacement cost, then $R_{pv} = b \cdot Qn_b(1 - S)[(1 + g_1)/(1 + k)]\dfrac{n}{(NR+1)}$

Where, S = Per unit Salvage cost of replaced battery.

g_1 = Inflation rate of battery replacements.

n_b = No. of battery cells.

Q = Total *Ah* capacity of battery.

$B\,Q\,n_b$ = Battery cost.

NR = No. of battery replacement over *n* years.

Here the term *NR* denotes the number of battery replacements over, *n* years. Finally, the total life cycle cost (TLC) becomes

$$TLC = IC + X_{pv} + R_{pv} \quad ...(5.13.6)$$

5.14 THERMOPHOTOVOLTAIC (TPV) ENERGY CONVERSION

Thermophotovoltaic Energy Conversion converts the radiant energy of a high temperature body (emitter) directly into electricity using a *PV* cell. *TPV* has a number of attractive features like fuel versatility (nuclear, fossil, solar, etc.), quiet operation, low maintenance, low emession levels, light weight, high power density modularity and co-generation of heat and power. While this concept is an old one, *TPV* is experiencing renewed interest due to recent advances in low band-gap *PV* cells. Low band-gap *PV* cells are required to work with emitters at mangeable temperatures (1300 to 1800K). In principle, the efficiency of a *TPV* system could be significantly improved if the spectrum of the emitter could be tailored to match the peak response of the *PV* cell. Spectral control can be achieved by either controlling the spectrum of the emitted radiation ('selective emitter') or by employing spectral filters to return unwanted radiation to the emitter. Achieving high efficiencies requires very high performance in terms of the spectral selectivity/reflectivity, bandwidth and the angular distribution with either a selective emitter or a spectral filter. Recently, the control of spectral emissivity by physically structuring the emitter has been examined by several researchers—a new and powerful approach relying on thin film and microelectronic fabrication of Photonic Crystals.

5.14.1 Photonic Crystals

Photonic crystals are structured materials where the refractive index is modulated in one, two or three dimensions. This modulation of refractive index is responsible for the photonic density of states spectrum. Many properties involving the interaction of radiation and materials, such as spontaneous emission or spectral emissivity were once thought of as intrinsic properties of the material. However, these properties are in fact a function of the photonic density of states and are subject to engineering through physical sturcturing of the materials. It is, in fact, possible to design structures where there are no photonic states for a specific frequency range, i.e., a photonic bandgap. Such structures are said to exhibit a full photonic band-gap if there are no photonic states in any direction and for any polarization. The spectral emissivity is zero within a full photonic band-gap.

Practical *TPV* energy conversion systems require high emitter temperatures of around 1800°K, to obtain high efficiencies and power densities. This calls for a refractory material to be chosen

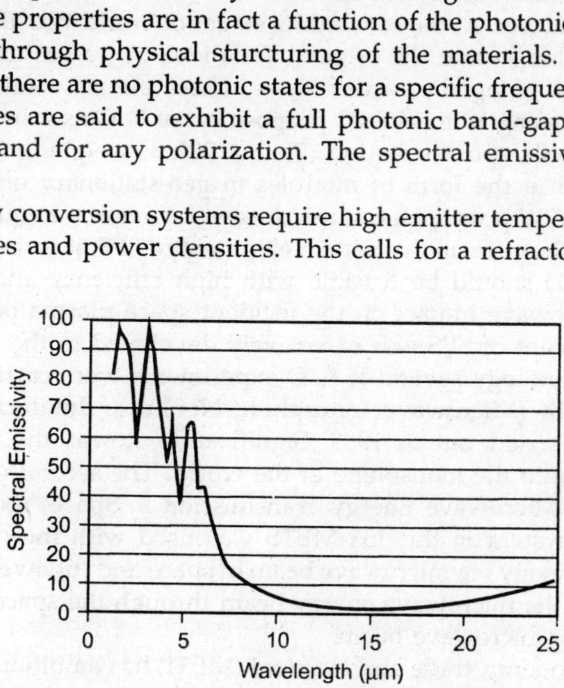

Fig. 5.14.1: Spectral Emissivity of Tungsten Three-dimensional Photonic Crystal

as the emitter where Tungstan is a natural choice. Besides, the capability to withstand high temperatures, it offers the advantage of a wide refractive index contrast, which determines the magnitude and bandwidth of the photonic band-gap. Also, Tungsten is a familiar material in the microelectronics environment and can be easily deposited and patterned using commonly available tools in the microelectronics industry. Principally, these crystals are charecterised via reflectance and transmittance measurements from which absorptivity and emissivity can also be estimated. One such experimental curve corresponding to spectral emissivity of Tungsten three dimensional crystal is shown in Fig. 5.14.1. These crystals exhibited an enormous band-gap around 8 to 20µm and attenuation factor (30dB per unit cell at 12µm). Results of *TPV* parameter estimates using different emitters with a 1800°K radiator and a GaSb *PV* cell is shown in Table 5.14.1.

Table 5.14.1: Results of TPV Calculations using Various Emitters. Model used on 1800K Radiator and GaSb Photovoltaic Cell

Emitter	*Radiant flux (W/cm²)*	*Electric power (W/cm²)*	*Efficiency (%)*
Blackbody	56.9	7.2	12.6
Photonic crystal	14.2	3.8	26.9
Microstructured W	19.6	4.4	22.4
Erbia/yttria	16.6	2.6	15.5

5.15 SOLAR POWER STATION/SATELLITE (SPS)

The idea of radio power transmission was first conceived by Tesla about a century ago. However, the first practical use of radio waves was intended for the transmission of information and not for transmitting electrical power. At the concluding stages of World War-II, high power microwave technology became available so that the original Tesla idea of transmitting electrical power was rekindled with microwave being the carrier wave. These efforts resulted in the idea of solar power satellite (SPS) according to the initial proposition of P. Glaser in 1968.

The SPS is a gigant solar power station (5km × 20km, weighing 50,000 tons) using Si or GaAlAs solar cells placed in the form of modules in geo-stationary orbit (above 36,000km). It typically generates 5 to 10GW power. Here a microwave carrier at a frequency of 2.45GHz is used for energy transmission. Thus, for an overall development of SPS, it is imperative that microwave power transmission (MPT) should be feasible with high efficiency and high safety alongwith scientific analysis of microwave impact on the incident space plasma environment.

The NASA/DOE concept of SPS was extensively developed in the late 1970's. Towards the development of MPT technology several R & D experiments were carried out in Japan. Out of these one is called MINIX (Microwave Ionosphere Nonlinear Interaction Experiment) rocket experiment which was carried out in 1983. Significantly it was the first microwave energy transmission experiment into the ionosphere of the world. The second one was titled ISY-METS (International Space Year-Microwave Energy Transmission in Space) rocket experiment in 1993. An active phased array system in this ISY-METS was used with the objective of (a) verifying energy transmission technology via microwave beam in space and (b) investigation of the nonlinear plasma effects caused by the microwave energy beam through the space plasma and its counter effects experienced by the microwave beam.

Japan's ministry of economy, trade and industry (METI) has ambitious plans to launch a giant solar power station by 2040. Currently METI is also planning to launch a satellite capable of generating one million kilowatts of electrical power per second. Significantly, this is equivalent to the output of a nuclear power plant. The SPS will be placed into a geostationary orbit about

36,000kms above the earth's surface. This satellite will have the special feature of two solar power generating wing panels, each measuring three kilometers with a 1000 meters diameter power transmission antenna between them.

For all terrestrial modules and panels, the question of receiving sun rays does not arise at night time. Even during day time a solar photovoltaic panel is confronted with reduced power generation depending on clearness index of the sky. But in space, SPS is always capable of generating electrical power even at night. Further such a power production scheme would benefit the environment significantly compared to thermal power plant as there is no emission of green house gases from the SPS. On the other hand, the size of the receiving antenna of an SPS will have to be several kilometers in diameter and its probable location will be either a desert or sea. This implies a higher provision to be reserved towards the distribution and transmission costs of the electricity ultimately feeding into a consumer. This economic hurdle is further constrained as the current evaluated cost of generating electrical power through an SPS is almost 2.5 times compared to the costs of electricity generated by a thermal or a nuclear plant.

EXERCISE

1. What are Photovoltaic devices? Outline very briefly their principle of operation.
2. (a) Sketch a neat cross section of a silicon solar cell.
   (b) What is the necessity of (i) *AR* coating, (ii) Encapsulation?
   (c) Point out two particular merits of single crystal silicon solar cell.
3. (a) What are the factors on which R_s of a solar cell depends?
   (b) Identify the factors on which Fill Factor of a cell depends.
   (c) Three solar cells of identical short circuit current and open circuit voltage gave the following results from their *V-I* characteristics.

	Value of R_s	Value of R_{sh}
Cell–1	5Ω	200Ω
Cell–2	2.5Ω	100Ω
Cell–3	10Ω	1000Ω

   Which one of these cells is expected to yield the best efficiency? Give reasons in support of your choice.
4. (a) Draw equivalent circuit of a solar cell and deduce the relation $V_{oc} = AKT/q \cdot \ln [(I_{sc}/I_o) + 1]$, where the symbols have their usual significance.
   (b) Try to draw the equivalent circuit of a PV module containing 5 identical solar cells connected in series.
5. (a) What is the maximum power delivered by a solar cell having a short circuit current density of 500 mA/cm^2, open circuit voltage = 0.4 volts and a fill factor (FF) equal to 40%? Assuming an effective area of 4 cm^2, what is the efficiency under an illumination of 0.6 SUN?
   (b) A *PV* module operated at STC yields a V_{oc} = 18 volts at an ambient temperature of 40°C, what is the open-circuit voltage of the module?
6. (a) Give a sketch of the block diagram of a *PV* system.
   (b) Enumerate the role played by the power conditioner.
   (c) Classify all possible *PV* systems in use and give as many examples as possible from each category.
   (d) Give one example where *PV* systems are useful for generating employment on a community basis.

7. (a) What types of failure mechanisms may be encountered by—
   (i) Cells in a *PV* module?
   (ii) *PV* modules in a panel configuration?
   (b) What is the difference between a *PV* panel and a *PV* array?
8. (a) Three different modules S_{11}, S_{12} and S_{13} gave the following datas on I_{sc}, V_{oc}, I_m and V_m when *V-I* characteristics were drawn at 0.7 SUN. Estimate the *RPL* value of the S_{1P} (formed by parallel combination of S_{11}, S_{12} and S_{13}) configuration from the set of data.

Module No.	I_{sc} (Amp)	V_{oc} (volts)	I_m (Amp)	V_m (volts)
S_{11}	1.97	21.6	1.66	16.0
S_{12}	1.88	27.6	1.615	16.9
S_{13}	1.87	21.6	1.49	17.16

   (b) If *RPL* value of a series or parallel string of modules is negative by estimate, what is your inference?
9. A *PV* pump, installed as a part of the National Drinking water mission in Rajasthan, pumps water from a depth of 48 metres. The solar cells are made of single crystal silicon and the array consists of 24 modules. (Each cell is a square area 10.4 cms × 10.4 cms). It is also given that the efficiencies of the inverter and the pump-motor set are 85% and 45% respectively. Estimate the rate at which the *PV* pump discharges water at noon (Global Radiation incident normally on the modules = 945 W/m^2). Assume density of water = 0.996 gm/cc.
10. The monthly coverage of daily insolation on the titled *PV* array at Baripada is 5.14 KWh/m^2/day during the winter months when the daily load demand is 38.3 KWh. Assuming overall system efficiency to be below 5%, estimate (i) Array area, (ii) Number of 30 peak watt PV modules needed for constituting the array.
11. The voltage current characteristics of a solar cell shows a current of 1.25mA and 1.0 mA corresponding to forward biasing of 400mV and 375mV respectively under dark condition. Estimate (i) the diode quality factor and (ii) the reverse saturation current of the device. Also, determine the open-circuit voltage from the estimates made in (i) and (ii) with a short circuit current of 40mA/cm^2. Deduce all necessary steps used for estimation.

ANNEXURE-I

Dependence of V-I Characteristics of PV Cell/Module on Insolation and Temperature: V-I characteristics of *PV* module depends on insolation as well as temperature. With respect to standard test conditions, having insolation 1 sun (== 1000W/m^2) and temperature 28°C, the current and voltage changes with the change in insolation according to the following relationship.

$$\Delta I = I_{sc}\,[E/1000 + a^* (T - 28) - 1] \quad ...(1)$$

where *E* and *T* are insolation (W/m^2) and temperature (°C) in non standard condition and ΔI is the change in current when translated to STC.

$$I_{sc} = \text{Short circuit current, } \alpha^* = \alpha/I_{sc} = \text{Normal current temperature co-efficient}$$

$$= 0.0014/°C.$$

Therefore the current at non standard condition (I') with respect to that at standard condition (I, say) can be given by

$$I' = I + \Delta I \quad ...(2)$$

Table 6.1.2.1: Physical and Chemical Properties of Coconut Shell

Moisture content	14.52 (wt %)
Volatile Matter content	55.91 (%)
Ash content	2.95 (%)
Fixed carbon content	26.62 (%)
Carbon	51.05 (%)
Hydrogen	5.70 (%)
Oxygen	42.58 (%)
Nitrogen	0.46 (%)
Sulphur	0.21 (%)
Gross calorific value	17.95 MJ/Kg.

6.1.3 Bio-fuels

The energy stored in dry bio-mass like wood and straw can be easily obtained by direct combustion although these dry materials can also be converted to liquid and gaseous fuels by a variety of techniques that will be presented in the next section. It calls for considerable energy to drive off the water and moisture content in wet bio-mass. As a result, the fuel value of the bio-mass is diminished. It is usually more appropriate to convert wet organic matters directly into premium fuels by a wet process called digestion or fermentation. Significantly, refined bio-fuels can be used to replace conventional fuel systems with the advantage that (a) very little or no modification of conventional systems are required, (b) compatibility with current patterns of energy use, (c) improvement of economic potential (d) refined bio-fuels have high calorific values, (e) storage for long periods without degradation is possible. Some alternative liquid fuels have been considered in sec. 6.12.

6.2 BIO-MASS CONVERSION TECHNOLOGIES

A wide variety of conversion techniques are in use for manufacturing bio-fuels from bio-mass. Techniques like digestion and fermentation are simple and well understood. Gasification techniques have been thoroughly demonstrated through a number of pilot plants and have already reached the stage of commercialization. Each of the bio-mass resources like wood, dung, vegetable waste can be treated in various ways to provide a series of useful bio-fuels. Bio-mass conversion can be of any one of these forms:

(a) Direct combustion such as wood waste and bagasse (sugarcane refute).
(b) Thermochemical conversion; and
(c) Bio-chemical conversion.

6.2.1 Direct Combustion

Direct combustion of wood to generate heat and then to generate steam happens to be the primitive energy producing methods for cooking, space heating (comfort in winter) and several small scale industrial processes. In developing countries like India, poor households are compelled to depend on fuel wood as many of them can not afford conventional fuels like kerosene, LPG and soft coke. Next, cellulosic wastes and residues are similarly used in agro based industries. Application of bagasse for firing boilers is a world-wide practice in sugar industry. Another important option is the co-firing of fossil fuels with cellulosic and wood related bio-mass.

However, continued dependence on fuelwood has the adverse impact on forest cover, which is certainly not a welcome trend. One way to cope up with this problem is to keep the efficiency of fuelwood stores high. In this context, the National Programme on Improved Chullas (NPIC) has been introduced by the MNES in December, 1983. This has helped to conserve fuelwood and minimize grrenhouse effect with the direct benefits percolating to the men and women in the lower income groups. Non-government organizations(NGO) has installed Improved Chullas (IC) in community kitchen of hotels, hospitals and army camps. An efficient cook store is expected to: (a) consume less fuel, (b) minimize smoke and drudgery and (c) ensure quick ignition and facilitate slow, sustained and controlled fire. This is also clarified through Table 6.2.1.1. A view of traditional chulla along with two Improved Chullas (IC) are presented in Figs. 6.2.1.1, 6.2.1.2 and 6.2.1.3 respectively. A later section 6.10 has been devoted to bio-mass stoves where the aforesaid features of stoves have been considered in details.

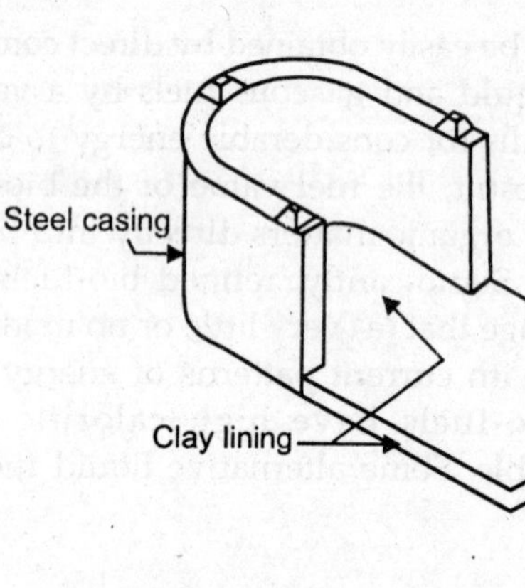

Fig. 6.2.1.1: Traditional Chulha

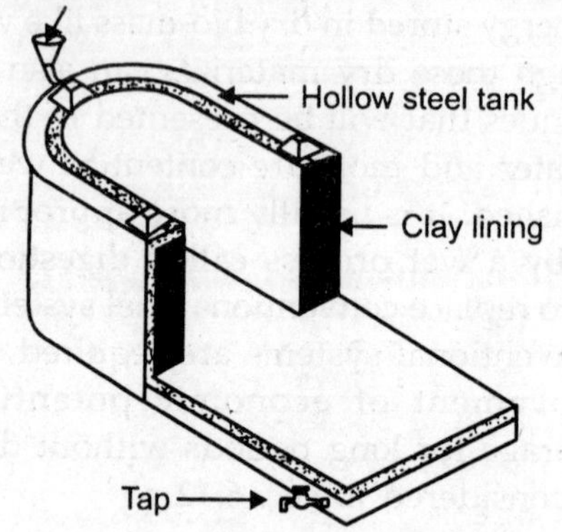

Fig. 6.2.1.2: Improved Chulha with Water Jacket

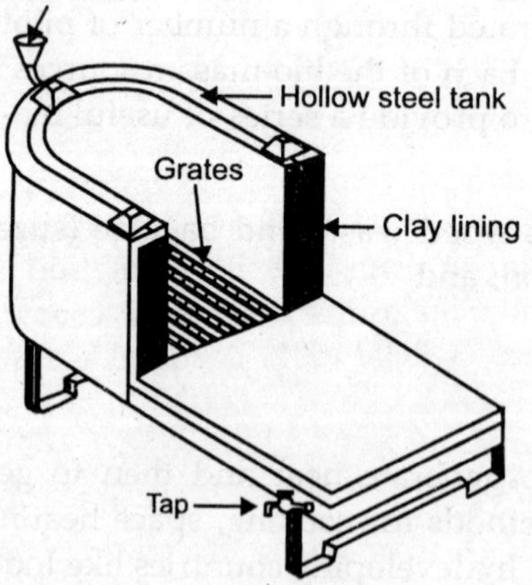

Fig. 6.2.1.3: Improved Chulha with Firegate

Table 6.2.1.1: Improved Chullas

	Traditional Chulla	*Improved Chulla (IC)*
• Wood needed/annum	2000 to 2500 Kg	1000–1500 Kg.
• Manufacturer	Individuals & state agencies	Small scale units
• Life Cycle	3 years	Five years
• Savings generated		(a) 8.7 Kg. of dung cake (b) 77 Kg. of agro wastes (c) 1.3 Kg. of coal (d) 1.8 litres of kerosene per year.

6.2.2 Thermochemical Conversion

This involves two forms—gasification and liquefaction. In the former step, the bio-mass is heated with limited oxygen to produce low BTU gas (LBG), medium BTU gas (MBG), Substitute natural gas (SNG) and some light fuel oils. MBG's have two to five times the energy density of low BTU gases, making transportation of these feasible over short distances through pipeline. Alternatively, it is reacted with steam and oxygen at high pressure and temperature to produce medium heating values gas. The latter may be consumed as fuel directly or used in liquefaction by converting it to methanol (Methyl Alcohol CH_3OH) or Ethanol (Ethyl Alcohol CH_3CH_2OH). Alternatively, these can be converted to high heating value gases.

6.2.3 Bio-chemical Conversion

The two wellknown forms being the *Anaerobic Digestion* and *Fermentation.*

6.2.3.1 Anaerobic Digestion

This is the microbial digestion of bio-mass, which has been exercised principally with animal manure. But it is applicable to other sources of bio-mass also. This process requires a low temperature upto 65°C and a high moisture content of at least 80%. It generates a gas consisting of CO_2 and methane (CH_4) with impurities such as hydrogen sulphide. This combination can be ignited directly or upgraded to SNG by removing the CO_2 and the impurities. The residue is usually a protein rich sludge that can be used as animal feed and liquid effluents that are biologically treated by standard techniques and returned to the soil. The underlying principles in this technology has been clarified with reference to bio-gas generation in 6.5.

6.2.3.2 Fermentation

This is a process of breakdown of complex molecules in an organic compound under the influence of a ferment such as yeast, bacteria, enzymes, etc. Fermentation is a well established and widely used technology for the conversion of grains and sugar crops into *Ethanol.* This can then be combined with gasoline to produce *Gasohol* (90% gasoline + 10% ethanol).

This is normally capital intensive technology, which demands high input energy. A substitute carbohydrate polymer called lingo cellulose is being considered as a cheaper alternative to bio-mass source.

6.3 PYROLYSIS

This involves an irreversible chemical conversion of bio-mass into char, pyrolytic oils and fuel gases through a process of heating in oxygen free or low oxygen environment. The process is similar to distillation of coal to produce synthetic gases, tars, oils and cokes. The gases produced are a mixture of hydrogen, methane, lower hydrocarbons carbon dioxide and carbon monoxide. The liquids are oily while the solid product is similar to charcoal. Process efficiency depends on process temperature, (maintained below 600°C for ease of operation and prevention of excess H_2 production), feedstock used and air/fuel ratio used in combustion. Improvement in conversion efficiency can be advanced by using more sophisticated charcoal kilns made of brick, concrete or metal. Below 600°C, there are generally four stages in the distillation process:

(a) *100 to 120°C* – The input material dries with moisture passing up through the bed;
(b) *275°C* – The output gases are mainly N_2, CO and CO_2; Acetic acid and methanol distill off;
(c) *230 to 350°C* – Exothermic reactions drive off complex mixtures of chemicals (aldehydes, esters, etc.) with CO_2, CO, CH_4, C_2H_6 and H_2;
(d) *Above 350°C* – All volatile matters are expelled with a higher proportion of H_2 remaining with CO, charcoal and ash residues.

Tars and pyroligneous acid, called condensed liquids may be separated. On treatment, chemical products like methanol (CH_3OH) may be obtained. Traditional charcoal preparation can also be conceived as pyrolysis where the vapours and gases are not collected. When pyrolysis is aimed at the production of maximum amount of secondary fuel gas, the process is known as gasification. Various sections of a small scale pyrolysis unit are depicted in Fig. 6.3.1., where vertical top loading devices are usually most convenient. Here the fuel product is more clean and transportable compared to original bio-mass.

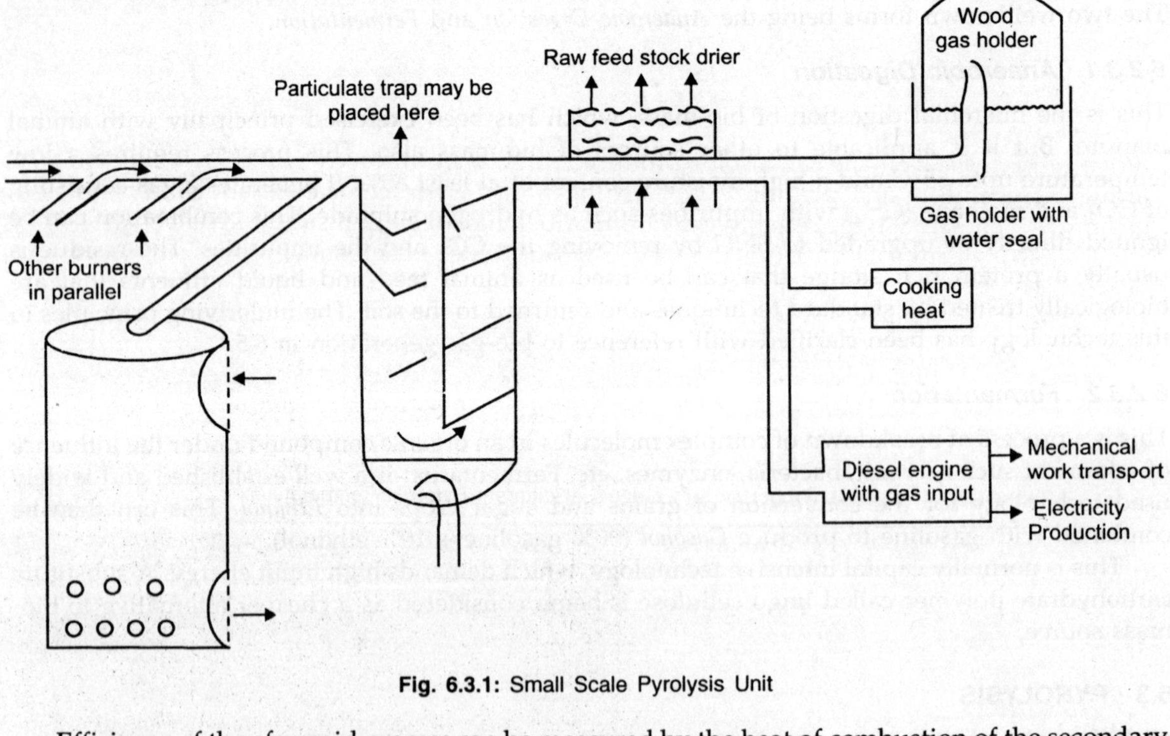

Fig. 6.3.1: Small Scale Pyrolysis Unit

Efficiency of the aforesaid process can be measured by the heat of combustion of the secondary fuels produced, divided by the heat of combustion of the input feedstock used. Typical efficiency values of a series of classified gasifiers will be presented in the forthcoming relevant section.

Recent studies have demonstrated that *fast* or *flash pyrolysis* can reduce the yield of solid charcoal to about 5% or less by weight with a proportionate increase in the yield of light liquids and gases. The inherent advantages of charcoal when used as solid fuel (briquetts) lies in its low sulphur content and higher calorific value compared to wood, ability to create a smokeless hotter flame and environment friendly nature.

6.3.1 Liquefication

Lianic yields are maximized by rapid heating of the feedstock to comparatively low temperatures. The vapours are condensed from the gas stream and these separate into a two phase liquor—

(A) An aqueous phase containing a soup of water soluble organic materials like acetic acid, methanol and acetone.

(B) Non-aqueous phase consisting of oils and tars.

Usually, it is profitable to upgrade this crude resources to premium fuels by conventional refining techniques.

Pyrolysis can be carried out also in presence of limited quantities of (i) oxygen (gasification), (ii) water (steam gasification) and (iii) hydrogen (hydrogenation). Only a brief outline of aforesaid processes is presented as a detailed description with schematic included in the section on bio-mass gasifiers.

(i) *Gasification:* One technique to maximize gas yield is to blow small quantities of air or oxygen into the reactor vessel and to increase the temperature above 1000°C. This causes a part of the feed to burn. Typical characteristics of fuel-gas from air blown gasifiers are: (a) Low calorific value around 5 MJ/m^3, (b) May contain upto 40% of inert N_2 gas, (c) Overall yields of 80 to 85% is expected.

Fuel gas from oxygen fed system has a higher calorific value of 10-20 MJ/m^3. This can be consumed or converted into methanol by standard catalytic process with yields around 50%.

(ii) *Steam gasification:* Methane can also be produced directly from woody matter by treatment at high temperature and pressures with hydrogen gas. Hydrogen can be added or more commonly generated in the reactor vessel from carbon monoxide and steam. Recent analysis suggests that steam gasification is the most efficient route to methanol production, where net energy yields vary between 55 to 60%.

(iii) *Hydrogenation:* Using a temperature of 300 to 400°C and 100 atmospheres, carbon monoxide and steam react with cellulose to produce heavy oils which can be separated and refined to premium fuels. It has been observed that operational condition have a major influence on the relative production of the reaction products. This is briefly presented in Table 6.3.1.1.

Table 6.3.1.1: Relative Proportion of Products at different Pyrolysis Temperatures

Pyrolysis temp (°C)	*Char (wt %)*	*Liquid (wt %)*	*Gases (wt %)*
400	42.14	35.85	22.01
500	32.02	39.56	28.42
600	29.81	29.53	40.66
700	25.46	20.47	54.07

6.4 PHOTOSYNTHESIS

The most important and inevitable chemical reaction on earth starts with Sun's radiation being intercepted and absorbed by the green pigment chlorophyll in the plant. Here, solar energy is stored in the plant in the form of chemical bond energy. This is photosynthesis, a phenomenon which illustrates biological conversion of solar energy into energy rich compounds (sugars and starches). These compounds can be burnt to produce steam required to drive turbines in thermal power stations. Thus, selective energy plantations can theoretically be a source of renewable energy depending on the photosynthesis efficiency and overall efficiency of conversion of sunlight to electricity.

Since the aforesaid conversion efficiencies are very low (of the order of 1% and even less) as compared to usual photovoltaic efficiencies, this process of photosynthesis has lacked serious consideration towards commercial utilization. Further the reaction is complex and yet not fully digested by scientists. Here, water and carbon dioxide molecule brakes down to form a carbohydrate with the release of pure oxygen. Written in the language of chemistry,

$$CO_2 + H_2O + \text{light} + \text{Chlorophyll} \rightarrow (H_2CO)_6 + O_2 \text{ (sugar)} + \text{Chlorophyll}$$

Or, $$6CO_2 + 12H_2O \rightarrow C_6H_{12}O_6 + 6H_20 + 6O_2$$

Visible light (wavelength range of 400°A to 700°A) is absorbed by the green chlorophyll which becomes activated and passes its energy to the water molecules. This range of light is called photo-synthetically active radiation (PAR) which constitutes about 40 to 45% of incident solar radiation. A hydrogen atom is then released and reacts with the carbon dioxide molecule to produce H_2CO and oxygen. It may be recollected that H_2CO is the basic molecule forming carbohydrate. It is stable at low temperature and breaks at high temperature through the following endothermic reaction.

$$H_2CO + O_2 \rightarrow CO_2 + H_2O + 112 \text{ Kcal/mole}$$

6.5 BIO-GAS GENERATION

This is essentially a mixture of the following gases:

Carbon dioxide 30 to 40%; Methane 55 to 65%;

H_2, H_2S and $N_2 < 10\%$;

It can be produced from the decomposition of animal, plant and human waste. It is a clean but slowly burning gas having calorific value between 5000 to 5500 Kcal/Kg. Significantly enough, the ingredients constituting the bio-gas can retain their value as a fertilizer when they are returned to the soil (as the by product in a typical bio-gasifier system). This is the reason why this fuel is a valuable renewable energy resource. It has assumed the popular name '*GOBAR GAS*' in the Indian scenario due to its origin linked with the cow-dung or cattle waste. But, a host of other biomass sources like crop residues, garbage kitchen wastes, paper and sugarcane wastes, sea wood and water hyacinth can also be effective for its generation. Any cellulose organic material of animal or plant origin which is easily bio-degradable has the potential to generate bio-gas.

Bio-gas is produced by digestion, pyrolysis or hydrogasification. Digestion is a biological process that occurs in absence of oxygen and in presence of anaerobic organisms at ambient pressure and temperatures of 35 to 70°C. The container in which this process takes place is a digester.

6.5.1 Anaerobic Digestion

Concerning the science of micro-organisms, there are living creatures which are microscopic in size and invisible to the naked eye. These are the different micro-organisms called bacteria, fungi virus, etc. Once again bacteria fall into two different groups, depending on their oxygen requirement. Those which grow in the presence of oxygen are called aerobic while anaerobic ones grow in absence of oxygen. The former produces CO_2, NH_3 and small amounts of other gases along with a decomposed mass and evolution of heat, while the latter produces CO_2, CH_4, H_2, traces of other gases along with decomposed mass. In a bio-gas plant, the option is to use anaerobic digestion as it produces methane (CH_4). In the wet process of producing bio-gas, this technique is speeded up by using a thermally insulated air tight tank with a stirrer unit and heating system. Bio-gas accumulates in the digester tank above the slurry and can be piped off continuously.

It will be worthwhile to highlight the advantages of the aforesaid process, which are as follows:

(a) *Calorific Value of Gas:* One of the by products has a calorific value which provides the necessary energy to produce steam or hot water;

(b) *New Sludge Yield:* The conversion process of organic mass to methane and CO_2 results in a small quantity of excess sludge;

(c) *Low Running Cost:* As there is no provision for oxygen/air in an anaerobic system, its running cost is hardly 25 to 30% of the corresponding aerobic system;

(d) *Nitrogen content:* The resultant sludge has often a nitrogen content rendering it as a very useful fertilizer;

(e) *Environment Friendly Nature:* If bio-gas can be a substitute for coal in certain boilers, it will definitely minimize air pollution. Further, the chemical compounds which are responsible for odour are broken down (the only slight odour resulting from H_2S). Since the system is enclosed, the odours are largely contained.

6.5.2 Factors Affecting Bio-gas Generation

The following factors are responsible for controlling the quantity and quality of bio-gas:

(i) *pH/Hydrogen Ion Concentration:* Along with a constant supply of bio-gas, the pH range has to be monitored at regular time intervals. A few ideal pH values are the following:

Type of waste/bio-mass	*Optimum pH range*
I. Digestion of sewage solids	7 to 7.5
II. Digestion of animal/plant wastes	8 to 8.5

Besides, in case I, the digester is expected to be fully active with micro-organisms so that bio-digestion will be very efficient. The pH ranges between 4 to 6 and 9 to 10 are regarded as acidic and alkaline respectively. These ranges are avoided as these pH values are detrimental to methane production organisms.

(ii) *Temperature:* For satisfactory amount of methane generation from fixed volume of slurry (say cattle dung), it has been experimentally observed that the optimum temperature range is 35 to 38°C with fall in bio-gas production as temperature decreases below 20°C. This can be evidenced by casting a glimpse at the temperature dependence of the period of digestion (Fig. 6.5.2.1) for a 90% anaerobic digestion of sewage sludges. It can be seen that there are two different micro organisms – Mesophilic and Thermophilic that are predominant in the two temperature ranges. The optimum mesophilic and thermophilic temperatures are 35°C and 55°C respectively. In Indian climates the time required for digestion may be lowered following the mesophilic bacteria activity. The other organism is seldom used as it calls for heating of storage tanks to high temperatures.

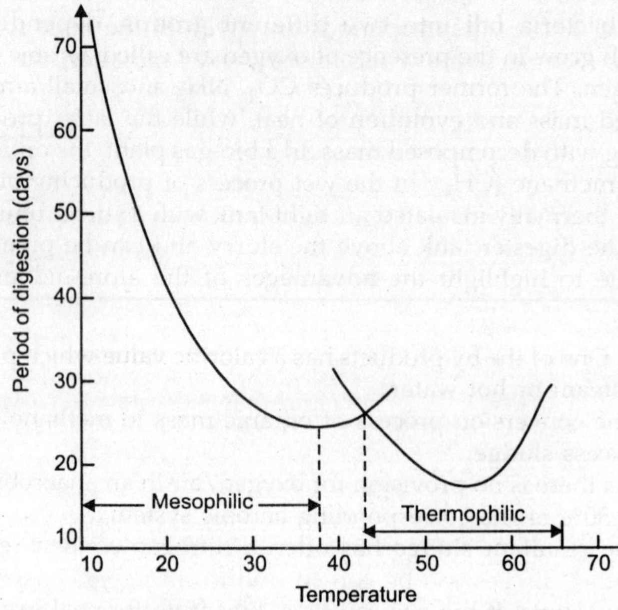

Fig. 6.5.2.1: Effect of Temperature on Digestion

Besides ambient temperature control, other weather conditions influencing the generation of bio-gas are wind velocity (chill factor), directly available sunlight corresponding to optimum temperature and type of food given to cattle (Gobar gas generation).

(iii) *Loading Rate:* This is the amount of raw material (usually Kg. of volatile solids) fed into the digester per day per unit volume. As an example, most municipal sewage treatment plants operate at a loading rate between 0.5 to 1.6 Kg/m^3/day. Higher rates are possible, but care has to be exercised to ensure that pH value goes low enough (accumulation of acids) leading to cessation of fermentation.

(iv) *Total Solid Content (TSC):* The raw cow-dung fed into a Goabr Gas Plant usually contains moisture in the range of 80 to 82%. The balance of 18 to 20% is termed as total solids. The optimization of total solid content is also necessary for digesting the feedstock at a faster rate and also in deciding the mixing ratios with other sources of bio-mass such as crop residues, wood, plants, etc.

(iv) *Carbon to Nitrogen Ratio in the Imput Wastes:* It is important to understand here that the bacterial elements present in the wastes uses up carbon 30 times faster than they use up nitrogen. This is why the fixation of carbon to nitrogen ratio in the right proportion is essential along with monitoring of temperature, pH, etc.

Favoured by optimum values of pH, temperature, etc. a carbon to nitrogen (C/N) ratio of 30: 1 will permit digestion to proceed at an optimum rate. If there is too much carbon in the raw wastes, nitrogen will be used up first and carbon left over, making the digester slow down quickly. On the other hand, if nitrogen is present in excess, the carbon is exhausted soon leading to total ceasation of fermentation. Further, the formation of undesired NH_3 by the excess nitrogen can kill the growth of bacteria, essential for generating methane.

(v) *Retention Time/Rate of Feeding:* The time period during which feedstocks are retained inside the digester is the retention period. This naturally depends on the type of feedstocks used and the optimum temperature range. Table 6.5.2.1 shows the retention periods for different sources of bio-mass. The minimum periods are given in the table for an average Indian Climatic conditions (Temperature, wind velocity, solar radiation, etc.).

Table 6.5.2.1: Retention Period

Materials	*Retention period*	
Cow & Buffalo dung	50 days	
Pig dung	20 days	Assuming an average
Poultry droppings	20 days	ambient of 30°C.
Night soil	30 days	

6.6 SITE SELECTION

The factors that are to be considered while selecting the site for a bio-gas plant, are—

(i) *Distance:* The distance between the plant and the site of gas consumption should be less in order to achieve system economy and minimization of gas leakage, for example, this optimum distance is 10 metres or less for a plant of capacity of $2m^3$.

(ii) *Run off Water:* Care has to be exercised to prevent the flow of run off water during the monsoon- intercepting ditches can be constructed as an integral part of the plant.

(iii) *Exposure of Sunlight:* The selected site should be favoured by insolation so as to maintain a temperature of 15°C to 30°C for bio-gas generation at a good rate.

(iv) *Distance from Wells:* As the seepage of fermented slurry may polute the well water, a minimum distance of 15 to 20 metres from the nearest well must be adhered to.

(v) *Space Requirements:* Sufficient space must be available for day to day maintenance. As a thumb rule, 10 to $15m^2$ of area is needed per m^3 of gas produced.

(vi) *Availability of Water:* Plenty of water must be available as the slurry uses a solid concentration of 7 to 9%.

6.7 DIGESTER DESIGN PROCEDURES

Shape of the digesting tanks used in practices are diverse and no particular shape can be recommended as convenient. There has to be a cover for retaining the gas. This may be fixed one or floating one. For operation of the digester in the thermophilic range, maintenance of nearly constant temperature is a vital criterion (Fig. 6.5.2.1).

A number of thumb rules for bio-gas production from cow dung in digesters are:

(a) Ratio of dry dung to water in wet dung is 4 : 1.

(b) 1 Kg. of wet dung is mixed with 1 Kg. of water to obtain the slurry.

(c) A digester reserves 10% of its volume as the gas space above the slurry. Therefore, total volume of digester can be 1.1 times the volume of slurry.

(d) At 15°C, 1 Kg. of dry dung produces 0.186 m^3 of gas. However, the exact digester sizing steps are worked out in next section 6.7.1.

6.7.1 Digester Sizing

The energy (E) available from bio-gas digester is

$$E = n_c H_b \cdot V_b \quad ...(6.7.1.1)$$

Where n_c is the combustion efficiency of the burner. (60%), H_b represents the heat of combustion per unit volume (calorific value of bio-gas 20 MJ/m^3 at 10 cm water gauge pressure, 0.01 atmosphere) and V_b the volume of bio-gas.

or, E can be written as $E = n_c H_m F_m V_b$ ...(6.7.1.2)

Where H_m stands for the heat of combustion of methane gas (56 MJ/Kg, 28 MJ/m^3 at STP) and F_m is purely a fraction depicting the methane content in the bio-gas produced. Once again V_b can be expressed as:

$$V_b = Cm_o \quad ...(6.7.1.3)$$

Where C is the bio-gas per unit dry mass of whole imput and m_o is the mass of the dry imput. The volume of fluid (V_f) is $V_f = m_o/P_m$...(6.7.1.4)

Where P_m is the density of dry material in the fluid. Finally, the volume of digester (V_d) is simply

$$V_d = V_f \cdot t_r \quad ...(6.7.1.5)$$

With t_r representing the retention time in the digester tank.

Example: Estimate the volume of a family bio-gas digester and the power available from the digester, suitable for the output of ten cows with a retention time of 20 days at an ambient temperature of 30°C.

Assume the following:

(i) Dry matter consumed per day = 1 Kg.
(ii) Bio-gas yield = 0.24 m^3/ Kg.
(iii) Efficiency of burner = 60%.
(iv) Proportion of methane in bio-gas = 80%.
(v) Heat of combustion of methane = 28 MJ/m^3.

Solution: Mass of dry imput (m_o) = 1 Kg/day × 10 = 10 Kg/day.
Using relation (6.7.1.4), fluid volume (V_f) = m_o/P_m = 10 Kg/day/50 Kg/m^3 = 0.20 m^3/day.
From relation (6.7.1.5) V_d = Digester volume = 0.2 × 20 = 4.0 m^3.
Volume of bio-gas (V_b) = Cm_o = 0.24 × 10 = 2.4 m^3/day.
Energy available from the digester = $E = n H_m F_m V_b$.
∴ E = 0.6 × 28 × 0.8 × 2.4 = 32.25 MJ/day = 8.80 KWh/day.

6.8 TYPES OF BIO-GAS PLANTS

There exists numerous models of bio-gas plants like:

(a) The continuous and batch types,
(b) The Dome and the Drum types,
(c) Different Drum types.

In the type (a), there is a single digester in which raw materials are charged regularly and the process goes on without interruption except during maintenance schedules. Advantage of such plants lies with respect to continuous gas production, smaller digestion chambers and lesser period of digestion and problems compared to batch type. In batch type plants, there is a need of fermented slurry to start the digester. There is a possibility of a direct change to the acid phase in absence of a fermented slurry, which is likely to affect the formation of CH_4. Also, the plant is capital intensive compared to continuous ones.

In type (b), the two models that are commonly employed are called: (i) Floating gas holder plant and (ii) Fixed dome digester plant.

The gas holder is separate from the digester in a floating gas holder plant while the two are combined in case of fixed dome digester plant. The Khadi Village Industries Commission Model (KVIC) follows the floating gas holder design. Here the plant shown in Fig. 6.8.1. contains steel

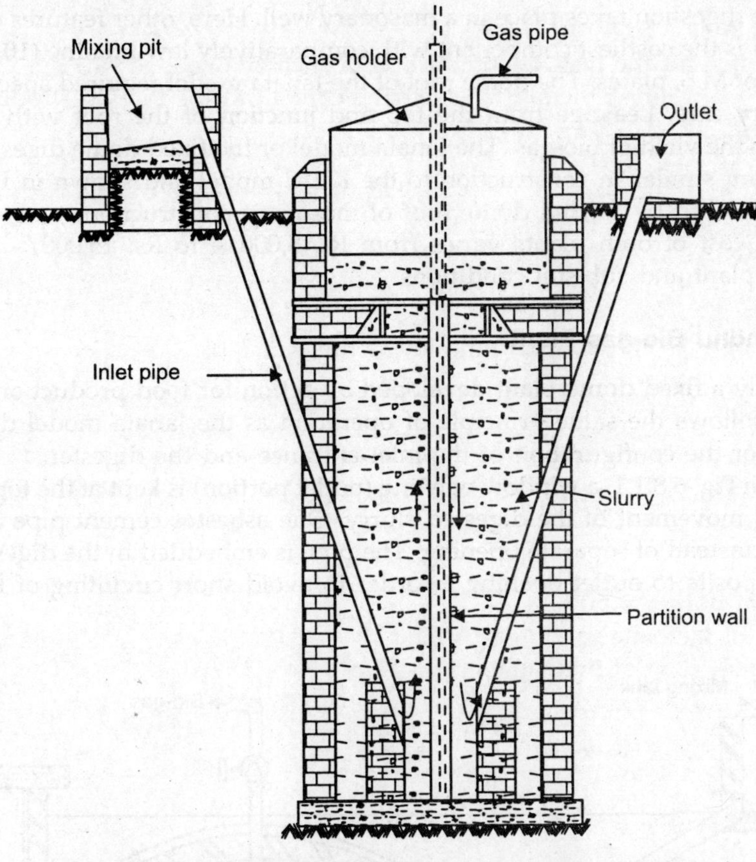

Fig. 6.8.1: Common Circular Digester with Floating Gas Holder and no Water Seal (India). (KVIC digester)

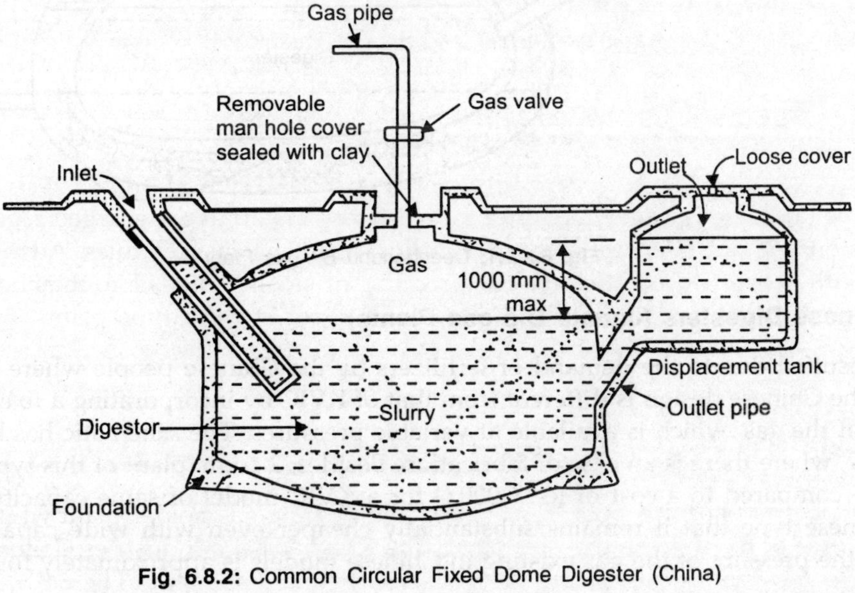

Fig. 6.8.2: Common Circular Fixed Dome Digester (China)

drum where the digestion takes place in a masonary well. Here, other features of the KVIC model is that the drum is the costliest component with comparatively low lifetime (10 years) and the gas holder is made of M.S. plates. The dome roof of the Janata model required specialized design and skilled masonary care. Leakage from the top and junction of the roof with the digester wall, causes a drop in the yield of bio-gas. The Janata model or the fixed dome digester (Chinese Plant) is drum less plant similar in construction to the KVIC model and shown in Fig. 6.8.2. Here the steel drum is replaced by a fixed dome roof of masonary construction.

The overall cost of both plants varies from Rs. 5,000/– to Rs. 15,000/– depending on the capacity of the plant and sub-soil conditions.

6.8.1 Deenbandhu Bio-gas Plant

This is essentially a fixed dome plant developed by action for food production, New Delhi. This low cost plant follows the same principle of operation as the Janata model described in earlier section except for the configuration of the inlet entrance and the digester.

As seen from Fig. 6.8.1.1, a window opening (outlet portion) is kept at the top of the foundation for the outward movement of the digested slurry. One asbestos cement pipe of 15 cm diameter is used for inlet instead of separate opening. The pipe is embedded in the digester wall at a fixed position just opposite to outlet opening in order to avoid short circuiting of fresh material and digested slurry.

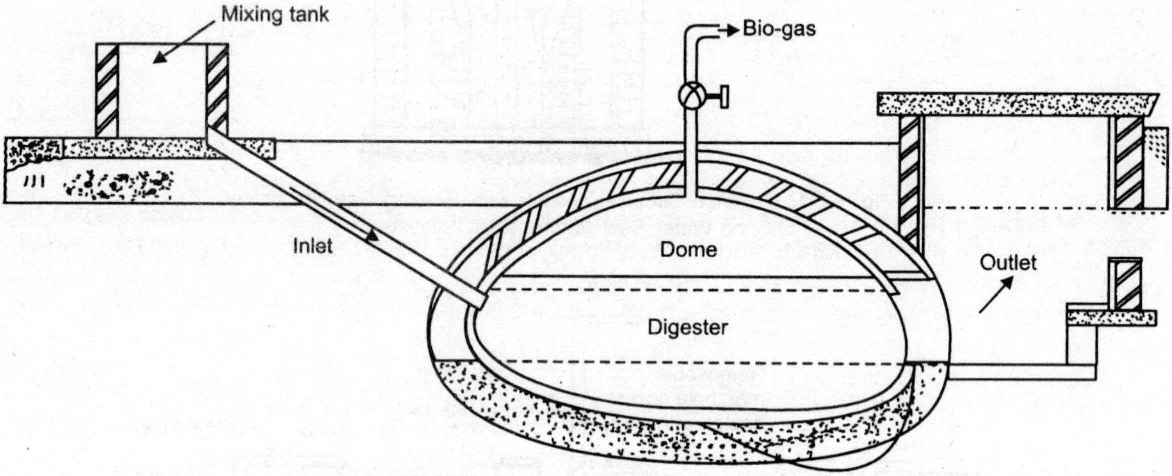

Fig. 6.8.1.1: Deenbandhu Bio-gas Plant

6.8.2 Chinese Digesters (Janata Bio-gas Plants)

This has resulted due to the demand of fertilizers by the Chinese people where bio-gas is a by product. The Chinese design is different from that of KVIC by incorporating a fixed dome for the collection of the gas, which is available at variable pressures. The schematic has been presented in Fig. 6.8.2, where there is an ease of fabrication. Besides, 2 cu.m. plant of this type costs around Rs. 3,000/- compared to a cost of Rs. 6,000/- for a KVIC model of same capacity. It is an edge of the Chinese type that it remains substantially cheaper even with wide capacity variations. Generally, the pressure of the gas existing in Chinese models is approximately four to five times that in KVIC plants. Planning, Research and Action division (PRAD) Lucknow has designed lower cost and popular Janata plants by modifying the design of Chinese plants.

6.8.3 Maintenance Procedures for Bio-gas Digesters

A few techniques for maintaining the bio-gas production from the aforesaid plants are:

(A) *Insulation:* Normally the external surface of the digester is insulated using different materials like mineral wool, aluminium cladding, fiber glass, straw, etc. This is why Janata type plants are usually fabricated below the ground level. In the floating dome design, 54% of the heat losses occur from the top of the gas holder, which is prevented by using paddy straw mat and other agricultural residues.

(B) *Handling of Effluent Slurry:* This is a major problem if the operator is not having sufficient open space or compost pits to get the slurry dry. For a typical domestic plant, 200 litres capacity oil drums can be used to carry this effluent to the fields but this will involve additional human/animal labour or use of diesel auto vehicles.

(C) *Methanogenic Activity:* During winter, as the temperature falls, there is a decrease in the activity of methogenic bacteria and subsequently lowering of the gas production rate. Methods used to overcome this temperature problems are circulation of hot water obtained either from solar water heater or IC engine. Care has to exercised that the temperature does not exceed 60°C, so as to permit the existence of mesophilic bacteria (refer Fig. 6.5.2.1 of section 6.5.2) other practices include addition of various nutrients for bacteria and covering the bio-gas plant by straw bags during night hours.

(D) Addition of Urea fertilizers in large quantity is to be avoided as toxicity of ammonia may cause a decrease in gas production.

(E) pH and volatile fatty acids play an important role in anaerobic digestion and should be controlled within the appropriate ranges.

(F) Leakage of gas especially from Janata type bio-gas plants is a very common problem. Detection of leakage points and repairing is of utmost necessity. During repairing there should be no gas inside the gas holder and the stop cock remains open till repaired points are dried.

6.8.4 Performance Prediction of Bio-gas Plants

Given a fixed amount of feed stock, the bio-gas production depends on:- (a) Retention time, (b) Loading rate, (c) Volatile solid concentration in feedstock and (d) Slurry temperature. The slurry temperature depends upon thermal environmental conditions of a particular place. All these parameters except thermal environment conditions can be controlled and maintained at a desired level at a particular location. Since it is not easy to compare the different design procedures for maximizing efficiency under varying environmental conditions, there is a need for a performance prediction procedure or algorithm.

6.9 BIO-MASS GASIFIERS

It is an equipment which can gasify a variety of bio-mass such as wood waste, agricultural waste like stalks, roots of various crops, maize cobs, etc. In fact, a solid fuel in a gasifier is converted by a series of thermochemical processes like drying, pyrolysis, oxidation and reduction to producer gas. When atmospheric air is the gasification agent, typical composition of the gas obtained from wood gasification on volumetric basis is given in Table 6.9.1.

Table 6.9.1: Typical Results of Wood Gasification

Carbon monoxide	18 to 22%
Hydrogen	13 to 19%
Methane	1 to 5%
Heavier Hydrocarbons	0.2 to 0.4%
Carbon dioxide	9 to 12%
Nitrogen	45 to 55%
Water vapour	4%

Producer gas consists mainly of carbon monoxide, hydrogen and oxygen and can be used for generation of motive power in either dual-fuel engines or modified diesel engines. Engines operating on spark ignition systems can be fully run by this producer gas while those using compression ignition systems can replace 60 to 80% of diesel oil. In several industrial oil fired boilers, the gas can be burnt directly.

Composition of the dry producer gas having net heating values for various types of bio-mass are presented in Table 6.9.2. In addition to the components depicted in the table, the product gas contains light hydro-carbons and tar which increases net heating values by 0.05 to 0.4 MJ/Nm^3. Typically wood gas contains about 50-100 gm/Nm^3 tar and that for straw is about 10 to 40 gm/Nm^3.

Table 6.9.2: Composition of Producer Gas

Fuel	*Moist (%wt)*	*Particle Size (mm)*	*Ash (%)*	*CO %v*	*H_2 %v*	*CH_4 %v*	*CO_2 %v*	*HHV MJ/Nm*
MSW* briquettes	8	0-35	11.8	17.8	14.6	2.6	11.0	4.7
Crushed waste+ Wood chips (45%)	29/42	0-35	16.8	17.9	14.8	1.9	13.0	4.5
Forest residue	52	0-30	4.2	23.1	14.9	2.3	9.9	5.3
Barley straw	29	0-100	5.7	18.9	17.6	2.3	13.8	5.1
Wheat straw	16	0-70	4.1	14.9	11.3	2.9	14.8	4.1
Car scrap yard waste	16	10-60	19.4	22.5	19.0	1.4	9.0	5.4
Sod peat	52	-	-	16.1	16.7	2.0	14.3	4.5
Wood chips	43	-	<1.0	29.0	15.4	1.6	6.8	5.9
Rice Husk	14	-	23.3	17.9	16.5	2.4	13.5	4.9
Rice straw	12	-	18.5	19.2	17.7	2.7	14.2	5.2

*MSW—Municipal Solid Waste; HHV—Higher Heating Value; Expected values

6.9.1 Classification of Gasifiers

Normally the gasification of solid fuels containing carbon is accomplished in an air sealed closed column under slight vacuum relative to ambient pressure. The fuel column is ignited at one point and exposed to air blast. The gas is drawn off at a different location. Depending on the aforesaid positions of air inlet and gas withdrawal with reference to the fuel bed movement, the gasifiers that are operated till date are shown in Table 6.9.1.1.

6.9.1.1 Updraft Gasifiers

This type of gasifier is, in principle a refractory levied shaft furnace. Solid fuel is fed through the hydraulically operated lock hopper placed on top of the reactor usually equipped with a scraper or a rod discharger. Corresponding to a 5MW size gasifier, the approximate volume of fuel bin is about 300 m^3. This fuel is transported to the fuel feeder of the gasifier using a conveyor. Rate of fuel feeding is controlled by a level sensor in the gasifier, which keeps the fuel layer within the gasifier at a standard level regardless of the output of the boiler plant. The fuel is gasified with air, which is blown into the gasifier through a rotating grate placed at the bottom. The gasification air amount is adjusted by the temperature or pressure of the boiler water stream. The special feature of the gasifier is the possibility of regulating the gasification temperature – an important factor while using fuels with low ash melting/sintering temperature. The regulation is conducted by humidifying the gasification air in a scrubber or by using steam injection. Fuel bed moves down as more and more ash is removed from the bottom by the use of hydraulically rotated mechanical grate. The producer gas generated drift upwards against the fuel flow to exit at the upper part of the gasifier. During operation, different functional zones develop in the gasifier, which determines the gas composition and properties. These zones are labelled in Fig. 6.9.1.1.1.

Table 6.9.1.1: Bio-mass Gasifiers

Direction of gas flow	*Type of Bed used*	*Capacity of the gasifiers*
Up draught	Fixed Bed	(a) Small size < 10 KW
Down draught	Fluidesed Bed	(b) Medium size – 10 to 50 KW
Cross draught		(c) Large size – 50 to 300 KW
		(d) Very large size – 300 KW

Applications:

(a) Providing shaft line power to agriculture pumps and processing m/cs like thrashers, straw choppers.

(b) Providing shaft line power of rural industries like saw mills, carpentary and fabrication shops and small rice mills. Decentralised source of electrical energy in rural health centres, Milk chilling centres.

(c) Shaft power requirements to various industries like dairy, oil mill, ceramics and poltry industries. Mining operations as wood processing units and well drilling. Electrification small village.

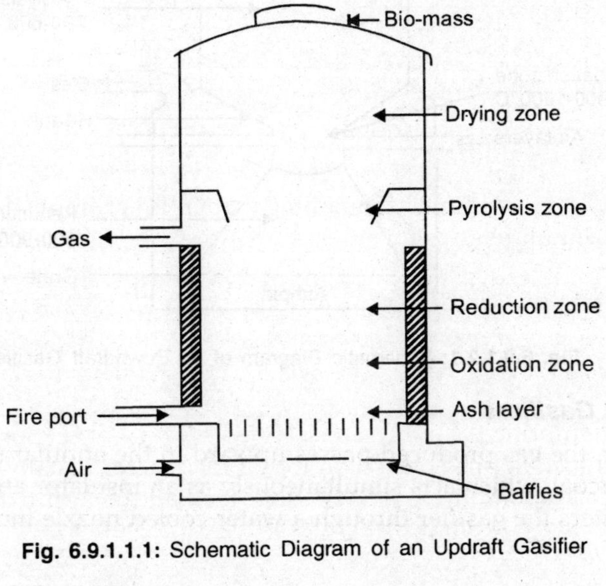

Fig. 6.9.1.1.1: Schematic Diagram of an Updraft Gasifier

Several applications of updraft gasifiers include:

(a) Retrofilling oil/gas fired boilers with gasifiers;
(b) Connection possible with large oil/load fired boilers;
(c) Capacities ranging from 1 to 5 MW.

Some of its prominent advantages are the following:

(a) Use of different feedstock/mixture (wood, woodwaste, bagasse, straws, husks, etc.);
(b) Different particle size varying from few mm upto 100 mm can be used;
(c) Producer gas having calorific value in the range 4 to 7 MJ/Nm^3;
(d) Higher thermal conversion efficiencies;
(e) Low ash carryover;
(f) Gasification process can be completely automated due to easy control over the parameters of interest.

6.9.1.2 Down Draft Gasifier

A schematic labelled diagram is presented in Fig. 6.9.1.2.1. where air enters through radial tyers near the top of the fire box. The partial combustion occurs in the zone in front of one tyer openings. The heat of combustion pyrolyzes the fuel immediately above and the hot gases exit through the grate after proceeding downward through the fire bos constriction. As the raw gases pass through a high temperature zone, most of the unburnt pyrolysis product can be cracked into gaseous hydrocarbon thus yielding a relatively clean gas. This ability of the downdraft gasifier renders it highly compatible with engine applications. Modifications to this category through R & D work have led to throatless gasifiers (as throat poses problem of fuel movement) and also fluidized bed gasifiers.

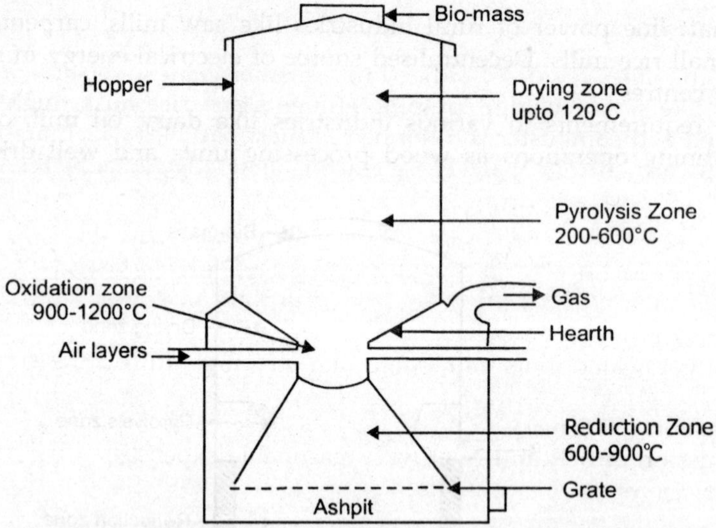

Fig. 6.9.1.2.1: Schematic Diagram of an Downdraft Gasifier

6.9.1.3 Cross Draft Gasifiers

In this type of gasifier, the gas produced passes upward in the annular space around the gasifier that is filled with charcoal, which acts simultaneously as an insulator and a dust filler. As shown in Fig. 6.9.1.3.1, air enters the gasifier through a water cooled nozzle mounted on one side of the firebox. The gas is produced in the horizontal zone in front of the nozzle and passes through a vertical grate into the hot gas port on the opposite side. Due to the short path length for the gasification reactions, this type of gas producer responds most rapidly for changes in gas production. However, this is seldom used in actual practice.

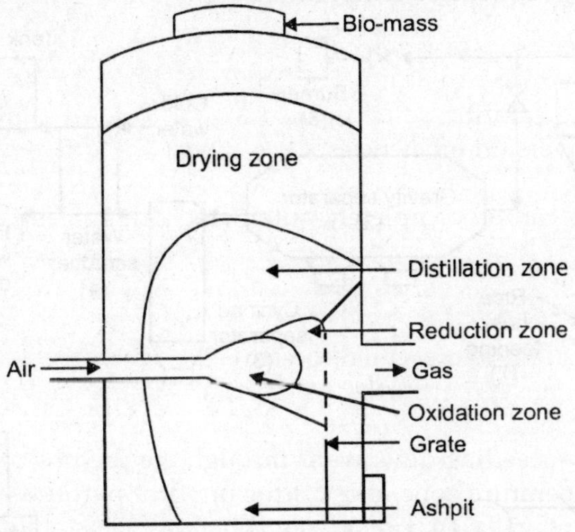

Fig. 6.9.1.3.1: Schematic Diagram of a Crossdraft Gasifier

6.9.1.4 Fluidised Bed Gasifier

At this stage it is easy to appreciate that the performance, of a fixed bed gasifier will depend on fuel properties such as chemical analysis, volatile contents, calorific value, size distribution and ash contents. On the other hand, a fluidized bed gasifier is most versatile as it is able to handle bio-mass sources like sewagecludge, pulping effluents and rice husk (high ash content). At the heart of a fluidized bed combuster is a hot bed of inert particles which are held in suspension – fluidized by an upward current of air. These reactors generally contain either inert material (sand) or reactive material (limestone or catalyst). The bed material is kept in fluid state by the rising column of gas. Principal advantages of these gasifiers are:

(A) Fuels with calorific value from 800 to 8000 Kcal/Kg can be used;
(B) Bio-mass with high moisture and ash content do not pose problems;
(C) Normal operating temperature of fluidized bed is in between 750°C to 950°C, so that ash zones do not get heated to its initial deformation temperature, thus preventing clinkering or slagging;
(D) High combustion efficiency;
(E) Reduced emission of SO_2 and N_2O with reasonable expenses;
(F) Simplicity of operation compared to that of an oil fired boiler.

Schematic diagrams of a Rice-husk gasifier is shown in Fig. 6.9.1.4.1.

Here, the main components of the system are (1) Reactor, (2) Gravity dust seperator, (3) Cyclone seperator, (4) Wet scrubber, (5) Moisture seperator, (6) Blower, (7) Rice husk feeder, (8) Burner, (9) 5 KVA DG set. The reactor (made of SS-310) is 150 mm in diameter and 1000 mm in height. A perforated plate is used at the bottom for uniform air distribution. Provision is made for fuel feeding at two locations namely 300mm and 600mm height from the distributor plate. The reactor as well as the pipe lines are well insulated to contain heat loss and also to avoid tar condensation. Ash drain holes are provided above the distributor plates.

To start the operation, the gasifier was filled with sand bed to a required height and heated using the ignition electric coil. On attainment of requisite temperature, the rice husk was fed from the top and fluidising air was sent from the bottom through a blower. It took 45 minutes to reach the steady state i.e. the attainment of consistent gas quality. The gas composition should be investigated as a function of reactor temperature.

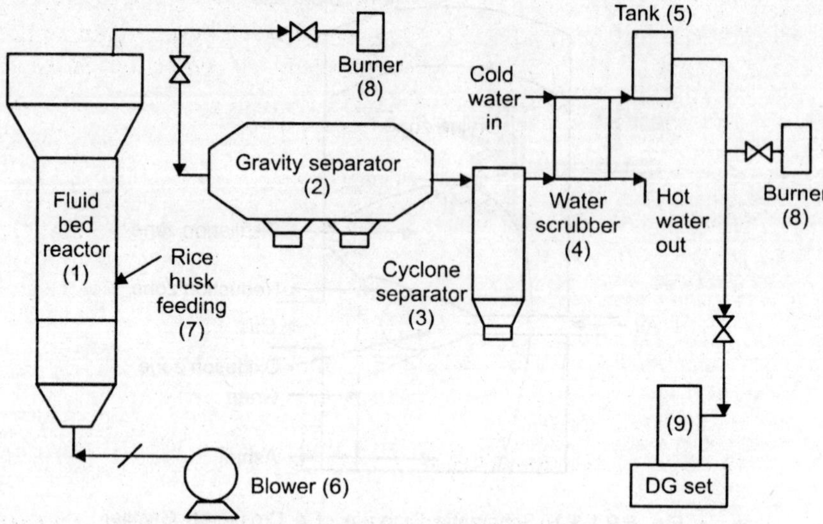

Fig. 6.9.1.4.1: Schematic Diagram of Fluidised Bed Rice Husk Gasifier

The aforesaid system has been in operation at the Anna University, Chennai through funds forwarded by Swedish International Development Agency and the principal results obtained are summarized below:

(A) At a reactor temperature of 800°C, the gas composition obtained contained 20% CO and 15% H_2;
(B) The calorific value ranged between 800 to 1000 Kcal/mm;
(C) The gas was fed into the DG set and the specific fuel consumption (SFC) rate was noted against the varying DG set load. A 50 to 60% diesel replacement could be achieved on an average.

6.9.1.5 Pressurised Gasification for IGCC

Pressurised gasifiers suitable for higher power capacities (above 75 MW) are developed in both fixed and fluid bed types. Some of the attractive features of this mode of operation are (a) High reaction rates, (b) High carbon conversion, (c) Higher inbed sulphur removal, (d) Smaller reactor configuration, (e) High scale up potentials, (f) Compatibility with gas turbines in the IGCC mode. This type of gasifiers derives the principal benefit from the high efficiency of the combined cycle technology of solid fuel to combustible gas converting process. Figure 6.9.1.5.1 presents the schematic of the gasifier while Fig. 6.9.1.5.2. depicts the schematic of an IGCC plant.

The gasifier system of Fig. 6.9.1.5.1 is a pressurized fluidized bed reactor developed by carbona for bio-mass gasification utilizing IGI's U-gas and Renugas experiences. The fuel feeding is done by a lock hopper system and a screw feeder. 10 to 15% of the air from the gas turbine compressor is extracted for this purpose and fed into the gasifier after passing through the booster compressor/heat exchanger system. A small amount of steam is also used in the gasifier

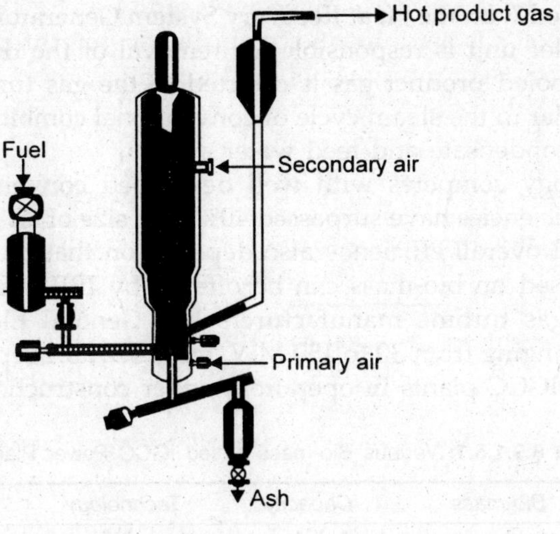

Fig. 6.9.1.5.1: Schematic of Pressurised Gasification

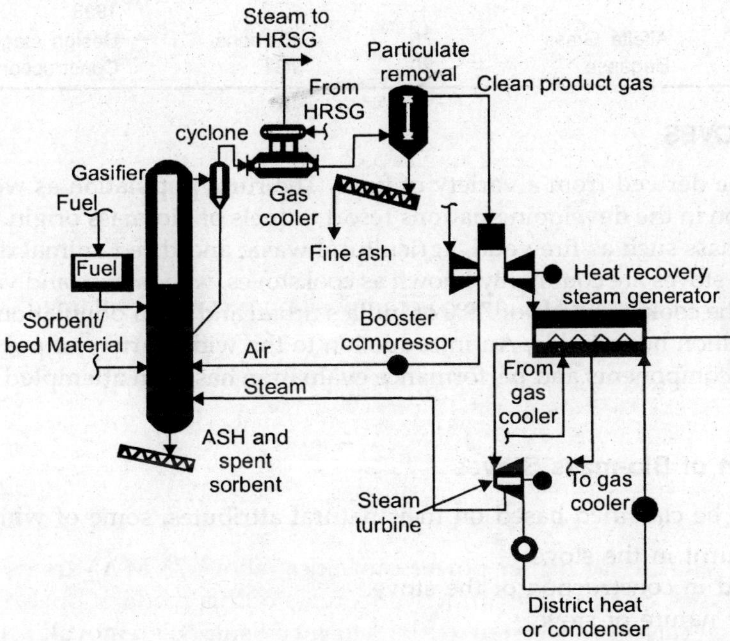

Fig. 6.9.1.5.2: Schematic of IGCC Process

for better control of temperature. The product gas contains combustible components like carbon monoxide (9 – 12%) and methane (5 to 10%) by volume besides inert gases like CO_2, N_2 and water vapour and is characterized by a typical low calorific value of 4 – 6 MJ/Nm^3.

As can be seen from Fig. 6.9.1.5.2, the product gas treatment of the simplified IGCC includes the steps of gas cooling and gas clean-up. The product gas existing in the gasifier is cooled from free board temperature to 450 – 500°C. In a fire tube type gas cooler, the saturated steam generated by the gas cooler is integrated with the Heat Recovery System Generator (HRSG) in the combined cycle. A ceramic candle filler unit is responsible for removal of the dust content of the cooled product gas. Finally, the cooled product gas is directed to the gas turbine. Steam cycle of this TGCC plant is, in fact, similar to the steam cycle of conventional combined cycle plants including HRSG steam turbine and condensate and feed water system.

The aforesaid technology competes with well developed conventional power generation systems where the plant efficiencies have surpassed 40%. The size of an IGCC plant is determined by the gas turbine size and overall efficiency also depends on that of gas turbine.

Today IGCC plants based on bio-mass can be offered by IBIL/Carbon with a commercial guarantees from leading gas turbine manufacturers like General Electric Westinghouse and Mitsubishi for plant sizes ranging from 30 to 150 MW. Table 6.9.1.5.1 represents the current status of various bio-mass based IGCC plants in operation/under construction.

Table 6.9.1.5.1: Various Bio-mass based IGCC Power Plants

Location	*Bio-mass*	*Capacity*	*Technology*	*Status*
Sweeden	Wood waste	6	Ahlstrom	Operation since 1996
Finland	Wood	8	Carbona	Start-up in 1998
U.S.A.	Alfalfa Grass	75	Carbona	Design stage
Hawall	Bagasse	20	IGT	Construction

6.10 BIO-MASS STOVES

Cooking energy can be derived from a variety of fuels. The rural population as well as a section of the urban population in the developing nations resort to fuels of bio-mass origin. Thus, cooking stoves that burn bio-mass such as firewood, agricultural waste and dried animal dung are called bio-mass stoves. These stoves are commonly known as cookstoves, woodstoves and various chulhas. These stoves cater to the cooking needs of 75% of India's urban and rural population – constituting approximately 130 million households. An introduction to the wide variety of possible design of bio-mass stoves; their components and performance evaluation has been attempted in the sections to follow.

6.10.1 Classification of Bio-mass Stoves

These stoves can best be classified based on their natural attributes, some of which are—

(A) The fuel being burnt in the stove.
(B) The material used in construction of the stove.
(C) Fixed or portable nature of stove.
(D) Number of pots or pans the stove can heat at a time.

In the 1st category, a stove that can burn wood can also be used to burn agricultural wastes and animal dung. Similarly, a stove that uses charcoal as fuel can also operate with fossil fuels such as coal and coke. Thus, the possible alternatives in this category are—

(i) The stoves for woody bio-mass capable of burning wood, animal dung and agricultural waste.
(ii) Stoves that are capable of burning saw dust, rice husk and such powdery bio-mass (Mukunda et al.).
(iii) Charcoal stoves that are suitable for fuels that burn without a flame such as charcoal or coke.

It is obvious that the fuels consumed in a biomass stove depends on the locally available resources in that region during that season. A stove designed for best performance using a given fuel does not depict same performance when the fuel is changed.

Materials for construction of stoves used in a region depends on the locally available materials for construction such as masonry, unfired clay, fired clay, metallic sheets or cast iron. Fixed stoves in India employ one of the first two materials. Mudstoves can be fabricated by housewives themselves in their kitchen. Ceramic and metallic stoves, on the other hand are light weight and portable.

These have considerably lower thermal mass and they store less thermal energy in their bodies. On the contrary, the fixed stoves can absorb a lot of thermal energy of combustion of the fuel.

Stoves can be fabricated to heat a single pan or multi pan with a single combustion chamber. In case of multi pan stoves, passages are provided for the hot fuel gases to pass under all the pans. Single, two and three pot designs are common in traditional stoves as well as in improved versions. Most multi pan stoves are fixed, but portable multi pan stoves of metals/ceramics are also available.

6.10.3 Components of Bio-mass Stoves

Although several types of bio-mass stoves are traditionally available and many more versions developed through the improved cookstoves programmes, generally the basic components of a bio-mass stove can be identified easily and their basic functions defined.

The most important part of any bio-mass stove is the combustion chamber or the firebox. This part of the stove houses the fuel bed which burns in the presence of air. The highest temperature in the stove occurs in this part. Many stoves contain a second firebox to burn the volatile matter released from bio-mass due to pyrolysis. The geometry of the combustion chamber obviously influences the combustion efficiency.

Air inlet to the combustion chamber is the next important component whose size and dimensions can be crucially controlled for efficient performance of the combustion chamber. Some improved supplies air to the fuel bed, where the fuel is pyrolysed and burnt and the secondary air let which supplies air to the region above the fuel bed to complete the combustion of the volatile products of pyrolysis.

Most improved stoves use grates for improving the combustion quality and rate. Grate is a perforated frame of metal bars or other convenient material on which the solid fuel is burnt. The area of air passage openings in a grate has been typically optimized as 25 to 30% of the total grate area.

Chimney is a vertical pipe attached to the stove to carry the combustion products out of the kitchen. With reference to a stove, the chimney has to provide the draught for the suction of the inlet air. Due to free convection, hot gases from the combustion chamber rise into the chimney, giving rise to the draught. The height and diameter of the chimney can thus be fixed up based on the size and power output of the stove. Materials commonly employed for chimneys are clay and concrete pipes, cast iron, masonry, bamboo, ferromud, etc. Portable stoves are usually chimneyless and chimneys are employed only in fixed stoves.

6.10.4 Performance Evaluation and Testing

Performance testing of bio-mass stoves to determine their thermal efficiency and emission levels is a complex task in view of the variety of designs of stoves, fuels burnt in them, sizes and types of vessels used, ambient conditions, ventilation levels and available wind speed at cooking site.

In this section, testing procedure for thermal efficiency of stoves has been elaborated with respect to various other involved process efficiencies.

Thermal efficiency of bio-mass stoves is an overall efficiency (n_o) which is a product of several process efficiencies.

Thus, $n_o = n_c.n_t.n_p.n_f$ = Heat absorbed by food/Energy potential of bio-mass.

Where n_c, n_t, n_p and n_f represents respectively the combustion, heat transfer pot and control efficiency. Out of these, n_c depends on the fraction of energy potential of bio-mass fuel that is released by combustion process.

Thus, $$n_c = \frac{\text{Heat generated by combustion}}{\text{Energy potential of the fuel}}$$

Poor mixing of air with fuel is likely to reduce the value of n_c, n_t is the fraction of heat generated by combustion that is actually transferred to the pot. The remaining heat is lost either to the stove body or to the ambient air through flue gases and radiation.

Thus, $$n_t = \frac{\text{Heat transferred to food from pot}}{\text{Gross heat imput to the pot}}$$

Finally, there is a fraction of heat transferred from the pot to the food and this amount of heat is actually absorbed by food.

Thus, $$n_f = \frac{\text{Heat absorbed by food}}{\text{Heat transferred to the food from pot}}$$

The aforesaid partial efficiencies are measurable in controlled laboratory conditions only.

Another significant figure of merit is the Specific Fuel Consumption (SFC), which is the mass of fuel in grams burnt to cook 1 Kg. of a particular dish of food.

$$\text{SFC} = \frac{\text{Gms of fuel consumed}}{\text{Kg of food cooked}}$$

This is more often a realistic measure of efficiency for various gasifiers also.

6.11 ENERGY PLANTATION

Plant materials with appropriate fuel value offers a renewable source of liquid fuel and organic chemicals. When these land plants grow by capturing the solar energy, the ultimate plantations are called, 'Energy Plantations'. They promise to serve as a solar energy storage device in which the accumulation energy can be released at temperatures comparable with those achievable from fossil fuels. In fact, more energy than that can be derived from the sum total of all existing thermal power plants in the country can be derived from this renewable source. Thus it is imperative that these plantations has to be cultivated and managed to provide substantial amounts of usable fuel continuously throughout the year at costs competitive with other fuels. It is a common observation that wild or unmanaged vegetations vary greatly in from and quantity between widely separated sites to be a reliable and manageable source of fuel.

6.11.1 Choice of calorie plants and their management

This naturally depends on the climate and soil where the energy plantations are to be located. Lands primarily meant for agricultural crops cannot be earmarked for cultivating fuels on a large scale.

However the calorie crops require adequate moisture, a moderate growing season and a few centimeters height of the top soil capable of retaining moisture. Large hectres of land do exist in our country with some of these favourable conditions where a particular species of calorie crop density has a typical value of 12,000 to 24,000 trees/hectre. Plantations are harvested initially after one to three years and subsequently on two to four years cycle. In some cases of experimental farming, bio-mass more than 25 tons/ha/year has been achieved. This signifies that a plantation of 50 million hectares can yield fuel equivalent of 3,000 to 4,000 trillion kilocalories. Since the annual average insolation in India is over 400 calories/cm^2/day, the sunshine conversion rate can be considered equivalent to collecting and storing between 100 to 175 trillion Kcal/ha/year. Thus, it is possible to achieve a solar energy conversion rate of 0.9% for optimizing plant material production for fuel value by methods like double cropping, mix species planting of a crop rotation. Technology for these operations are already available in the existing sectors of our agricultural economy. It is very significant to note that the estimated heat equivalent of the energy plantations is lower than the current cost of hydrocarbon fuels but slightly higher than the cost of coal in areas close to the coal mines. Some such species of high yield rate are presented in Table 6.11.1.

Table 6.11.1: Common Indian Species

Species	*Yield (m^3/ha/yr)*	*Calorific value (Kcal/Kg.)*
Eucalyptus Camalulensis	7–10	4800
Leucaena leucocephala	30–40	4200–4800
Acavia Auriculiformis	10–12	4800–4900
Albizia Lebbek	05	5200
Prosopis Juliflora	5–6	4000

6.11.2 Proposed Plants for Energy Plantation

Some of the plants to be highlighted in this section has been exemplified in Table 6.11.1. The plant/species that deserves special mention are:

Casuarina: This is a large ever green tree whose botanic name is Casuarina equisetifolia. Its principal uses are as wood, pulp, wind break and in control of erosion. These species grow well in warm, subtropical and tropical climates and can be planted from sea level upto 1500 metres. These plants have also tolerated rainfall as high as 5000 mm and as low as 200 to 300mm. When allowed to grow, these can attain heights of 50 metres with trunk diameter of 30 cms.

Eucalyptus: These plantations show remarkable early growths on favourable sites and attains nearly 60 to 70 m in height and 2.3 metres in diameter with a straight trunk as long as two thirds of its total height. Almost all the 140 species are recognized as fast growing species for fuel wood. Typical production or yield rate is around 24.1 ton/acre/year.

Surghume: Principal advantage of this species is a high yield rate of 10 to 30 tons/acre and requirement of less water for normal growth (resulting in lower irrigation costs). These are also the source materials for production of alternative fuels such as ethanol.

6.12 ALTERNATIVE LIQUID FUELS

Alcohol and vegetable oil are now replacing petrol and diesel as transportation fuels in several countries. This process is likely to dominate as the era of oil price hike continues. One liquid bio-fuel is the hydrocarbon that can be derived from Euphorbia plantations, which has already received significant cultivation in desert areas.

Methanol and ethanol can be blended with unleaded petrol and burnt in modern internal combustion engines. Of course, minor engine modifications are necessary for pure alcohols and blends containing more than 20% alcohol. Care has to be exercised in pipes, pumps and fuel tanks as alcohols, particularly methanol contain traces of water and corrosive organic impurities. When alcohol percentage is as high as 98, the fuel has to be pre-heated, calling for a small electric heater in cold climates. This is why diesel engines have been modified for dual fuel use. Some of the notable features of alcohol fuels are: (a) They have a high octave rating but a lower calorific value than petrol, (b) Methane contains 25% less energy per gallon than ethanol and 50% less than petrol, (c) Alcohols burn with higher efficiency than petrol, (d) Alcohols burn at lower temperatures than petrol, (e) As these fuels are free from lead and sulphur, noxious emissions are significantly reduced.

Vegetable oils are processed from crushed seeds and nuts of sunflower, rapeseed, peanuts, palm, soya, etc. These can be blended with diesel fuel or used directly in diesel engines. Farmers in African and Latin American countries have long been using this fuel in their agricultural machinery.

Fuel gas have also replaced petrol or diesel. However, low and medium calorific value gases before transportation over long distances are often converted to more useful substances like methane and methanol.

Ethanol (C_2H_5OH) is normally prepared by certain micro-organisms from sugars under acidic conditions (pH 4 to 5). The most common microorganism, the yeast Saccharmonyces Cerevisiae is poisoned by C_2H_5OH concentrations greater than 10% so, higher concentrations upto 95% are produced by distilling and fractionating techniques. The different source along with the corresponding technologies are presented in the block diagram of Fig. 6.12.1. The improved production of ethanol has assumed special significance in view if its use in meter vehicles as gasoline extender (gasohol) or octane booster. Great care has to be exercised to ensure that these alcohol gasoline blends are free from water. Still, the prominent advantages are (a) cost of boiler conversion from oil to ethanol are modest compared with these required for coal and charcoal (b) the atmospheric pollution factor is significantly low.

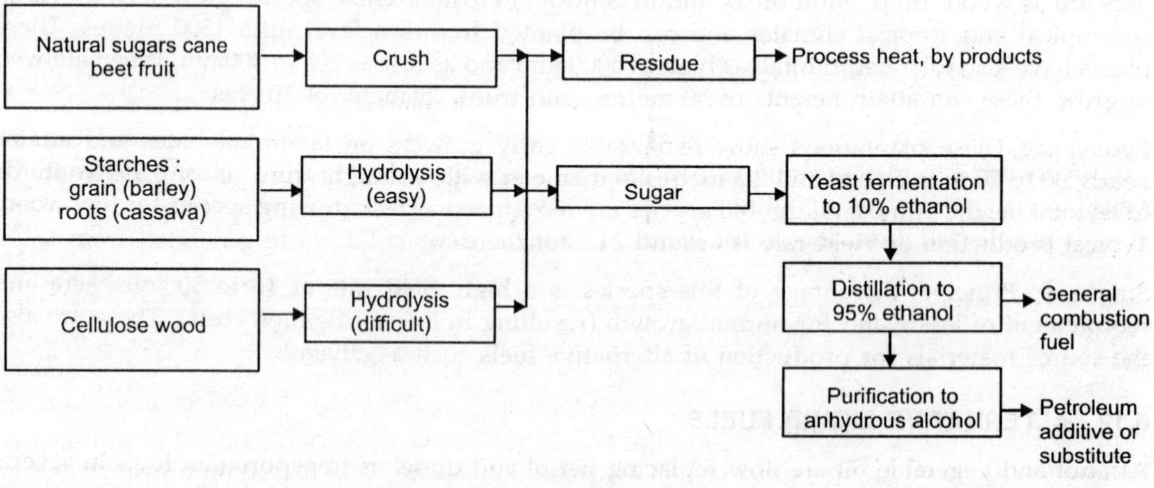

Fig. 6.12.1: Production Technologies for Alternative Liquid Fuels from Various Sources

6.12.1 Improved Chemical Processes

Several chemical processes have come up for refinement of bio-fuels, out of which one will be briefly presented here. About 50% of the while meat of coconut, copra, etc. is oil, that can be extracted by squeezing and rolling. This oil funds use in diesel engines with some modifications to the injection system. But the difficulties encountered are (a) toxic fumes, (b) deposits and (c) solidification below 23°C. Reaction of this oil with 20% by volume of ethanol or methanol produces volatile easter and glycerol. This modified product (easters) is a better diesel engine fuel where the glycerol is also a valuable by product. The energy density of 38 MJ/Kg. typically occurring in easters (produced by fermentation from natural sugars of gasification of wood wasters) compares favourably with that of petroleum in the range of 45 to 50 MJ/Kg.

6.13 BIO-METHANATION

This process applies in general to any kind of bio-degradables. Anaerobic digestion can occur at psychophilic range (below 20°C), mesophilic (20 – 45°C) and thermophilic range (45 – 60°C) of temperature. It is well known that bacterial activity varies at different temperature ranges. Mesophilic condition is best suited in our country and there are several experimental evidences that thermophilic digestion is not as fast as mesophilic one and the former also leads to production of bio-gas with lower methane content. The anaerobic digestion is well performed at neutral or slightly alkaline condition.

6.13.1 Application to Tannery Wastes

The process of bio-methanation reveals that solid wastes of tanneries undergo enzymatic hydrolysis. A mixed bacterial culture acts in symbiosis to digest the substrate where there are broadly two groups of micro-organisms such as acid farmers and methane farmers. The acidogenesis is brought about by acidifying the bacteria in a few hours. The methanogenic bacteria which grows in slightly alkaline condition are capable of producing methane following two different pathways viz. by using hydrogen and carbon dioxide according to two different reactions—

(i) CO + 4H (ii) CHCOOH

It further reveals that the accumulation of hydrogen in the digester will foster the transformation of pyruvic acid into propionic acid at the expense of acetic acid. This prepionic acid may only be transformed directly to methane. The necessity of passing through the acetic intermediary is a result of the action of specialized bacteria which has a slow growth rate. The solid substrates are thus converted to methane and carbon dioxide, keeping the H_2S content below 0.9% in the bio-gas.

It is collected from the recent literature that the daily solid waste generated from leather processing activities in Kolkata is approximately 175 to 180 metric tons having an energy potential of 2.3×10^6 MJ, which once again is equivalent to about 80 metric tons coal. Hides and skins are very rich in organic carbon content to the extent of about 50% on dry weight basis. The hydrogen content of the collagen is about 6.4% on dry weight basis. Another important feature is about 16 to 17% on dry weight basis presence of nitrogen. This is responsible for 30% of the digested solid wastes remaining as residue after energy recovery to serve as a useful source of nitrogen rich manure with an energy potential of 9.63 MJ/Kg. (dry residue). It is important to realize that bio-gas generated from tanneries contain 70 to 79% methane as compared to collections from cow-dung, piggery-waste, buffalo dung, etc. which is in the range of 50 to 55% only.

Depending on the characteristic of the waste generated from various tanneries, the scheme of bio-methanation may be suggested to consider characterize the recoverable energy potential of each solid waste after shredding to 6mm of untanned wastes—

(i) Anaerobiosis of solid untanned protein wastes
(ii) Anaerobiosis of dechromed protein wastes
(iii) Anaerobiosis of waste water sludge.

EXERCISE

1. Indicate briefly the reasons for considering bio-mass as
   (a) A renewable source of energy
   (b) An useful derivative of solar energy
2. Classify the principal sources of bio-mass with few examples of each category.
3. Mention some of the advantages and disadvantages of bio-mass for energetical use.
4. What is fermentation ? What are the constituents of (a) Gashol (b) Lingo-cellulose ?
5. What are the principal constituent of bio-gas ? What is its source ? Explain briefly the mechanism of anaerobic digestion with some of its merits.
6. Discuss critically the factors which control the yield of bio-gas.
7. What are the factors to be considered for selecting a site for a bio-gas plant ? Name only two common plants found invarious Indian sites. Which factors affect the sizing of a bio-gas plant?
8. Explain the process called photosynthesis. What are the necessary conditions ?
9. Explain the constructional details and working of KVIC digester.
10. What is a community bio-gas plant ?
11. What is meant by energy plantation ? Mention its advantages and disadvantages. List some of the main plants proposed for the purpose in India.
12. Write short notes on: (a) Wet and dry fermentation (b) Pyrolysis (c) Bio methonation from tannery wastes (d) Bio-mass stoves.
13. Estimate the volume of a bio-gas digester suitable for the output of four cows and also the power available from the digester. Retention time is 20 days, temperature 30°C, dry matter consumed is 2 Kg/day with a bio-gas yield of 0.24 m^3 per Kg. Assume burner efficiency to be 60% methane proportion to be 0.8 and heat of combustion of methane as 28 MJ/m^3 at STP. [Ans- 3.2m^3 ; 300W].

APPENDIX– I

Case I

'Ankur' Bio-mass Gasifier Installation in Ceramic Title Unit

The 'Ankur' WBG series gasifiers have been installed in March 2001 at M/s Classic Ceramics, manufacturers of glazed ceramic tiles. The gasifiers have been installed to provide thermal energy to their kilns. The system details are:

1. *Prior to Gasifier Installation*

Biscuits	: Kerosene fired
	: 1100°C operating temperature
	: 200 litres Kerosene consumption/day
Glazed tiles	: Kerosene fired
	: 1050°C operating temperature
	: 1800 litres Kerosene consumption/day

2. *With Gasifier*

Biscuits	: Gasifier WBG-300
	: Gas & Kerosene fired
	: 1100°C operating temperature
	: Kerosene saving of 1400 litres/day
	: Wood consumption 5200 Kg./day
Glazed tiles	: Gasifier WBG – 500
	: Gas & Kerosene fired
	: 1050°C operating temperature
	: Kerosene saving of 2000 litres/day (Part gas given to biscuit kiln)
	: Wood consumption 7800 Kg./day

3. *Overall Saving*

• Kerosene	: 1000 KL/annum
• Monetary Savings	: Rs. 80 lacs/annum
• Total Investment	: Rs. 30 lacs
• Payback Period	: Four to five months

4. *Highlights* : Such high temperature applications have been successfully implemented for the first time.

Trouble free un-interrupted operation in excess of 1000 hours.

'Ankur' Bio-mass Gasifier Installation for Steam Generation

1. *Prior to Gasifier Installation*

Boiler capacity	: 4.0 MT/hour steam
Furnace oil consumption	: 6.0 KL/day

2. *With Gasifier*

Gasifier model	: WBG – 500
Furnace oil consumption	: 3.0 KL/day
Wood consumption	: 12.0 MT/day

3. *Overall Savings*

Furnace oil savings	: 3.0 KL/day
Monetary savings	: Rs. 10,000/- day
Total investment	: Rs. 30 lacs
Payback period	: One year

4. *Highlights* : Gasifier designed to operate on totally automatic control mode.

Fully automatic duel-fuel burner in use.

Mechanised charging of biomass through skip charger.

Automatic nozzle closure for shutdown; variable frequency based blower for Hi-low mode etc.

Case II

'Ankur' Bio-mass Gasifier Installation at Rice Mill in West Bengal

Our 'Ankur' FBG series gasifier has been installed in June 2001 at Sree Gopal Rice Mills, Kachripatty Village, P.O. Bolpur, Birbhum Dist. West Bengal. The gasifier replaces the diesel consumed in DG set used by the rice mill to the extent of about 70 – 75% using rice husk as feedstocks.

1. *Prior to Gasifier Installation*
   DG set model : Kirloskar, SL90, 125 KVA – 2 Nos.
   Diesel consumption : 30 litres/hour
   Operating hours : 10 to 20 hours/day
   Fuel cost / hour : Rs. 600/hour
2. *With Gasifier*
   Gasifier model : FBG – 120
   Diesel consumption : 8 litres/hour
   Rice husk consumption : 110 Kg./hour
   Fuel cost / hour : Rs. 225/hour
3. *Overall Savings*
   Diesel savings : 22 litres/hour
   Monetary savings : Rs. 375/hour
   Total investment : Rs. 8.0 lacs
   Payback period : 2150 hours of operation
4. *Highlights* : Trouble free operation since the system was commissioned in mid 2001.
   Engine performance has been certified by KOEL, the OEM, on dual-fuel operations.

Case III

'Ankur' Bio-mass Gasifier Installation for Annealing of Steel Tubes

An 'Ankur' Bio-mass Gasifier has been installed in January 2002 at M/s. Patson Industries, Por, near Vadodara. The gasifier is used to provide thermal energy for a tube annealing furnace, with a view to completely replace the furnace oil with gas.

1. *Prior to Installation of Gasifier*
   Furnace oil consumption : 12 – 15 litres/hour
   Furnace temperature : 1000°C approx.
   Fuel costs : Rs. 120 – 150 per hour
2. *With Gasifier*
   Gasifier model : WBG – 60 in Hot Gas mode
   Furnace oil consumption : Nil
   Wood consumption : 60 Kg./hour
   Furnace temperature : 1000°C approx.
   Fuel costs : Rs. 90/- per hour
3. *Overall Savings*
   Furnace oil savings : 90 KL/annum
   Monetary savings : Rs. 4.0 Lacs/annum

Capital investment	:	Rs. 3.75 lacs
Payback period	:	One year
4. *Highlights*	:	Only oil burner replaced by gas burner. No other modification on the furnace. Very clean operation as against large amounts of black smoke generated with furnace oil.

SUMMARY

Bio-mass in the Indian context primarily means field residues of crops, waste products of crops processing, forestry and agro-based industries residues. Separate table deploys the availability and use of various terms of agro-wastes along with selective cultivation of energy efficient plantations. Attractiveness of bio-mass as a fuel for power generation derives from its many similarities to fossil fuels. It is easy to understand that equipments designed for fossil fuels could easily be adopted for bio-mass resources with practically very little modification, adjustments and major transitional economic costs. This is why bio-mass based power generation can be regarded as a true replacement technology and in the current Indian scenario this renewable energy component is meeting over 70% of the energy needs of the rural areas covering 70% of the population. In this context, the various types of bio-gas digestors have been described and a comparative evaluation of their design and performance also highlighted. A detailed classification along with description of bio-mass gasifiers have been included with particular emphasis on rice-husk gasifier. It has to be in mind that rice-husk is the second largest available agro-industrial residue, next only to Bagasse accounting for nearly 22% of the agro by products for the country. Currently rice-mills are the single largest consumers of rice-husk consuming upto 43% of the husk generated. Necessary technical features of various models of ANKUR bio-mass gasifiers in use in different industrial scenario has been presented in the Appendices. As the era of oil price hike continues with fresh threats of yet another Gulf war, alternative liquid fuels are fast replacing petrol and diesel in several countries. A brief coverage of the organic materials along with improved chemical processing and Bio-methanation have also been included to appreciate the full potential of some of the prominent liquid fuels.

It may be appreciated that if used for power generation, surplus bio-mass resources in India can save over 70 million tonnes of coal every year. Further, 100 million tonnes of CO_2 emission into the atmosphere can be avoided. Bio-mass projects are ideally suited for developing countries like India considering the magnitude of surplus bio-mass available. MNES has undertaken an ambitious project on *'Bio-mass Resource Atlas for India'*, which seeks to integrate data obtained from field level studies on bio-mass assessment. This, in future, is expected to create a bio-residue map of India for use in power generation.

REFERENCES

1. Wu Chuangzhi, Development and Commercialisation of bio-mass gasification in China, Proc. of International Seminar on Decentralised Energy: Options and Management, Feb 22-24, Kolkata, India, 2002.
2. G. Chen, et al., Decentralised Energy production from bio-mass gasification using circulating fluidized Bed Technology, pp. 74-80, Proc. of International Seminar on Decentralised Energy: Options and Management, Feb 22-24, Kolkata, India, 2002.

3. Li Jung Jing, et al., Assessment of Bio-mass Resource Availability in China, China Environmental Science Press, Beijing, 1999.
4. Rijal, Critical Evaluation of the improved cooking stove programme in Nepal, Renewable Energy Review Journal, 9(1), 1987.
5. R.V. Siemons, Identifying a role for bio-mass gasification in rural electrification in developing countries: The Economic Perspective, bio-mass and Bio-energy, pp. 271-285, 20(2001).
6. G.D. Rai, Non-conventional Energy Sources, Dhanpat Rai and Sons, New Delhi, 1998.
7. K.M. Mital, Non-conventional Energy Sources – Their Principles, Progress and Prospects, Wheeler Publising, 1997.
8. B. Ghosh, et al., Possibilities and potential for industrial grade power from bio-mass gassification, pp. 116-122, Proc. of International Seminar on Decentralised Energy: Options and Management, Feb 22-24, Kolkata, India, 2002.
9. D.C. Elliot, Issues value added Products from bio-mass, Progress in Thermochemical conversion, pp. 1186-1196, 2001.
10. P.D. Grover, Status of bio-mass briquetting in India, Wood Energy News, Regional Wood Energy Development Programme in Asia, 14(2): October (1999).
11. S.S. Chandel, et al., Evaluation of Wood-stove Programme in Himachal Pradesh, Wood Energy News pp. 105-116, vol. 2, no. 2, Dec' 1992.
12. M. Kinoshita Charles, et al., Power Generation Potential of bio-mass gasification systems, Energy 123(3), pp. 88-99, 1997.
13. J. Mathur, et al., Development of Mini bio-mass gassifier for converting waste to Energy, pp. 463-467, Proc. of NREC-2002, Coimbatore, India, Jan'2003.
14. Y.P. Gupta, Experimental Investigations on a domestic fuel pyrolyzer using Paddy Straw, SESI journal 8(2), pp. 79-84, 1998.
15. L. Fugbemi, et al., 'Pyrolysis products from different Bio-masses: Application to the thermal cracking of Tar', Journal of Applied Energy, vol. 69, pp. 293-306, 2002.
16. R. Warnecle, 'Gassification of Bio-mass: Comparision of Fixed Bed and Fluidised Bed Gassifier', J.Bio-mass and bio-energy, vol. 18, pp. 489-497, 2000.
17. S. Ramaswamy, et al., Estimation of Bio-mass Resource: Gandhigram Experience: pp. 533-537, Proc. of NREC-2002, Coimbatore, India, Jan'2003.
18. Mathur and Sharma, Internal Combustion Engines, Dhanpat Rai and Sons, 1998.
19. A. Sadesh, et al., Optimization of Cold Storage Systems powered by bio-mass energy for Rural Area Applications, World Renewable Energy Congress VII, WREC-2002, Cologne, Germany.
20. P. Ahuja, et al., A model for primary and heterogeneous secondary reactions of Wood Pyrolysis, Chemical Engg. Tech., 19, pp. 272-282, 1996.
21. A.K. Jain, et al., Availability and Charecteristics of Paddy husk as renewable sources of energy, Journal of Agricultural Engg., vol. 34(1), 1997.
22. Mohan Sudhir and J.R. Meshram, bio-mass based power generation-Prospects in India, Energy Management, Jan-March, 1998.
23. Anonymous [1999a], Annual Report, Ministry of Non-conventional Energy Sources, New Delhi, 1998-1999.
24. S. Kumar, et al., Implementation of night soil based large size bio-gas programme in Maharasthra, Proc. of 21st NSEC, New Delhi, Dec'1997.
25. Y.P. Anand, Study of the systems of management in a Railway Complex, Ph.D. Thesis, Indian Institute of Technology, New Delhi, 1989.
26. G. Nagarajan, et al., Ethanol in glow plug assisted DI Diesel Engine, Proc. of NREC-2002, pp. 498-503, Coimbatore, India, Jan'2003.
27. K.C. Pandey, et al., Critical Review on Vegetable Oils as substitute fuels for diesel engines, Proc. of the XVI National Conference on IC Engines and Combustion, Calcutta, Jan'20-22, 2000.
28. S. Muthu, et al., Energy from bio-mass – Prospects in India, Proc. of International Seminar on Decentralised Energy: Options and Management, p. 203, Feb 22-24, Kolkata, India.

29. K.K. Moorhead, et al., Water hyacinth growth in anaerobic digester effluents, Biological Wastes, 34, vol. 2, pp. 91-99.
30. S.T. Ramesh, et al., Bio-mass as energy source – Energy recovery from water hyacinth by anaerobic digestion, pp. 468-473, Proc. of NREC-2002, Coimbatore, India, Jan'2003.
31. A.M. Reddy and M.K.Reddy, 'A Bio-mass Power Project in Operation' IREDA NEWS, pp. 29-31, vol. II, no. 4, Oct-Dec'2000.
32. H.B. Mathur, Ethanol – A Bio-Solar Fuel Substitute, IREDA NEWS, pp. 39-41, vol. II, no. 4, Oct-Dec'2000.
33. American Petroleum Institute, Use of Alcohol in Motor Gasoline – A Review, Publication no. 4080.
34. A.K. Dhussa and R.C.Tewari, Waste to Energy in India, Bio Energy News, pp. 12-14, March'2000.
35. A.K. Dhussa, et al., Energy Recovery from Municipal Solid Waste: Potential and Possibilities, Bio Energy News, pp. 7-8, March'2000.
36. K.B.K. Reddy, Financing of Bio-mass power/Cogeneration Projects – An Indian Perspective, IREDA NEWS, vol. 12, no. 2, April-June'2001.
37. A.K. Jain, Hydrogen fuel from Bio-mass Resources, SESI Journal, 10(2), pp. 107-116, 2000.
38. R.C. Pal, et al., Performance Monitoring of TERI Mark – 4D Model Bio-gas Plants in Rural House hold, SESI Journal, 10(2), pp. 95-105, 2000.
39. A.K. Kalia and S.P.Singh, 1m^3 Deenabandhu Bio-gas plants for hilly region, Bio-gas forum, 60, pp. 18-20, 1995.
40. A.K. Kalia, Status of Bio-gas plants in District Kangra, Proc. of 24th NREC-2000, pp. 344-348, Nov'2000.
41. A.K. Kalia, Bio-gas as Source of Rural Energy, Energy Sources, 22(1), pp. 67-76, 2000.
42. C. Palaniappan et al., Renewable Energy Applications to Industries, Narosa (1998).

7

Wind Energy Conversion

INTRODUCTION

Wind is only a tiny fraction of 5 million tonnes of matter being converted every second by the sun and radiated as energy. Indeed, in selected windy locations, this non-conventional energy source has the capability of reducing the electrical energy deficits to a significantly large extent. Accordingly, wind power could become the most economical new source of electricity and generate fresh sources of employment in the next decade. The wind energy generating capacity at the turn of the century (2000) is around 3500 MW and growing @ 35 percent/annum. The World Watch Institute, in a recent analysis, has pointed to India as the new wind super power as the installed capacity in India has already exceeded 1000 MW.

In order to appreciate the genesis of wind nature, this Chapter, therefore, starts from the basic principles of realizing flow of wind from the solar radiation. Next the estimates of power in the wind motion and corresponding estimates of mechanical and electrical energy are derived. The machines that convert wind energy to electrical energy have been taken up next as Wind Energy Conversion Systems (WECS). A detailed classification of WECS have also been attempted with an emphasis on the related merits and demerits. A separate section has been devoted to the performance analysis of WECS to cover the vital parameters like power co-efficient, tip-speed ratio and solidity from the designer's point of view. Finally, a few recent reports of wind power generation in India and abroad have been reviewed with the help of few useful figures and some relevant tables.

7.1 BASIC PRINCIPLES

Flow of wind originates from two natural factors (i) the absorption of solar energy on the earth's surface and in the atmosphere and (ii) the rotation of the earth about its axis and its motion around the sun. As the incident radiation from the sun strikes the different atmospheric layers of air, these layers gets heated. As a result, there is expansion and upward movement of air. These layer is replaced by colder layers and this mechanism continues so long as solar radiation continues to heat the various air layers in an uneven way. At night the process is reversed because the air cools down more rapidly over the land surface resulting in breezes (off-shore) in the reverse direction. Air in motion relative to a surface possesses energy. The well known example is that of a wind mill which operates either by slowing down the wind velocity or capturing some of its energy when it converts wind energy to mechanical energy. The average wind speed at a particular site and height varies from time to time, place to place and season to season. Obviously, the power extracted from the wind machine depends on time of the day, place and the season of the year. Further, the wind energy density at selected sites is also very low calling for deployment of large rotor areas to obtain significant electrical energy. Despite this intermittent nature, wind velocity patterns at any particular site remains remarkably constant year by year (on a seasonal basis). Average wind speeds have been observed to be greater in hilly and coastal areas compared to inland values. Significantly, winds blow more consistently and with greater strength on the surface of water where there is less surface drag. Generally wind speeds at a site increases with height above the surface and are found to be typically 20% to 25% greater at a height of 10 metres from the surface. At a height of 60 metres they may be even 60% higher due to a reduction in the drag effect of the earth's surface.

Besides the intermittent and dilute nature of wind energy, there are two major sources of noise resulting from a wind machine- mechanical noise and aerodynamic noise. There is also a possibility that electromagnetic signals (Radio, Television, Cellphone, etc.) may be scattered in the vicinity of wind energy turbines/systems leading to interference or distortion of TV and high frequency channels. But, these problems can certainly be tackled to a significant extent by a group of electrical and communication engineers.

7.2 RECENT DEVELOPMENTS

Despite all the aforesaid disadvantages, it has to be born in mind firmly that harnessing of wind energy does not cause any air pollution. In fact, the generation of wind power is also a step forward against the unwanted environmental phenomenon of global warming and acid rain. With the rapid strides of telecommunication engineers into the technology and application of microprocessors, installed WECS all over the world have become fully automatic and reliable. Significant amounts of electrical energy have been generated from wind energy in U.S.A., U.K., Netherlands and India. In U.S.A. alone, there are more than 18,000 wind turbines of various capacities, the largest being in California, utilizing the wind power for agriculture, residential, commercial water supply and industrial purposes. Besides these, WECS installations are characterized by low maintenance cost, lower gestation periods and minimum transmission loss. A typical wind power project, on date, is expected to cost around Rs. 3.5 crore/MW with generation costing around Rs. 2.25/kWh, (comparable with unit power cost from a thermal power plant, gestation period around one year with an operation and running cost of 2% only.)

India provides great opportunity for generation of electricity through wind power as it has been revealed through wind-mapping that several sites in the country are gifted with the optimum wind velocity for operating WECS. Ministry of non-conventional energy sources (MNES) has completed a nation wide wind mapping programme a few years back where 58 cities were identified for wind-farms with a total potential of 4000MW. Through another comprehensive wind resource assessment survey, 53 cities have been identified (Fig. 7.2.1) for having wind speeds above 18 kmph.

MNES, accordingly, gave the necessary impetus to the grid quality power generation by wind turbine technology in the 7th National Five Year Plan (NFYP) (1985-90). Since then wind power generation has picked up a substantial momentum from industries, entrepreneurs and businessman. As a result, 'Grid Quality Wind Power Generation' emerged rightly as the thrust area by the beginning of the 8th NFYP (1992-97). Significantly, 95% of the installed capacity till date have come from the private sector. BHEL has entered into WEG production in collaboration with M/s. Nordex of Denmark. As a result, the joint venture has specialized in construction of 'Lattice Tower Structures', which has an edge over conventional tubular towers. An overview of prominent wind sites in India have been mapped in Fig. 7.2.2.

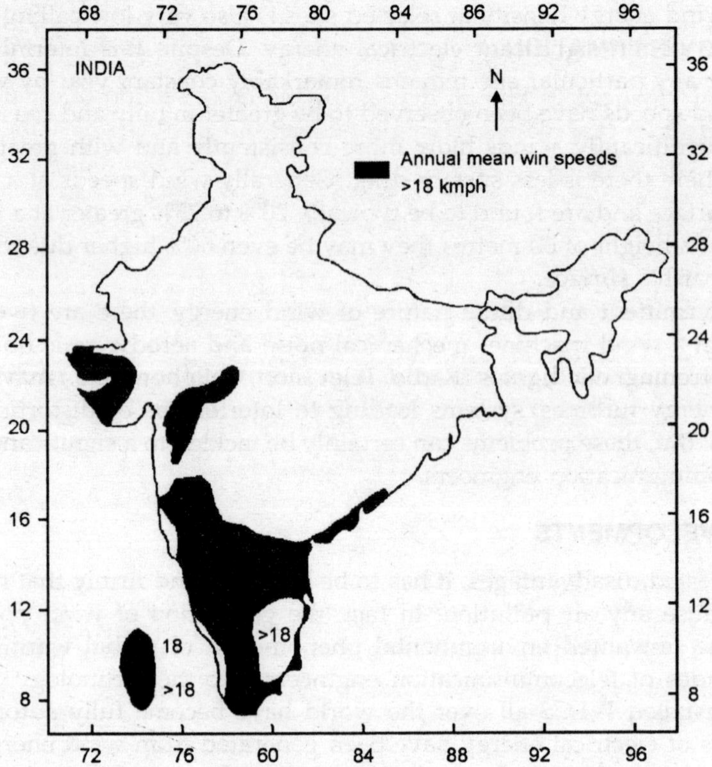

Fig. 7.2.1: Map of India Showing Areas (Shaded) with Mean Annual Wind Speed exceeding 18 kmph

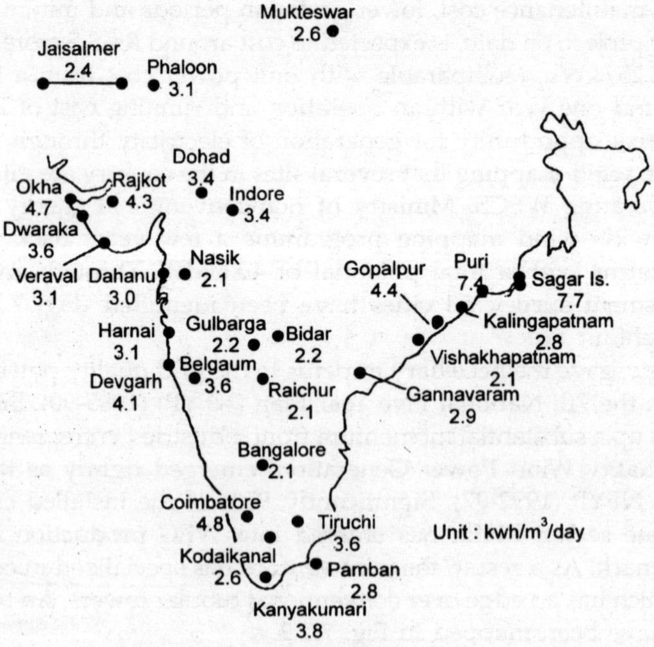

Fig. 7.2.2: Prominent Wind Sites in India

7.3 WIND ENERGY ESTIMATION

From the concept of kinetic, it is known that a flowing cross-section of air possesses energy by virtue of its velocity. A device like a sail or propeller (part of a WECS) which slows down due to interaction with the mass of a moving cross-section of air essentially extracts a part of the flowing kinetic energy to mechanical energy. If the incoming rate of kinetic energy of a flowing wind stream is KE_i, then

$$KE_i = m\left(\frac{V_i^2}{2}\right) \quad \text{i.e. } P_{tot} = m\left(\frac{V_i^2}{2}\right) \qquad ...(7.3.1)$$

where V_i is the incoming wind velocity and m is the mass flow rate, given by the continuity equation

$$m = \rho A V_i \qquad ...(7.3.2)$$

where it is easy to detect that symbols ρ and A represent incoming wind density and its cross-sectional area respectively.

$$\therefore \quad P_{tot} = \frac{1}{2}\rho A V_i^3 \qquad ...(7.3.3)$$

It is clear from (7.3.3) that total power of a wind stream is directly proportional to the cube of its flowing velocity, wind density and the intercepted cross sectional area. Thus, small increase in wind speeds can register a significant increase in the power extracted. But, the situation is not that simple. Power available also varies depending on the density of air which can be assumed as equal to 1.225 Kg/m^3 at sea level. However, a designer has to tolerate a variation of 10 to 15 percent depending on the season and time of the year contributing to temperature, pressure and humidity variations. Considering the example of horizontal axis aeroturbines with circular diameter D, having $A = \frac{\pi D^2}{4}$

Eq (7.3.3) becomes

$$P_{tot} = \frac{\rho}{2}\cdot\frac{\pi}{4}\cdot D^2V^3 \qquad ...(7.3.4)$$

Eq (7.3.4) depicts that the total power available varies according to the square of the diameter of the intercept area i.e., square of the rotor diameter. Thus, for a fixed wind velocity at a site, there will be a four times increase in P_{tot} if the rotor diameter is doubled. This can be clearly observed from the combined effects of wind speed and rotor diameter on the available wind power in Fig. 7.3.1. When a rotor diameter of 40 metres is chosen in lieu of 20 metres at a site where the wind speed is 10 metres/sec., the available power shoots upto 1 MW from a value of 0.25 MW.

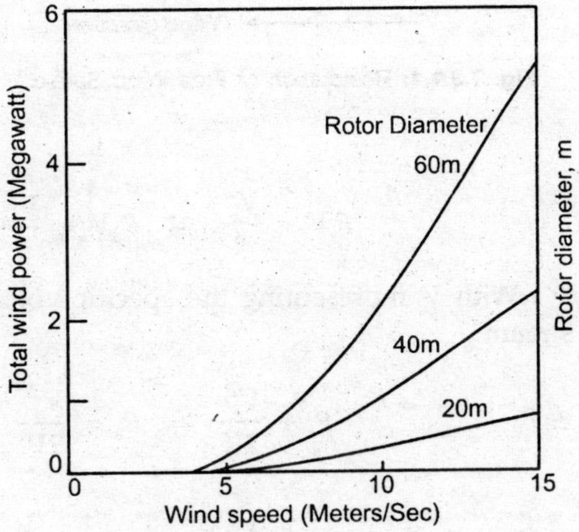

Fig. 7.3.1: Dependence of Wind-rotor Power on Wind Speed and Rotor Diameter

7.3.1 Maximum Power

Obviously, the physical design of a wind turbine are such that only a fraction of the available wind power can be converted into useful wind power. As the free wind stream interacts with a rotor, the wind transfers part of its energy into the rotor and the speed of the wind decreases to a minimum in the rotor wake. Subsequently the wind stream regains energy from the surrounding air and the free wind speed is restored at a sufficient distance from the rotor [Fig. 7.3.1.1.]. Let us assume that the wheel of such a turbine has thickness a-b such that the incoming wind pressure and velocity, far upstream of the turbine are P_i and V_i respectively. Consequently, let the exit wind pressure and velocity far downstream of the turbine be represented by P_e and V_e respectively. Such pressure and velocity profiles of a wind moving through horizontal axis propeller type wind turbine is depicted in Fig. 7.3.1.2. Since kinetic energy has to be extracted by a turbine, $V_e < V_i$ where it is reasonable to expect that the pressure and temperature changes are very small compared to ambient. Thus the density of incoming air between *i* and *a* of the aforesaid thermodynamic system may be assumed constant. So, if no heat or work are added or removed between *i* and *a*, then the general energy equation contains only the kinetic and flow energy terms as

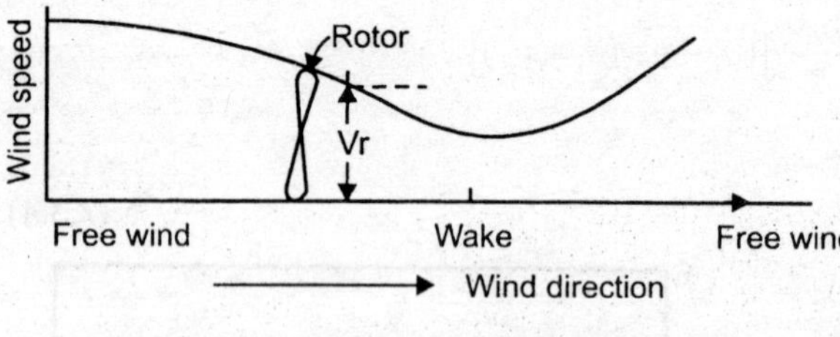

Fig. 7.3.1.1: Restoration of Free Wind Speed

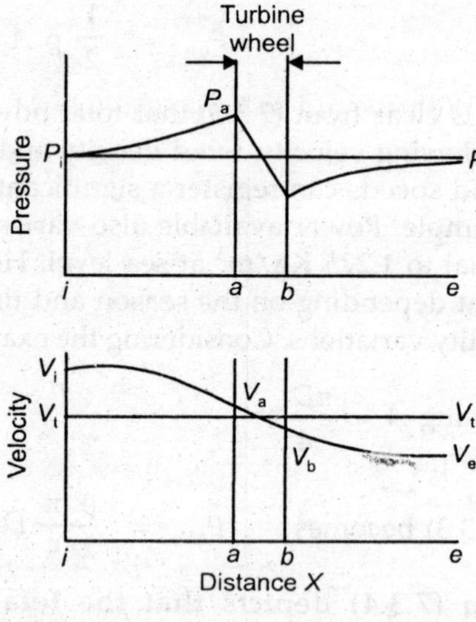

Fig. 7.3.1.2: Pressure and Velocity Profiles of a Wind moving through a Horizontal-axis Propeller-type Wind Turbine.

$$P_iV + \frac{V_i^2}{2} = P_aV + \frac{V_a^2}{2} \quad \text{...(7.3.5)}$$

With *V* representing the specific volume (volume per unit mass in m^3/Kg) of flowing air stream.

$$\therefore \quad P_i + \frac{V_i^2}{2V} = P_a = \frac{V_a^2}{2V}$$

so that

$$P_i + \rho\frac{V_i^2}{2} = P_a + \rho\,\frac{V_a^2}{2} \quad \text{...(7.3.6)}$$

where the density of air $\rho = \frac{1}{V}$ has been used. In an exactly similar way, for the exit region from b to e,

$$P_e + \rho\frac{V_e^2}{2} = P_b + \rho\frac{V_b^2}{2} \qquad ...(7.3.7)$$

Now, combining eqns. (7.3.6) and (7.3.7)

$$P_a - P_b = \left\{P_i + \rho.\left(\frac{v_1^2 - v_2^2}{2}\right)\right\} - \left\{P_e + \rho.\left(v_e^2 - v_b^2\right)\right\} \qquad ...(7.3.8)$$

It can be realized from the Fig. 7.3.1.2. that at the exit end of the turbine the wind pressure is restored to its ambient value, i.e.,

$$P_e = P_i \qquad ...(7.3.9)$$

Also, the wind velocity within the turbine (v_t) does not change with the blade width from a to b with the total transit, then

$$v_t = v_a = v_b \qquad ...(7.3.10)$$

With the aid of eqns. (7.3.9) and (7.3.10) results as

$$P_a - P_b = \frac{\rho}{2}.(v_i^2 - v_e^2) = \frac{\rho}{2}\cdot[(v_i - v_e)(v_i + v_e)] \qquad ...(7.3.11)$$

Now, the axial force (F_x), in the direction of wind stream on the turbine results from the pressure difference in (7.3.11) and the projected area (A) perpendicular to the stream.

Thus, $$F_x = (P_a - P_b)\,A = \frac{\rho A}{2}\cdot[(v_i - v_e)(v_i + v_e)] \qquad ...(7.3.12)$$

Since the mass flow rate (m) is given by

$$m = \rho A v_t \qquad ...(7.3.13)$$

F_x is thus also equal to the change of momentum of the aforesaid mass flow rate in eq. (7.3.13). In accordance with Newton's second law,

$$\therefore \quad F_x = (m\,dv) = \rho\,A\,v_t(v_i - v_e) \qquad ...(7.3.14)$$

where dv represents the change in velocity from v_i to v_e.
Comparison of (7.4.12) and (7.4.14) yields

$$v_t = \frac{(v_i + v_e)}{2} \qquad ...(7.3.15)$$

Now, in the total thermodynamic system of Fig. 7.3.1.2. the changes in potential energy, internal energy and flow energy are zero. Also, no heat is added to or rejected from the system. Thus, the general energy equation depicts only the steady flow work (W) in terms of kinetic energies as

$$W = (KE)_I - (KE)_e = (v_i^2 - v_e^2) \qquad ...(7.3.16)$$

Rate of above work is the power, p, given by

$$P = m\frac{(v_i^2 - v_e^2)}{2} = \frac{\rho A V_t}{2} \cdot (v_i^2 - v_e^2) \quad ...(7.3.17)$$

Combining this with eq. (7.3.15), the result is

$$P = \frac{1}{4\rho} A (v_i + v_e)(v_i^2 - v_e^2) \quad ...(7.3.18)$$

Now, there is an optimum exit velocity ($v_{e.opt}$) that is responsible for maximum power (P_{max}). This is obtained by differentiating P in eqn.(7.3.18) with respect to v_e for a known v_i and putting $\frac{dP}{dV_e}$ = 0 which yields

$$3v_e^2 + 2v_i \cdot v_e - v_i^2 = 0 \quad ...(7.3.19)$$

$V_{e.opt}$ is solved by investigating a positive value of v_e in eqn. (7.3.19) which leads to

$$V_{e.opt} = \frac{v_i}{3} \quad ...(7.3.20)$$

Combining this solution in the eqn. (7.3.18) yields

$$P_{max} = \frac{8}{27} \cdot \rho A v_i^3 \quad ...(7.3.21)$$

7.3.2 Maximum Theoretical Efficiency

The fraction of the free-flow wind power that can be extracted by a rotor is called the power co-efficient of the turbine.

Thus, Power Co-efficient (η_{max}) = Power required at the wind rotor/Total power available in the wind.

$\eta_{max} = \frac{P_{max}}{P_{tot}} = \frac{16}{27}$ = 0.5926 ...(7.3.22) from a combination of (7.3.3) and (7.3.21). In evaluating the performance of a WECS, it is to be borne in mind that a wind turbine can only convert a maximum of 59.26% of the total available wind power.

Now, an ideal rotor with propeller type blades of aerodynamic design may have a power coefficient approaching a value of 0.59. But such a rotor would not be strong enough to withstand the stresses to which it is subjected when rotating at a high rate in a high speed wind stream. For the best practical rotors, a value of η_{max} ranging from 0.4 to 0.45 is used by the designers. When converted to electrical power, some of the rotor energy has to be sacrificed and the resulting coefficient of an aero-generator (i.e., electrical power generated/available wind power) is around 35 percent.

7.3.3 Axial Force and Thrust on Blades

It can be easily be understood that there are two types of forces operating on the blades of a propeller type wind turbine. Out of these, one is the circumferential force acting in the direction of wheel rotation that provide the torque. Other is the axial force acting in the same direction as that of the flowing windstream. The second force is responsible for an axial thrust, which must be carefully estimated and counteracted by the designer.

The circumferential force or torque T can be obtained as

$$T = \frac{P}{\omega} = \frac{P}{\pi} D.N \qquad \text{...(7.3.23)}$$

Where T has the dimensions of *Kgf* or Newtons, ω is the angular velocity of turbine wheel in metres/sec, and D is the diameter of turbine wheel in metres. N stands for the number of wheel revolutions/second.

Now, it is already understood that sectional power = $P = \eta_{max} \cdot P_{tot} = \frac{1}{2} . \rho A v_i^3 \eta_{max}$.

Thus, for a turbine operating at power P, the torque generated is

$T = (\eta_{max}/2) \{(\rho A v_i^3)/\pi DN\}$ ...(7.3.24) where area (A) can be written as $\pi D^2/4$,

so that $T = (\eta_{max}/8) \{(\rho D v_i^3)/N\}$ obviously

when $T = T_{max}$, $\eta_{max} = 16/27 = 0.593$

$$\therefore \quad T_{max} = (2/27)\,(\rho D v_i^3)/N \qquad \text{...(7.3.25)}$$

Here, the axial force or axial thrust presented in eqn. (7.3.12) is recollected so that

$$F_x = (\rho A/2)\,(v_i^2 - v_e^2) = (\rho \pi D^2/8)\,(v_i^2 - v_e^2) \qquad \text{...(7.3.26)}$$

Obviously, $\quad F_x = F_{x,max}$ when $v_e = v_i/3$

$$\therefore \quad F_{x,max} = (\rho \pi D^2 v_i^2)/9 = (4/9)\,\rho A v_i^2 \qquad \text{...(7.3.27)}$$

Eqn. (7.3.27) is an eye-opener to designer of WECS as large magnitudes of axial forces have to be tackled for large diameter machines. There is thus an upper limit of the diameter that has to be optimized from technical and economical considerations.

Example: Wind is flowing at 1 standard atmospheric pressure and 15° C temperature through a propeller type wind turbine with a velocity of 25 metres/sec. Assuming a turbine diameter of 60 metres and a turbine wheel revolution of 50 r.p.m, estimate:

(a) The maximum obtainable power density
(b) Torque at maximum efficiency
(c) Axial thrust

Solution:

For air, the universal value of gas constant is

$$R = 0.287 \text{ KJ/Kg.K and 1 atm} = 1.01325 \text{ bar} = 1.013 \times 10^5 \text{ N/m}^2.$$

So, $\quad$ Air density $= \rho = P/RT = (1.013 \times 10^5)/(287 \times 288) = 1.225 \text{ Kg/m}^3$

(a) Total Power = $P_{tot} = (\rho A v_i^3)/2$, $\quad$ Power density = $P_{tot}/A = (\rho v_i^3)/2 = (1.225 \times 25^3)/2$

= 9570.31 Watts.

(b) Torque at maximum efficiency $(T_{max}) = (2/27)\,(\rho D v_i^3)/N$

where D = 60 metres and N = 50 r.p.m. = 50/60 rev/sec.

$\therefore \quad T_{max} = (2/27)\,(1.225 \times 25^3 \times 60)/(50/60) = 102083.33$ Newtons.

(c) Axial thrust = $(\pi/9)\,\rho D^2 v_i^2 = \pi/9\,.\,1.225 \times 60^2 \times 25^2 = 6725250$ Newton.

Example: Wind velocity at the rate of 10 m/s at 1 standard atm pressure and 30°C temperature flows into a turbine. Estimate the value of reasonably obtainable power density in W/m².

Solution:

For air, R = 287 J/Kg.K and 1 atmosphere = 1.013 bar = 1.013×10^5 N/m²

∴ Density of air = $\rho = P/RT = (1.013 \times 10^5)/(287 \times 303)$ Kg/m³

∴ Power density = $P_{tot}/A = \rho v_i^3/2 = (10^3/2.)\ (1.013 \times 10^5)\ /\ (287 \times 303) = 582.2$ W/m²

Now, assuming Power Coefficient, $(\eta_{max}) = 40\%$

Reasonably obtainable power density = 0.4 × 582.2 = 232.9 W/m².

7.4 WIND ELECTRIC POWER GENERATION

Wind mills have been in use for converting wind power to mechanical energy and are traditionally the oldest gadget in the family of wind electric generators (WEG's). Traditional wind mills were used extensively in the middle ages to mill grain and to lift water for land drainage and watering cattle. WEG's are still used for these purposes today in some parts of the globe. But the main focus of attention now centers around their technical features to generate electricity. Power available from a windmill can be represented with the eqn. (7.3.3) described before. The power coefficient (C_p) for a windmill is the ratio of shaft power to the maximum power available in the wind and has also been represented in generator from eqn. (7.3.3) to eqn. (7.3.27). These are the strong dependence of power produced (shaft power) on the square of the diameter of the turbine wheel and the cube of the wind velocity. Obviously small fluctuations in wind velocity, which almost always occurs would mean large fluctuations in power. A glimpse of Fig. 7.3.1, representing the power-velocity characteristics of a wind turbine reveals that a decrease in wind velocity to approximately one half of its value would drop the available power by a factor of ten, assuming constant efficiency. Actually, a greater loss in power occurs as many machine losses are independent of wind velocity.

Since a WEG's wheel cannot be completely closed and due to spillage and other effects practical WEG's can achieve some 50 to 70% of the ideal efficiency, the actual power available (P) is

$$P = \eta . P_{tot} = (\eta A v_i^3 \rho)/2 \quad \text{...(7.4.1)}$$

Where a designer uses η (overall efficiency) values ranging from 30 to 40 percent for real turbines.

7.5 LIFT AND DRAG

The extraction of power and ultimately energy from wind depends on creation of certain forces and applying them to rotate a mechanism. The two primary mechanism for creating aforesaid forces are the lift and drag mechanisms.

By definition, lift forces act perpendicular to the air flow while drag forces act along the direction of flowing wind. The former occurs by changing the velocity of the air stream flowing over either side of the lifting surface. It is known that the pressure drops when the air flow is increased and the vice-versa. In principle, thus any change in velocity is responsible for a pressure difference across the lifting surface. By this difference of pressure, a force or airfoil is created on the high pressure side and more towards the low pressure side of the lifting surface. A good airfoil has normally a high lift to drag ratio of the order of 30:1. The lift force increases in magnitude as the angle formed at the junction of airfoil and air stream (angle of attack) becomes less acute. For efficient operation, the blades of a wind turbine needs to function with maximum possible lift and minimum possible drag. A lift is thus an efficient process at the cost of deflecting the air flow where as drag dissipates energy. Thus it is imperative that the design of every wind turbine specifies an optimum angle at which the aerofoil has to be set for achieving the maximum lift to drag ratio. Besides aerofoil, there exists another technique of blowing air through narrow slots in a cylinder so that it emerges tangentially. This is called a Thwait's slot. This also creates a rotation of the air flow suitable to produce a lift. Such techniques, however are limited in applications as they yield low values of lift to drag ratio.

Major research and development funds have gone largely to propeller type wind machines, to be discussed in details in next section. Some of these machines have been expanded to investigate other designs offering future technical and economical advantages. One such design is based on the magnus effect, demonstrated experimentally by Magnus in 1852. The effect showed that if a horizontal cylinder is rotated about its axis and moved through a still wind, a lift force is also produced. Similarly, if a stationary horizontal cylinder is rotated about its axis in a cross wind, it will experience a lift force. The effect is equally applicable to a vertical cylinder rotated about its axis in a cross wind. It will experience a force perpendicular to its axis causing it to move in a direction essentially perpendicular to that of the wind.

The magnus effect can be explained by first considering a cylinder in a non-viscous flow (Fig. 7.5.1). Since the cylinder has a length much greater than its diameter, flows around it can be considered to be two dimensional.

In Fig. 7.5.1 (a), a uniform stream of wind is flowing past a non-rotating cylinder. This results in streamlines as shown in the figure. The velocity at any point on the surface of the cylinder is given by

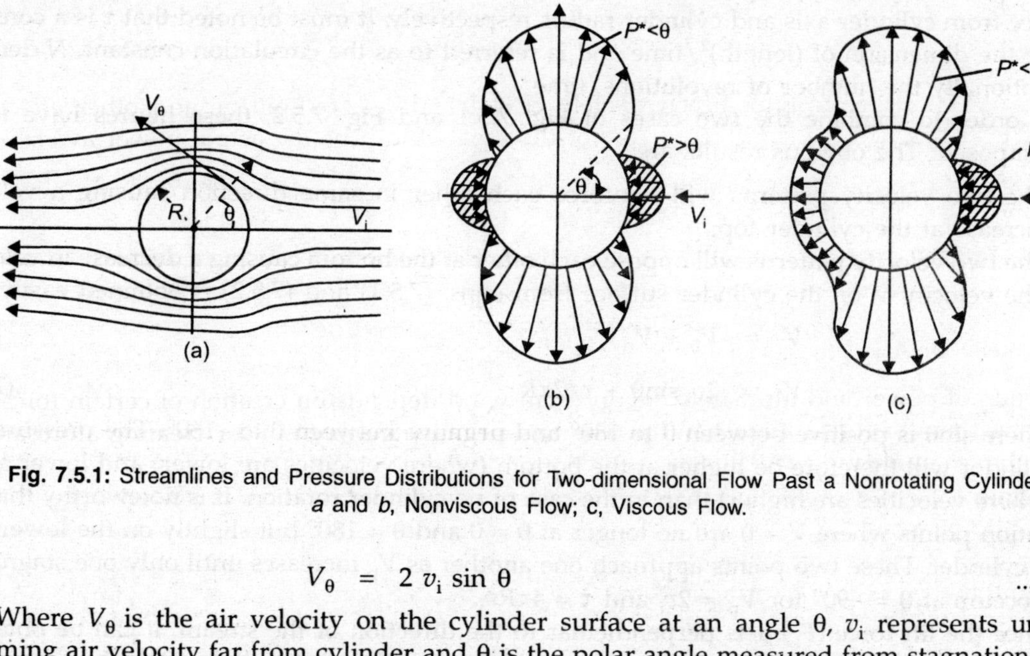

Fig. 7.5.1: Streamlines and Pressure Distributions for Two-dimensional Flow Past a Nonrotating Cylinder. *a* and *b*, Nonviscous Flow; c, Viscous Flow.

$$V_\theta = 2\, v_i \sin\theta \qquad \text{...(7.5.1)}$$

Where V_θ is the air velocity on the cylinder surface at an angle θ, v_i represents uniform incoming air velocity far from cylinder and θ is the polar angle measured from stagnation point. When Bernoullis eqn. is applied to any position θ on the surface of the cylinder,

$$P_i/\rho + v_i^2/2 = P_\theta/\rho + V_\theta^2/2 \qquad \text{...(7.5.2)}$$

Eqn. (7.5.1) and (7.5.2) gives $P_\theta = P_i + \frac{1}{2}\rho v_i^2 (1\text{-}4 \sin^2\theta)$ ...(7.5.3)

Or, $P^*_\theta = (P_\theta - P_i)/(\rho v_i^2/2) = 1 - 4\sin^2\theta$ ...(7.5.4)

where P_i is the pressure of incoming stream, P_θ is the pressure on the cylinder surface at θ and ρ is the constant air density. P_θ^* in eqn. (7.5.4) represents the non-dimensional pressure on cylinder surface at θ. The quantity $\rho v_i^2/2$ can easily be identified as the dynamic pressure of the undisturbed stream. The variation of P_θ^* with θ is demonstrated in Fig. 7.5.1(b), where no resultant force occur either parallel or perpendicular to the undisturbed stream. Thus, there is neither drag nor lift on the cylinder for the case of ideal non-viscous flow.

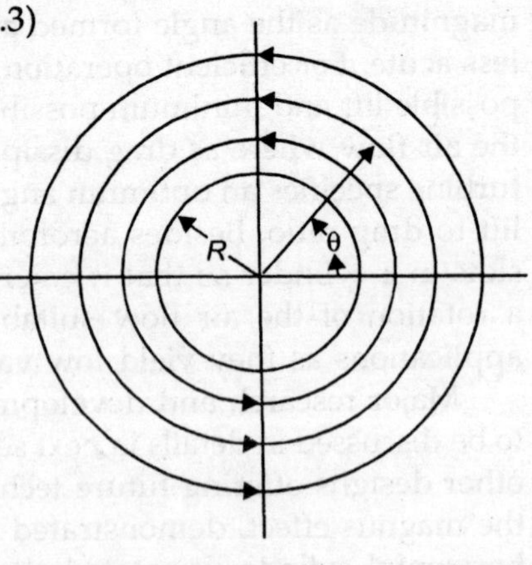

Fig. 7.5.2: Circulatory Flow around a Rotating Cylinder in an otherwise still Air

Now, we consider the case of the cylinder rotating about its axis but the air far from the cylinder is at rest this time. As seen in Fig. 7.5.2, a circulatory flow around the cylinder is produced due to its rotation where the velocity (V_r) is inversely proportional to the distance from the cylinder axis. Thus, $2\pi\, r.V_r = 2\pi\, RV_p = \tau$

Where $V_r = \tau/2\pi\, r$ ...(7.5.5)

and $V_p = \tau/2\pi\, R$ ...(7.5.6)

where V_p is the cylinder peripheral velocity equal to $2\pi RN$ with r and R representing the radial distance from cylinder axis and cylinder radius respectively. It must be noted that τ is a constant having the dimension of (length)^2/time and is referred to as the circulation constant. N denotes conventionally the number of revolutions/time.

In order to combine the two cases of Fig. 7.5.1 and Fig. 7.5.2, these figures have to be superimposed. The obvious results are:

(a) The two velocity patterns will reinforce each other in same direction causing a velocity increase at the cylinder top;

(b) The two velocity patterns will oppose each other at the bottom causing a decrease in velocity. The velocity V on the cylinder surface from eqns. (7.5.1) and (7.5.2) is obtained easily as

$$V = V_\theta + V_p$$

i.e., $$V = 2v_i \sin\theta + \tau/2\pi R \quad \text{...(7.5.7)}$$

where sinθ is positive between 0 to 180° and negative between 0 to –180°. The pressures on the cylinder will therefore be higher at the bottom (where velocities are lower) and lower at the top (where velocities are higher) than in the case of no cylinder rotation. It is noteworthy that the stagnation points where $V = 0$ are no longer at $\theta = 0$ and $\theta = 180°$ but slightly on the lower side of the cylinder. These two points approach one another as V_p increases until only one stagnation point occurs at $\theta = -90°$ for $V_p = 2v_i$ and $\tau = 4\pi R v_i$.

Since the lift force (F_L) acts perpendicular to the direction of the stream, it can be obtained by integrating the component of the pressure on the cylinder that is perpendicular to the free stream. If P represents the pressure on the cylinder surface at θ, their substituting the velocity value from eqn. (7.5.2) in place of P_θ,

$$P = P_i + \rho/2\,[v_i^2 - (2\,v_i \sin\theta + \tau/2\pi R)^2] \quad ...(7.5.8)$$

$$\therefore \quad F_L = \int_{-\pi}^{+\pi} - P \sin\theta \,.\, Rd\theta.H \quad ...(7.5.9)$$

where *H* stands for the length/height of the cylinder.
Now, aided by eqn. (7.5.2)

$$F_L = 2\pi\rho.R\,H\,V_p\,v_i = 4\pi^2\rho R^2 H\,N v_i \quad ...(7.5.10)$$

If C_L is the dimensionless lift coefficient, $P.v_i\,2/2$ is the dynamic pressure of the free stream and the projected area of the cylinder = A = 2RH,

$$\therefore \quad F_L = C_L.A.\rho.(v_i^2/2) \quad ...(7.5.11)$$

by combining eqns. (7.5.10) and (7.5.11),

$$C_L = 2\pi.\,V_p/v_i \quad ...(7.4.4.12)$$

It is obvious from eqn. (7.5.12) that lift coefficient for a non-viscous flow is directly proportional to the ratio of cylinder peripheral to the free stream velocities. However, experimental measurements on rotating cylinders in real fluids indicate that values of C_L are much lower than those given by eqn. 7.5.12. Choosing a speed ratio of V_P/v_i around 4 or 5, a maximum value around 9 to 10 can be obtained for C_L. This is clearly depicted in Fig. 7.5.3, which presents lift coefficient values versus values of V_p/v_i useful to the designer.

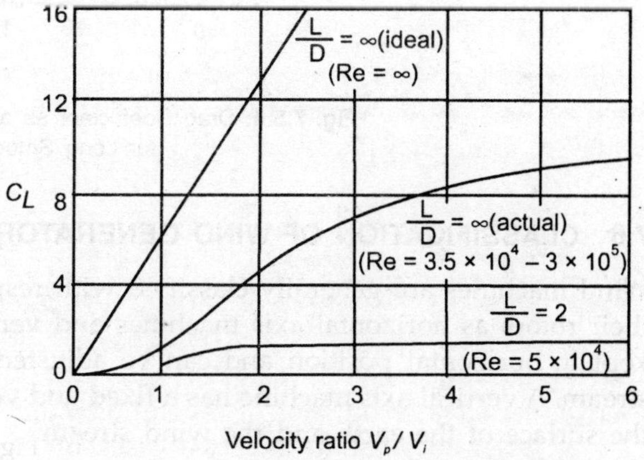

Fig. 7.5.3: Lift Coefficient of a Rotating Cylinder in an Air Stream as a Function of the Ratio of the Cylinder Peripheral and Free Stream Velocities

It has to be kept in mind that for a real fluid the imbalance in pressure (Fig. 7.5.3) results in a drag force in the direction of the free stream, given similarly to eqn. (7.5.11) by

$$F_D = C_D\,A\,\rho\,v_i^2/2 \quad ...(7.5.13)$$

where C_D is a dimensionless drag coefficient which is obviously a function of Reynold's number (*Re*) as

$$Re = (2Rv_i\rho)/\mu \quad ...(7.5.14)$$

where μ represents the viscosity of the fluid. The relationship between C_D and *Re* for long smooth cylinders is shown in Fig. 7.5.4. Obviously, the net force acting on the cylinder is the resultant force between the lift and drag forces.

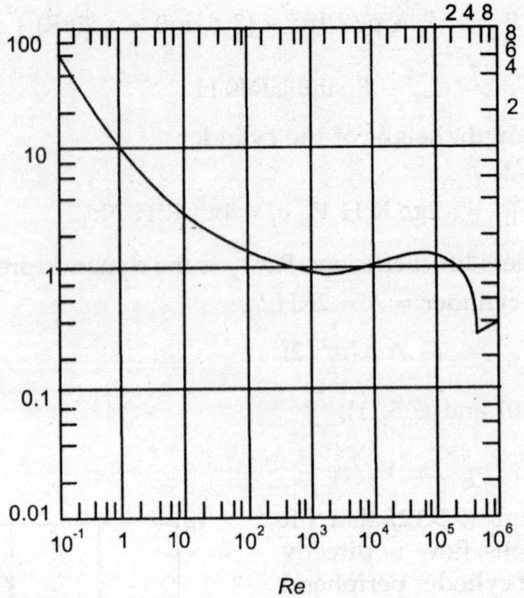

Fig. 7.5.4: Drag Coefficient as a Function of Reynolds Number for Long Smooth Cylinders

7.6 CLASSIFICATION OF WIND GENERATORS

Wind machines are generally classified with respect to the orientation of the axis of rotation of their rotors as horizontal axis machines and vertical axis machines. In the former, rotor axis is kept in horizontal position and can be adjusted so that it is parallel to the direction of wind stream. A vertical axis machine has a fixed and vertical rotor axis, which is perpendicular to both the surface of the earth and the wind stream.

7.6.1 Types of Rotors

Various types of rotors in use in WEG are: (i) Multi-blade type, (ii) Propeller type, (iii) Savonious type, (iv) Darrieus type. Out of these, the first two find use in horizontal axis machines while the other two are employed in vertical axis machines.

The multi-blade rotor consists of a number of curved sheet metal blades which increased in chord length while going outwards from the center. These blades, numbering from 12 to 20, are fixed at their inner ends to a circular rim. They are also connected near their outer edge to a second rim, to a fixed support. Diameter of these rotors vary from 2 to 5 metres (Fig. 7.6.1.1).

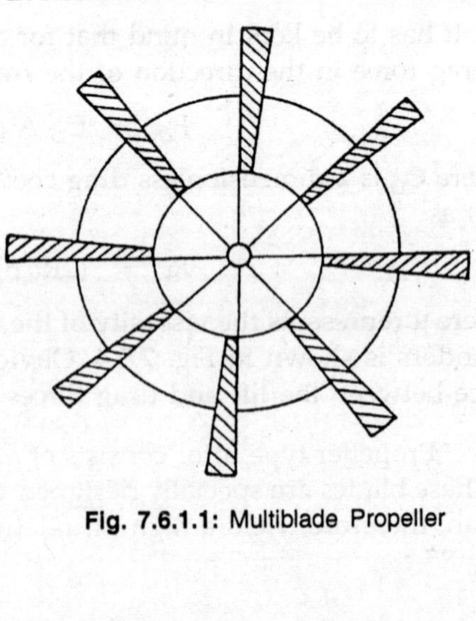

Fig. 7.6.1.1: Multiblade Propeller

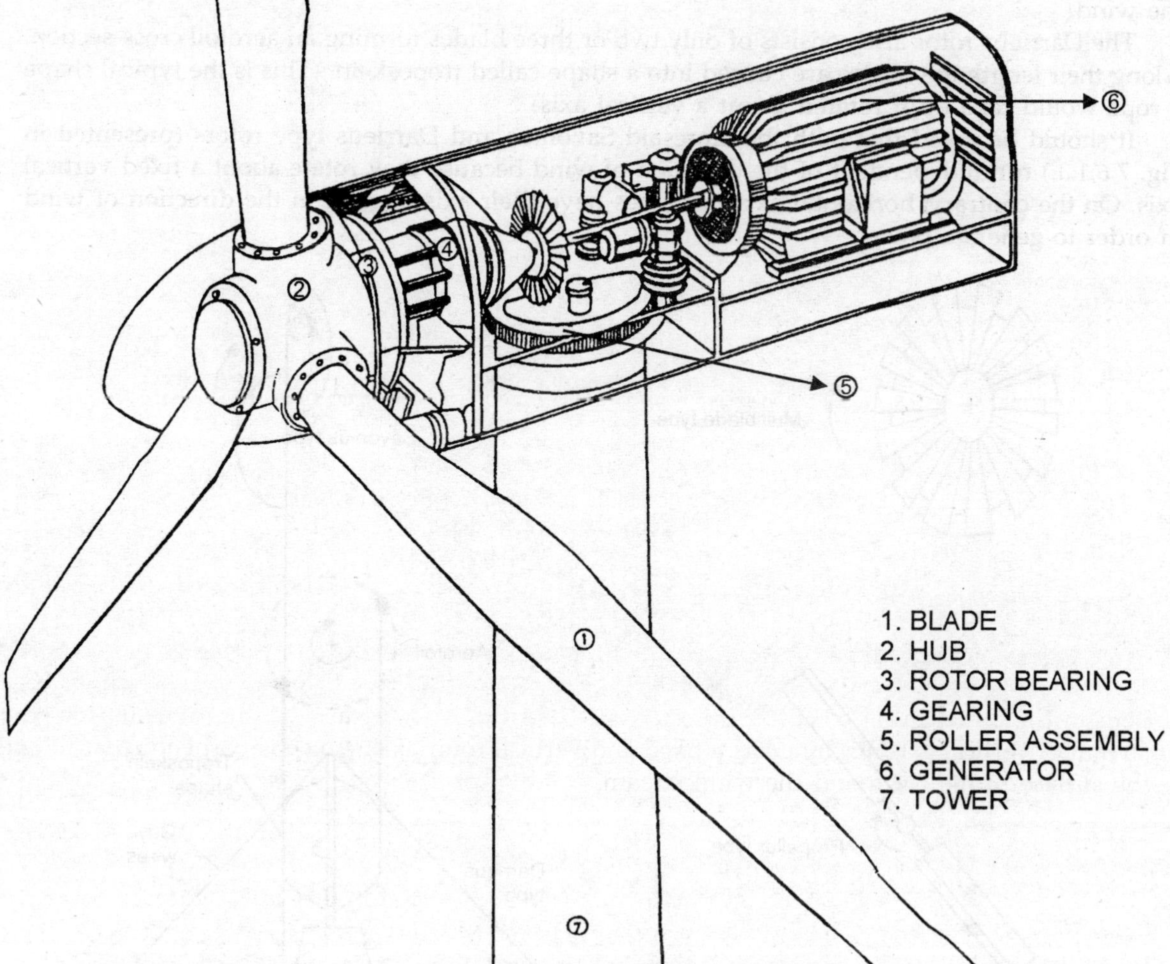

Fig. 7.6.1.2: Propeller Type Wind Machine

Propeller type rotor consists of only two to three blades made of glass fibre reinforced plastic. These blades are specially designed with aerofoil sections having high thickness to cord ratio and can, therefore, yield a high lift relative to the drag. Diameters of these rotors is in the range of 2 to 25 metres (Fig. 7.6.1.2).

To fabricate a Savonius rotor, a hollow elliptical cylinder is sliced into two pieces and each of these halves fixed to a vertical axis with a fixed gap to give rise to an S-shape. As a result, necessary torque is generated by the pressure difference between the two sides of the half facing the wind.

The Darrieus rotor also consists of only two or three blades forming an aerofoil cross-section. Along their length the blades are curved into a shape called troposkein. (This is the typical shape a rope would take up if rotated about a vertical axis).

It should be noted that both the aforesaid Savonius and Darrieus type rotors (presented in Fig. 7.6.1.3.) run independent of the direction of wind because they rotate about a fixed vertical axis. On the contrary, horizontal axis machines have their axis aligned in the direction of wind in order to generate power.

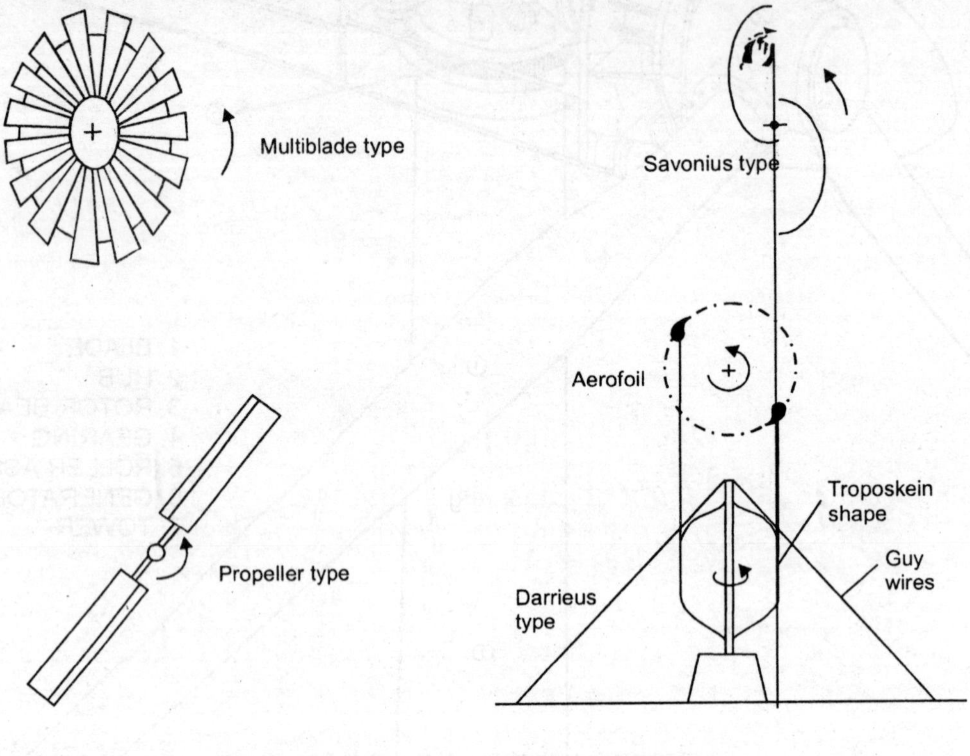

Fig. 7.6.1.3: Various Types of Wind Machine Rotors

7.6.2 Multi-blade Type Wind Machine

A simple sketch of this machine is shown in Fig. 7.6.1.4. It can be easily identified that the principal elements are the rotor, the wind mill head, the tail vane, the transmission system and the supporting structure (tower). An additional component like a water pump can be an integral part of the system if the machine is intended to pump water. The transmission system consists of a power shaft, a speed reducing gear drive, a crank shaft-connecting rod mechanism and a pump rod. The gear drive is intended to reduce the rotational speed by a factor of 3 to 4. The rotational motion is then converted to a reciprocating motion by the crank shaft-connecting rod mechanism, which in turn, is connected to the water pump through a pump rod. The center of the rotor, the machine head, is called the hub. The tail-vane, which is a simple cut out sheet metal attached to the back side of the machine head serves the purpose of aligning the rotor with the direction of wind flow. The tower, usually a steel truss construction, serves the very important purpose of keeping the rotor at an appropriate height from the ground where the wind speed is adequate.

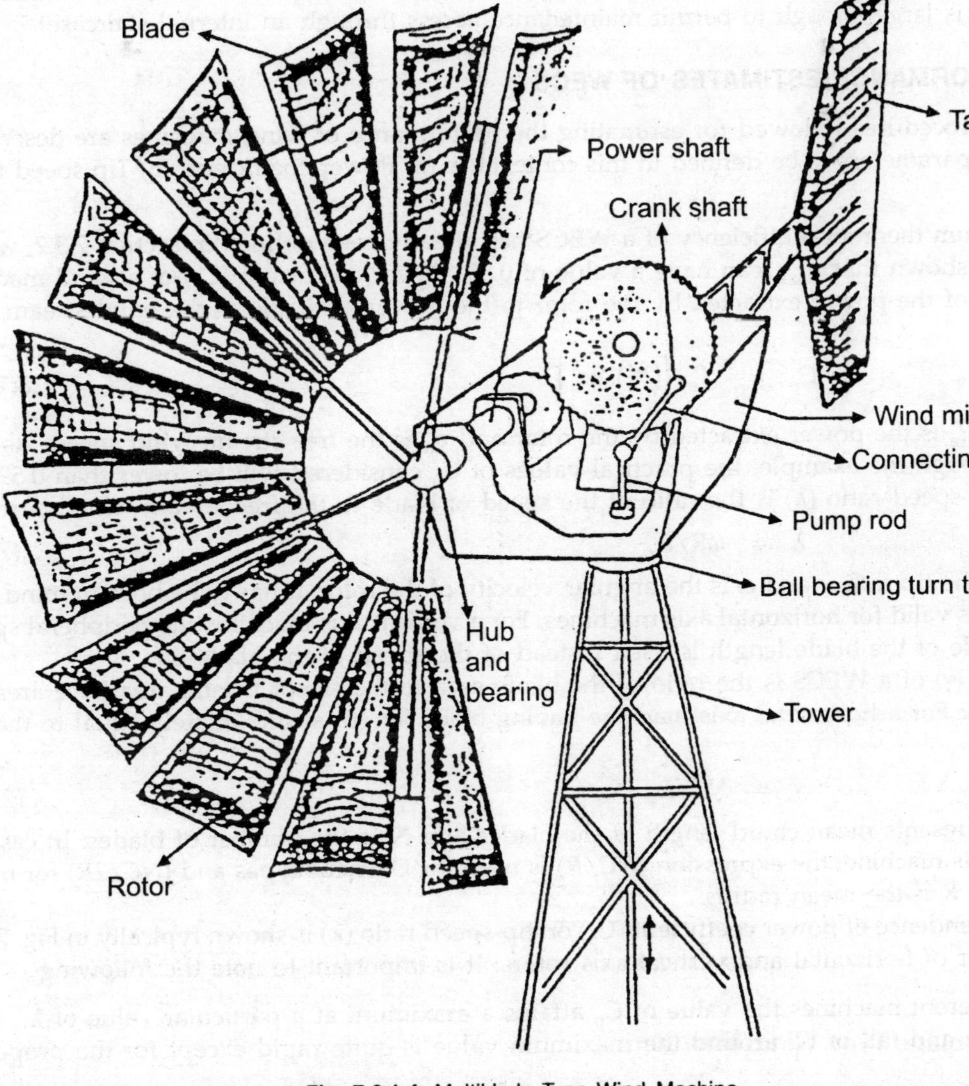

Fig. 7.6.1.4: Multiblade Type Wind Machine

7.6.3 Propeller Type Wind Machine

A cross-sectional view of this machine is presented in Fig. 7.6.1.2. which clearly identifies some of the main components like blade, hub, rotor bearing, gearing, roller assembly, generator and tower.

The power extracted from the wind by the blades is transmitted through the hub to a gear train and then on to a generator. Nacelle is a housing for accommodating gear train, generator and various control hardwares (including breaking mechanism and roller assembly linkages to the tower). The roller assembly permits rotation of the nacelle about a vertical axis for aligning the rotor with the direction of wind flow.

For small capacity machines ranging from fraction of a kilowatts to few kilowatts, the tower can be of steel truss design followed for the multi-blade type machines. For large capacity machines rated upto few hundred kilowatts, a tubular type tower design is preferred. (Here the diameter at the base is large enough to permit maintenance access through an internal staircase).

7.7 PERFORMANCE ESTIMATES OF WECS

Here, the procedures followed for estimating the performance of wind machines are described. The useful parameters to be defined in this context are (a) Power coefficient, (b) Tip speed ratio, (c) Solidity.

The maximum theoretical efficiency of a WECS has already been defined in section 7.3.2, where it has been shown that η_{max} can have a value of 0.59. Power coefficient (C_p) of a wind machine is the ratio of the power extracted by the rotor to the power available in the wind stream, i.e.

$$C_p = P_e/\{(\rho/2)Av_\alpha^3\} \quad \text{...(7.7.1)}$$

Where P_e is the power extracted by the rotor and v_α is the free stream wind speed. As will be seen through the example, the practical values of C_p considered will be lower than 0.59.

The Tip-speed ratio (λ) is the ratio of the speed of blade to the free stream wind speed

$$\therefore \quad \lambda = \omega R/v_\alpha \quad \text{...(7.7.2)}$$

where R is the tip radius and ω is the angular velocity of the rotor. It has to be born in mind that eqn. (7.7.2) is valid for horizontal axis machines. For a vertical axis machine, the peripheral speed at the middle of the blade length is used instead of the speed of the blade-tip.

Solidity (γ) of a WECS is the ratio of the blade area to the swept frontal area (face area) of the machine. For a horizontal axis machine having blade length approximately equal to the tip radius,

$$\gamma = NC/R \quad \text{....(7.7.3)}$$

where C represents mean chord length of the blades and N is the number of blades. In case of a vertical axis machine, the expression (NC/R) is used for Darrieus types and ($NC/2R$) for other types where R is the mean radius.

The dependence of power coefficient (C_p) on tip-speed ratio (λ) is shown typically in Fig. 7.7.1 for a number of horizontal and vertical axis rotors. It is important to note the following–

(a) For different machines the value of C_p attains a maximum at a particular value of λ.
(b) The rise and fall of C_p around the maximum value is quite rapid except for the propeller type.
(c) Highest value of C_p are noted with propeller type rotors.
(d) The multi-blade type and Savonius type are characterized by low values of λ (around one) while the propeller and Darrieus type are characterized by high values of λ ranging from 4 to 8.
(e) Ideal values of C_p (around 0.59) has been indicated as Betz limit in the figure.

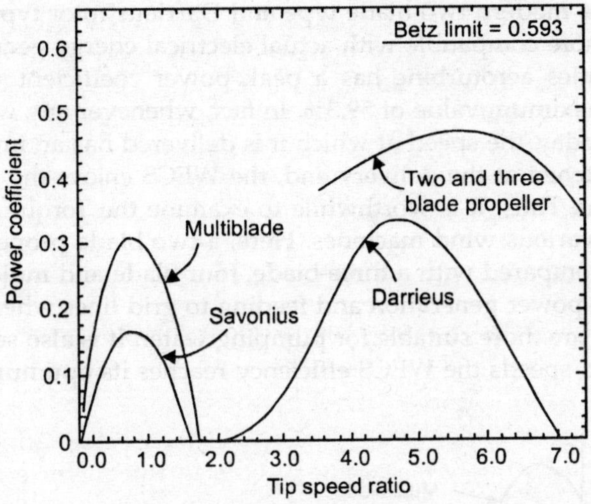

Fig. 7.7.1: Variation of Power Coefficient with Tip Speed Ratio for Various Types of Rotors

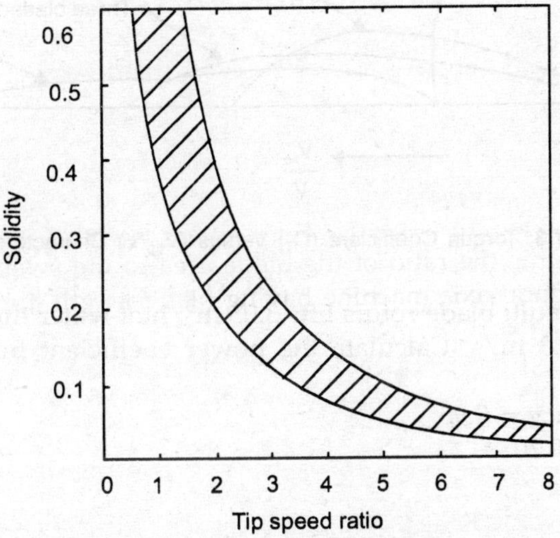

Fig. 7.7.2: Variation of Solidity with Tip Speed Ratio for Various Types of Rotors

Fig. 7.7.2 represents the variation of solidity (γ) versus the tip speed ratio (λ). Here, it is easy to observe the following trends:

(a) Rotors with low tip speed ratios have high value of solidity.

(b) Rotors with high tip speed ratios enjoy low values of solidity.

Thus, rotor of a WECS can be designed for a specific performance by interacting with typical characteristics given in Figs. 7.7.1 and 7.7.2 (with extrapolation, if needed).

It can be observed from Fig. 7.7.2 that there are large differences in the performances of various types of aeroturbine blades. The savonious and multiblade rotors are intended for low speed operation while the modern two blade type and Darrieus rotor type are intended for high speed operation, being more compatible with actual electrical energy generation. It is interesting to note that the two blades aeroturbine has a peak power coefficient which is the closest to theoretically permitted maximum value of 59.3%. In fact, whenever any wind power is delivered to the existing grid line today, the speed at which it is delivered has an important bearing. When these two factors are matched at the delivery end, the WECS enjoys the maximum utility as an energy conversion system. Thus, it is worthwhile to examine the Torque Coefficient (C_T) versus V_{tip}/V characteristics of various wind machines. Here, a two blade propeller has potentially the best performance when compared with a three-blade, four blade and multi-blade systems. These are the natural choice for power generation and feeding to grid lines where as multiblade WECS with high starting torque are more suitable for pumping water. It is also seen from Fig. 7.7.3. that only at intermediate wind-speeds the WECS efficiency reaches its optimum and power extracted follows a V^3 law.

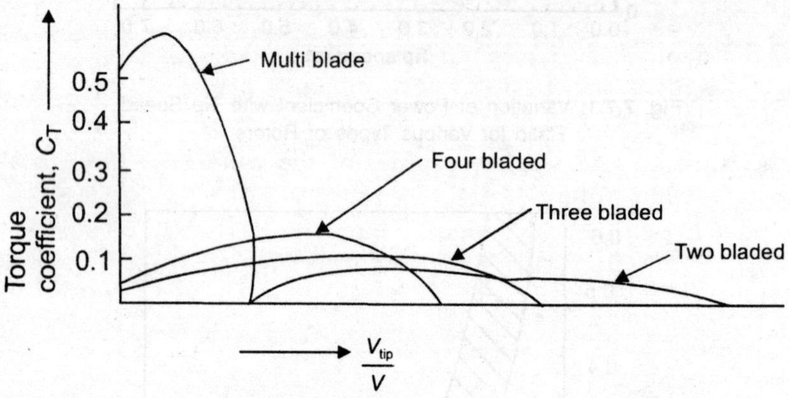

Fig. 7.7.3: Torque Coefficient (C_T) Versus (V_{tip}/V) Characterestics

Example: A wind mill with multi blade rotors lifts 3.03 m^3/h of water through a head of 28 metres when the wind speed is 3.3 m/s. Calculate the power coefficient for a rotor diameter of 4.5 metres. Assume,

(i) Transmission efficiency = 0.95
(ii) Pump efficiency = 0.70

Solution:

The density of water can be assumed as 996 Kg/m^3. Thus, the ideal power needed to pump water

$$= \text{Mass of water} \times \text{Accn. due to gravity} \times \text{Head of water.}$$

$$= 3.03 \times (996/3600) \times 9.81 \times 28 = 230.26 \text{ watts.}$$

$$\therefore \text{ Power required at rotor} = 230.26/\text{overall efficiency}$$

$$= 230.26/(0.95 \times 0.70) = 346.25 \text{ watts and}$$

$$P_{tot} = (\rho/2)A\, v_i^3.$$

where

$$\rho = \text{density of air} = 1.2 \text{ Kg/m}^3$$

$$A = \pi R^2 = 3.14\,(4.5)^2$$

$$\therefore \quad P_{tot} = (1.2/2) \times 3.14 \times (4.5)^2 \times (3.3)^3 = 1370.92 \text{ watts.}$$

$$\text{Power coefficient} = C_p = \text{Power Required at Rotor}/P_{tot} = 346.25/1370.92 = 0.252$$

Example: A multi-blade type wind machine (solidity = 0.5) uses 18 blades, each of length 1 metres. Considering a rotor diameter of 1.086 metres, what is the value of the mean chord length of each blade used?

Solution:

Let C metres be each chord length.

It is known that solidity = γ = (N × Area of blades)/R^2 where, area of blades = 1 × C. Radius of rotor R = 1.086 m and No. of blades = N = 18.

$$\therefore \quad 18 \times 1 \times C = 0.5 \times (1.086)^2 \quad \text{or, } C = 0.103 \text{ metres.}$$

EXERCISE

1. (a) Why is wind energy considered as a Renewable Energy Source? Mention its demerits as a source of energy.
   (b) List only the two major sources of noise from a wind machine.
2. Starting from the basic mechanism of wind flow, deduce the relation between power generated and velocity of wind. Sketch typical power velocity curves and offer your own comments.
3. (a) Sketch the pressure and velocity profiles of the wind flow through a horizontal axis propeller type wind turbine .
   (b) Starting from basic thermodynamics, deduce the necessary relation between incoming wind velocity and optimum exit velocity in order to derive maximum power.
4. Draw a neat sketch of multiblade type wind machine and label carefully the system components. Point out very briefly the role of (i) Crank shaft mechanism (ii) Tail-vane (iii) Tower.
5. What do you mean by: (a) Power coefficient (b) Tip-speed ratio (c) Solidity of a WECS ? Sketch neat curves to depict inter-relations between aforesaid parameters of interest to a designer. Offer your own comments on the nature of the curves by choosing typical designer values.
6. Following data give typical values of wind speed, rotor speed, generator speed and active power of one out of the four Wind Electric Generators installed at Fraserganj in West Bengal:

Date	*Time*	*Wind-speed*	*Rotor speed*	*Generator speed*	*Active power (KW)*
	10-35 am	5m/sec	25 rpm	1000 rpm	03
	10-40 am	4m/sec	26 rpm	1001 rpm	17
29.01.02.	10-45 am	4m/sec	25 rpm	1007 rpm	08
	10-50 am	4m/sec	26 rpm	1010 rpm	21
	10-55 am	4m/sec	26 rpm	1005 rpm	06

   Estimate the tip-speed ratio and power coefficient of the WEG at any two instant of time. Offer your own comments by comparing the results obtained at different time slots.

7. Wind flows at 1.03 bar pressure at a location through a turbine generator where ambient temperature is 28° C and incoming wind velocity is 28 m/sec. If the turbine is rotating at 50 rpm and its diameter is 65 metres, estimate the value of torque at maximum efficiency. Deduce necessary steps with necessary assumptions.

8. Fill in the gaps
   (i) __________ is an instrument for measuring wind velocity.
   (ii) Wind power project on date is expected to cost around Rs. _______.
   (iii) Generation of wind power costs around Rs. _________.
   (iv) Gestation period of installed wind power projects is typically ______ years.
   (v) Typical value of power coefficient for a WECS connected to grid line is around __________.

SUMMARY

This chapter initiates from the origin of wind flow and establishes the renewable nature of wind energy. Estimates of maximum power in wind motion and corresponding conditions imposed on the incoming and exit wind velocities have been analysed in details from the pressure and velocity profiles existing in a typical crossection of WECS. Procedure for the assessment of axial thrust on blades of wind machines have been outlined. The concept of Lift and Drag mechanism have also been explained with a suitable mathematical modelling leading to estimates of typical values of lift and drag co-efficients. Possible classification of WECS have been attempted on the basis of rotor design and the necessary figures of merit pertaining to any WECS have been introduced and then critically co-related from the actual designer's point of view.

The total installed capacity of WECS in India is around 1340MW and the rate of power generation increases in an exponential manner. This is why India has been executing fifth largest wind power programme in the world after Germany, U.S.A., Spain and Denmark. Out of this capacity, private sector wind power developers account for about 1278MW. Out of all the states, the highest share of around 812MW comes from Tamilnadu with an approximate number of 2800 WEG's in the state. Most of the wind farms in India have the ability to feed the power to the state electricity grids or to use it in various manufacturing units. The various state electricity boards have recently extended liberal facilities for wheeling, banking buy-back and third party sale of the power produced.

Recently, analyses have been complied to find out the actual extent of emission reduction and the saving of coal by the substitution of wind energy. This is quite significant in a developing country like India where the growing concern over the hazards of local, regional and global emissions from fossil fuel based power generation has assumed an alarming proportion. Table below shows the emission reduction by using of 1 unit of electricity and 1MW of electricity through wind energy.

Table

Sources of Emission	*Emission reduction for 1 unit of electricity (gms.)*	*Annual emission reduction for 1MW wind farm (Tonnes)*
Carbon di-oxide	800	2625–4025
Sulphur di-oxide	10	37.5–60
Nitrogen oxides	3.4	12.5–20

REFERENCES

1. M.M.El Wakil, Power Plant Technology, McGraw Hill, 1985.
2. Arora and Domkundwar, A course in Power Plant Engineering, Dhanpat Rai and Sons, New Delhi, 1998.
3. A. Mani and D.A. Mooley, Wind Energy Resource Survey in India, Allied Publishers, New Delhi, 1980.
4. List of Potential Sites for Wind Power Projects in the country, IREDA NEWS, vol. 11, no. 1, Jan-March '2000.
5. Annual report, Ministry of Non-conventional Energy Sources, Govt. of India, New Delhi, 1997-1998.
6. Annual Report, MNES, Govt. of India, 2000.
7. Annual Report, MNES, Govt. of India, 2001.
8. S. Jebaraj, et al., Wind Energy Programmes in India, Proc. of NREC-2002, Coimbatore, India, pp. 458-462, Jan '2003.
9. B. Rajsekhar, et al., Indian Wind Energy Programmes, Performance and future directions, IREDA NEWS, vol. 11, no. 1, Jan-March, 2000.
10. N.G. Mortensen, Wind Atlas Analysis and Application Program (WASP), vol. 1: Getting Standard, Riso National Laboratory, Denmark, p. 5.
11. B. Rajsekhar and F. Van Hulle, Assessment of Grid Integrational Problems of Wind Farms (ECN-C-97-045), Netherlands Energy Research Foundation Report, 1997.
12. Wind Energy Data for India, Anna Mani and D.A. Mooley, Allied Publishers Pvt. Ltd., 1983.
13. S.K. Philip, Stand-alone Wind Diesel Power Supply System for community application, Proc. of NREC-2002, Coimbatore, India, pp. 430-436, Jan '2003.
14. G.R. Bhagwatikar and W.G. Gandhare, Energy Conservation in Grid Connected Wind Power Projects, Proc. of NREC-2002, Coimbatore, India, pp. 417-422, Jan '2003.
15. G.R. Bhagwatikar and W.G. Gandhare, Power quality and harmonics generated by Grid connected windmills, Proc. of NREC-1999, Indore, India, pp. 334-337, 1999.
16. Wind Power Experiences and future Directions, Proc. of Workshop organized by TERI, New Delhi, India, July 23, 1999.
17. Centre for Wind Energy Technology (C-WET), List of Potential Sites for Wind Power Projects in the country, IREDA NEWS, vol. 13, no. 3, July to Sept., 2002.
18. Centre for Wind Energy Technology, Wind Pro, no. 63, pp. 15-16, Jan '2003.
19. Grid Stability and Expansion Challenges, Wind Pro, no. 63, pp. 17-18, Jan '2003.
20. J. Srikanth and D. Swamy, A Comprehensive Study on Socio-economic and environmental implications of Wind Energy Generation in Coimbatore, Tirunelveli and Kanya Kumari districts of Tamilnadu, Proc. of 26th NREC-2002, Coimbatore, India, pp. 392-397, Jan '2003.
21. C.S. Sinha, Cost of Renewable Electricity: Case of Wind Power Generation in India, , SESI Journal, vol. 6, no. 1, pp. 39-56, June 1996.
22. M.K. Deshmukh and S.S. Desmukh, Wind Power Generation option for ABT affected Power Producers, Proc. of 26th NREC-2002, Coimbatore, India, pp. 398-403, Jan '2003.

8

Hydro-energy Conversion

INTRODUCTION

The hydro energy conversion technology is not a new concept. This technology is being used since long back to convert hydro energy into useful mechanical energy or electrical energy. The most common hydro energy conversion system is the hydro electric power plant. The generated electricity can be used for irrigation or portable water pumping, lighting or health or education purpose etc. This provides power at cheapest rate. Rain falling upon the earth's surface has potential energy relative to oceans towards which it flows. This energy is converted into shaft work with the help of a hydraulic prime mover. Finally the electrical energy is obtained from an electric generator coupled to the shaft of the prime mover. The power developed at the shaft of the prime mover is given by

$$HP = (mh/75) \cdot \eta_h = (\rho Qh) \cdot \eta_h$$

where m is the mass flow rate of water through the prime mover in Kg/sec.

h is the height of fall in metres

η_h is the hydraulic efficiency of the prime mover

ρ is the density of water in Kg/m^3 and

Q is volume flow rate of water through the prime mover in m^3/sec.

The power available at the terminal of the generator in kW can be assessed by the following expression:

$$\text{kW (power)} = (\rho Qh/75) \cdot \eta_h \cdot \eta_m \cdot \eta_g \cdot 0.736$$

where η_m is the mechanical efficiency

η_g is the generator efficiency and

0.736 is the conversion factor from HP to kW.

Nearly 20% of the total world power is generated using hydro plant. According to the estimate of the World Power Organisation, the world hydro potential is about 5000 GW whereas only about 200 GW is harnessed. India's total hydro power potential is assessed as 84000MW at 60% load factor which corresponds to 1,60,000 MW installed capacity. Among them, only 14% has been harnessed.

Depending on the capacity of the unit, the hydro electric power plant may be of higher capacity or of lower capacity. Upto 1972, higher unit capacities and larger sites oriented hydro power plants were developed in view to produce very low cost energy to compete thermal power. Since fossil fuels are depleting at faster rate coupled with steady rise in oil prices, attention has been focussed on smaller size and smaller sites oriented hydro electric power plants. Apart from that, small hydro is environmentally benign, operationally flexible, suitable for peaking support to the local grid as-well-as for stand alone applications in isolated remote areas. The small hydro has been considered to be 'highest densities' resource among all the non-conventional renewable energy sources and it takes the leading place in generation of electricity throughout the world. The small hydro power is also covered in the renewable energy programme in India. Due to above mentioned reasons, the following sections and sub-sections have been discussed in relation to small hydro power (SHP) system.

8.1 SMALL HYDRO POWER SYSTEM (SHP)

The installed capacity of small hydro in the globe is around 50,000 MW against the estimated potential of 1,80,000 MW. The trend of utilization of small hydro power throughout the world is shown in Table 8.1.1. and 8.1.2. India has several rivulets, streams and rivers. It is having world's largest irrigation canal networks with thousands of dams. There are several monsoon fed, double monsoon fed as well as snow fed rivers and streams with perennial flows. In India, the estimated potential of small hydro is around 15,000 MW. The recent data up to March, 2001, created by MNES regarding the number of potential sites is 4096 and the aggregate capacity is 10,071 MW as shown in Table 8.1.3.

Table 8.1.1: Potential and Operation of Small Hydro Plants (in Twh/Year)

Countries	*Exploitable*	*In operation*
Industrialized	374	40.6
Developing	195	4.5
USSR	337	–
China	150	39.5
World	1156	84.6

Table 8.1.2: Number of Small Hydro Plants and Installed Capacity

Countries	*In operation*		*Under construction*		*Under planning*
	No.	GW	No.	GW	No.
Industrialized	16980	20.5	92	0.21	3541
Developing	846	1.7	125	0.31	674
USSR	–	–	100	0.22	–
China	58000	13.3	600	2.20	2000
World	75826	35.5	917	2.94	6215

The small hydro power stations can help for building the availability of decentralized power. Low head installations can be built everywhere in the mountain region, plains or even at the sea level. It can also be installed close to power consumption areas. In remote hilly areas like Ladakh or water logged areas as the Sundarbans in West Bengal, low head small hydro power plants are often the only source of energy. With the help of small hydro power, pumped irrigation and rural water supply can be accomplished. Rural electrification at house hold level can be done. Saw mills, grinding mills can be run with small hydro electricity. Reciprocating compressors or screw type compressors can be operated for cold storage or heat pump systems.

Table 8.1.3: State Wise Details of Identified Small Hydel sites up to 25 MW Capacity in India

Sl. No	*Name of the state*	*Identified Number* of sites	*Total capacity in* MW
1.	Haryana	22	30.05
2.	Himachal Pradesh	323	1624.78
3.	Jammu & Kashmir	201	1207.27
4.	Punjab	78	65.26
5.	Rajasthan	49	27.26
6.	Uttar Pradesh & Uttaranchal	445	1472.93
7.	Gujarat	290	156.83
8.	Madhya Pradesh & Chhatisgarh	125	410.13
9.	Maharashtra	234	599.47
10.	Andhra Pradesh	286	254.63
11.	Karnataka	230	652.61
12.	Kerala	198	466.85
13.	Tamil Nadu	147	338.92
14.	Bihar & Jharkhand	171	367.97
15.	Orissa	161	156.76
16.	Sikkim	68	202.75
17.	West Bengal	145	182.62
18.	Arunachal Pradesh	492	1059.03
19.	Assam	46	118.00
20.	Manipur	96	105.63
21.	Meghalaya	98	181.50
22.	Mizoram	88	190.32
23.	Nagaland	86	181.39
24.	Tripura	8	9.85
25.	Andaman & Nicobar Islands	6	6.40
26.	Goa	3	2.60
	Total	4096	10071.81

It has also been estimated that a small hydro power station of 1 KW installed capacity can prevent production of 200 tonnes of CO_2 over operating life of 25 – 30 years.

8.2 CLASSIFICATIONS OF SMALL HYDRO POWER SYSTEM

According the Central Electricity Authority and Bureau of Indian Standards, the small hydro power systems are classified depending on capacity and available head. The classifications are as follows:

(a) Depending on capacity—

Size	*Unit size*
Micro	upto 100 KW
Mini	101 to 1000 KW
Small	1001 to 6000 KW

(b) Depending on Head—

Type	*Head*
Ultra low head	bellow 3 metres
Low head	less than 30 metres
Medium head	between 30 to 75 metres
High head	above 75 metres

On the basis of heads available, the small hydro power systems are again classified as: (i) Independent scheme, (ii) Sub ordinate scheme.

In independent scheme, a stream flow is captured, regulated and developed for power generation only. The low head schemes are not observed to be economical for independent generation. High head or medium head schemes are considered. But the medium head schemes are noted to be more successful. These schemes have become successful in Himalayan territories states like, Jammu and Kashmir, Nagaland, Manipur and Himachal Pradesh etc.

In sub-ordinate schemes, the primary objective is irrigation or drinking water and secondary objective is power generation. These schemes are appropriate for those regions where there is extensive network of irrigation canals and the water can be stored in the form of reservoir. The schemes can be used for the sites of indogangetic plans (canals) and Peninsular India (reservoirs).

8.3 DIFFERENT COMPONENTS OF SMALL HYDRO ELECTRIC POWER SYSTEM AND THEIR DESIGN CONSIDERATIONS

A typical arrangement of a Small Hydro Electric Power System is shown in Fig. 8.3.1, with its different components. The basic components are as:

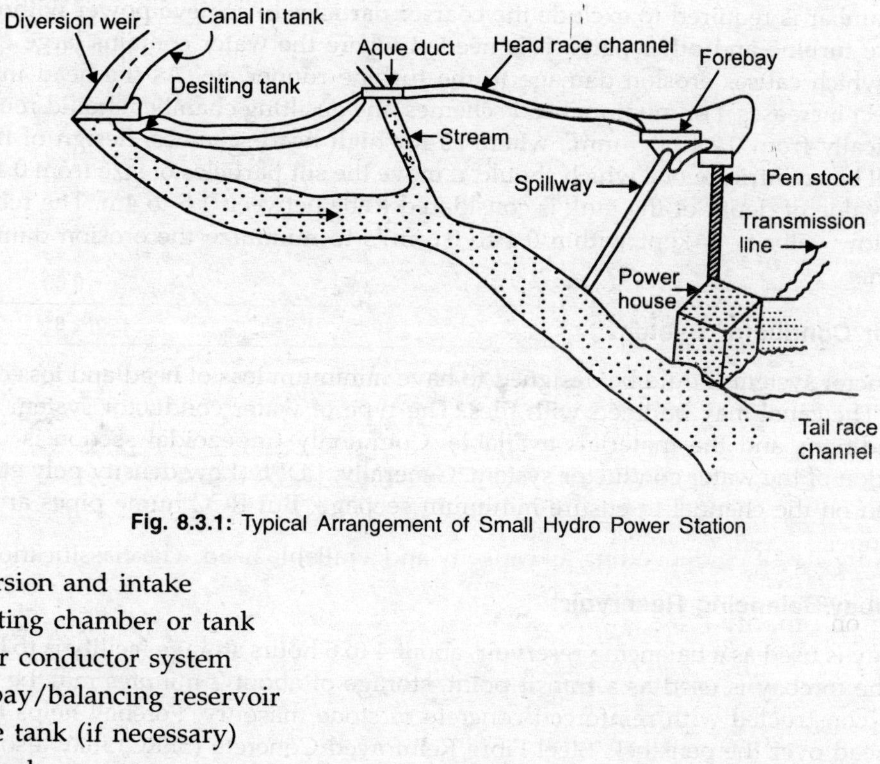

Fig. 8.3.1: Typical Arrangement of Small Hydro Power Station

(i) Diversion and intake
(ii) Desilting chamber or tank
(iii) Water conductor system
(iv) Forebay/balancing reservoir
(v) Surge tank (if necessary)
(vi) Penstock
(vii) Power house comprising of turbine, generator, protection and control system, dewatering system, drainage system, auxiliary power system, grounding, emergency and stand by power system, lighting and ventilation
(viii) Tail race channel

8.3.1 Diversion System

In an hydro-electric power generation system, dam, barrages, solid boulder structure and trench type weir are used to divert the required flow from the river bed or streams to the intake structure. But for small hydro-electric power generating system, solid boulder structure and trench type weir are considered. Boulder type is preferred when boulders are available in the river and the rock is encountered in the river bed within one meter depth, where as trench type is used where rock is not available in the river bed. The diversion weir should divert all the lean season flows and the structure should be safe during monsoon floods. Diversion structures in hill streams often face choking of intake. Underground structures may get damaged due to heavy boulders and bed loads. After all, this structure should be simple in construction and economical. Boulder weir is comparatively quite cheap. In the trench type, there will be a trapezoidal trough which will be located below the bed of the river. The top should be kept at the bed level of the river. The intake which is located at the end of the trench weir, is provided with a gate to control the release of water as per requirement. The horizontal trash back is kept at the bed level with a slope in the downstream direction. It helps in tapping the water and rolling down the boulders.

8.3.2 Desilting Chamber or Tank

Desilting chamber is required to exclude the coarser particles to achieve power without abrasion effects on the turbine and other parts. It is needed where the water contains large quantities of course silt, which causes erosion damage to the turbine runner etc. As the head increases, the abrasion effect increases. For medium head schemes, the desilting chamber should remove the silt of sizes typically from 0.2 to 0.5 mm., where as for high head schemes, design of the desilting chamber will be an intricate one which should remove the silt particles of size from 0.1 to 0.2 mm. The typical value of depth of the tank is considered to be between 1.5 to 4m. The recommended horizontal flow velocity is kept within 0.4 to 0.6 m/s to minimize the erosion damages to the turbine runner.

8.3.3 Water Conductor System

Water conductor system should be designed to have minimum loss of head and loss of water due to seepage. The canal may be lined with tiles. The type of water conductor system depends on the site conditions and the materials available. Commonly trapezoidal section is used for the channel section of the water conductor system. Generally, LDPE (Low density poly ethylene) film liner is given on the channel to ensure minimum seepage. But RCC hume pipes are suitable in steep hill slope.

8.3.4 Forebay/Balancing Reservoir

When forebay is used as a balancing reservoir, about 4 to 6 hours storage facility is to be provided. But when the forebay is used as a transit point, storage of about 2 minutes may be adequate. It is generally constructed with reinforced concrete of stone masonry. Forebay helps to provide a minimum head over the penstock. Steel Fibre Reinforced Concrete (SFRC) may also be used for forebay. It is having a better characteristics as compared to ordinary concrete. But its cost is high.

8.3.5 Surge Tank

Surge tank will be necessary for the water conductor conduit length of more than 5 times the head on the machine.

8.3.6 Penstock

Penstock is used to feed the water to the machine through it. In its hydraulic design consideration, diameter of the penstock is determined by considering its economic aspect also. During structural design of penstock, its thickness is assessed. In some cases, RC hume pipes are also used for penstocks. Penstock can also be made by SFRC. The materials for the penstock pipes may be mild steel and PVC depending on the design pressure. A bell mouth entry is preferred to ensure in reduction in head losses and a smooth entry of water from forebay tank into the penstock.

8.3.7 Spillway

The presence of a spillway is important as it does not allow the water level to rise and flood the area during the load rejection of the plant. Its crest is kept at the permissible water level. The channel or pipe can be used for the spillway.

8.3.8 Power House

In the power house, turbine, generator, control panels, auxiliary equipments etc. are kept. Therefore its building should accommodate them and it is usually constructed by RCC or stone masonry. Generally the height of the building is kept about 3 to 5m.

8.3.9 Tail Race

Tail race is a water channel or a cut and cover conduit. Through it, the water passes from the turbine outlet i.e. draft tube to the river. In the tail race, generally, the allowed maximum water velocity is considered to be 1 metre/second. The shape of the channel is usually trapezoidal or rectangular. It is constructed by stone masonry or brick masonry depending on the availability of the material locally with minimum cost.

8.4 TURBINES AND GENERATORS FOR SMALL HYDRO POWER SYSTEM

In small hydro power systems, usually the following types of turbines are used:

(i) Bulb or Tubular Turbine
(ii) Tube Turbine
(iii) Straflo Turbine

8.4.1 Bulb or Tubular Turbine

The configuration of bulb turbine system is shown in Fig. 8.4.1. In this system, the turbine and generator are enclosed inside the conduit through which water flows from head race to tail race. The encloser should be water tight in nature. Kaplan type or propeller type turbines are usually used. Generator is directly coupled to the turbine. Sometimes generator is connected through a set of gears to increase the speed of the generator. It reduces the generator size. Generator is usually kept at the upstream side. Turbine is located at the downstream end. The water is allowed to flow axially around the bulb through guide vanes and wicket gates to the turbine runner. Since the units are of package type, therefore its erection and dismantling take little time. Various sizes of various unit capacities of bulb turbines are presently available. A large number of these units are in operation throughout the world. These are suitable for comparatively low heads ranging from 3 to 18 m. The bulb turbines are available for capacities from 40 MW to 200 MW. Only lower capacity bulb turbines are used for small hydropower generation. The efficiency of this system is very high. Size is less and hydraulic losses are minimum for this type of turbines.

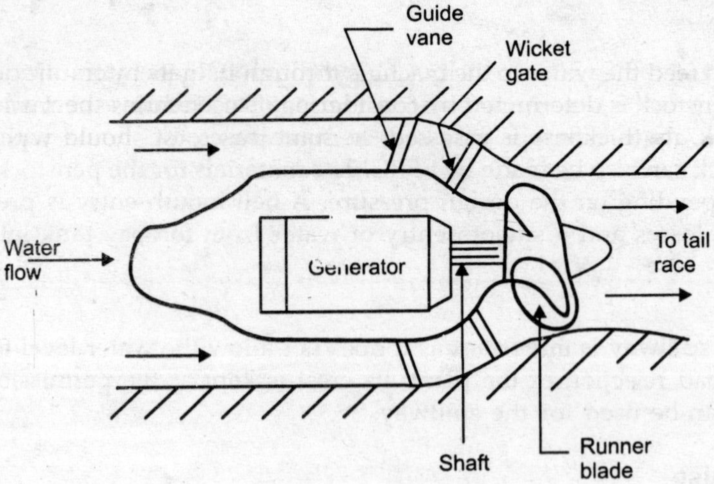

Fig 8.4.1: Bulb Turbine

8.4.2 Tube Turbine

The configuration of the tube turbine system is shown in Fig. 8.4.2. The turbine may be horizontal. It is kept in a tubular channel. The water comes from the head race to the runner block through wicket gate and guide vanes. Then water flows to the tail race. The flow is controlled by the wicket gates where as guide vanes help to direct water between the propeller blades. The generator is kept outside the channel in a pit and above the water level in the tail race. When the shaft is horizontal, the channel should be sloped downward beyond the turbine. If channel cannot be sloped downward, then turbine shaft may be sloped upward to have the same objective. This type is also known as "pit type". The main advantage of this system is the easy accessibility to the generator.

8.4.3 Straflo Turbine

In this design, the poles of the generators are mounted directly on the periphery of the turbine runner. Therefore, like the bulb type, there will be no rotor for the generator. The intricate sealing arrangements is required for this type to prevent water ingress into poles. It requires specialized technology. The water flow is always kept horizontal for this system. The merits of this type are the need of smaller size power house and the large range of power output with large range of head.

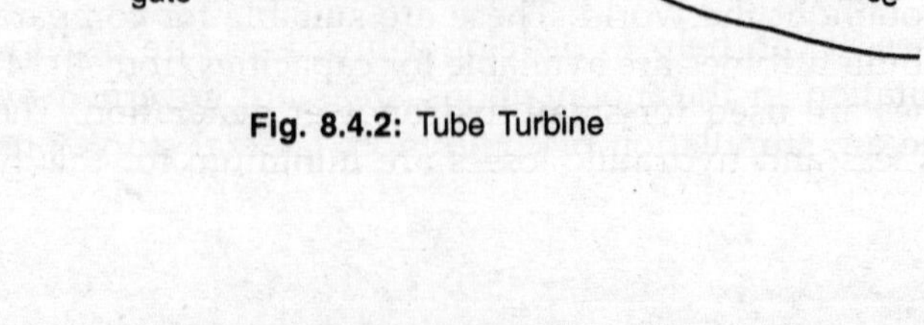

Fig. 8.4.2: Tube Turbine

8.4.4 Generators

Generally conventional hydro electric A.C. synchronous generators are used. For small hydro power systems, brushless excitation system is used. This reduces the maintenance costs. During the design of the generator rotor, it is kept in mind that rotor should withstand turbine runway speed i.e., speed attained on sudden load throw off conditions.

8.5 MINI AND MICRO-HYDEL POWER SYSTEMS

In India, there is high hydro potential with low head. The estimated low head hydro potential of this country is estimated to be about 20,000 MW. If properly planned and implemented, it is possible to commission a small hydro generating setup of 5 MW with a period of one and half year against the period of decade or two for conventional power plants. Several such sets upto 1000 KW each have been already installed in West Bengal, Bhutan, Arunachal Pradesh, Uttar Pradesh and Himachal Pradesh. China has developed lot of power through mini and micro hydel plants as small head power capacity in the country is estimated as 150×10^3 MW against 580×10^3 MW total. Tubular turbines are specially preferred for very low heads with large flow rates.

In recent time, in our country an innovative concept is being developed and implemented. This is the installation of light weight portable micro hydel sets in the hilly areas, particularly in the Himalayan region. This scheme helps to provide power to non-grid connected remote areas by using 'stand alone systems' ranging from 5 to 15 KW. Another major advantage of this scheme is that it can be easily installed and maintained by the local communities. At present 35 projects of portable micro hydel power generating schemes have been commissioned and those are generating power.

8.6 TECHNICAL FEASIBILITY CRITERION FOR SMALL HYDRO POWER SYSTEM

In our country, Hydrology and Trench type Weir Intake have been identified as important areas for the technical feasibility in formulation of small hydro electric power generating systems. These identical areas and their effects have been briefly discussed in this section.

8.6.1 Hydrology

Hydrology is the science which deals with rainfall and run-off. Water evaporates from plants, rivers, oceans and is carried with the air in the form of vapour known as clouds. When the water vapours in the atmosphere get cooled below its dew point temperature, it falls in the form of water (rain) or snow depending on the atmospheric temperature. This evaporation and precipitation is known as hydrological cycle. This cycle is a natural continuous process and therefore constitutes a perennial source of energy.

The altitude plays an important role in precipitation. The quantity of precipitation increases with increase in altitude upto a certain level and then decreases thereafter. Depending on local topography, the maximum precipitation varies from place to place. It is also noted that the average variability of precipitation generally increases with altitude. It is also identified that at higher altitudes, the maximum precipitation occur generally earlier than on the foothills.

It is very difficult to know the climatology of hilly or mountainous areas. However, recent aerial surveys and meteorological satellites are providing information on clouds, snow cover, glaciers, avalanches. It can help to develop a broad picture of climatology in the mountainous areas. The precipitation in the mountainous areas are governed by complex phenomenon like direct uplift of the air, stimulation of convection, general storm precipitation and spillover.

The direct lifting of the air leads to adiabatic cooling which produces rain. When the unstable air encounters a mountain slope, it gets lifted up beyond its instability level and causes convective precipitation. General storm precipitation occurs due to dynamic interactions which lead to convergence of air and cause consequent lifting. This type of precipitation can occur in mountaineous or plane areas. Spillover rainfall is the precipitation that falls on the leg side. It is formed on the windward side and is carried by the wind stream on the leg side. Finally, it reaches the ground on the windward side.

8.6.2 Trench Type Weir Intake

In hills, the streams experience heavy floods, boulder movements and river meandering. This trench type weir has simplified the intake design in hills. The length of the weirs is chosen to enable the passing of the peak flood discharge and rolling over the boulders. Usually the length of the weir is considered to be equal to the width of the stream.

In practice, the non-uniformity of discharge persists in the entire length and width area of the trench. To combat this, generally the width of the trench is kept more than the theoretical calculated value.

8.7 MERITS AND DEMERITS OF MINI, MICRO AND SMALL HYDRO POWER SYSTEMS

Merits:

(i) Mini, micro and small hydro electric power potential is more economical.

(ii) It takes the shortest time for developing an unit i.e., it has low gestation period.

(iii) It saves fossil fuels.

(iv) Running expenditure is almost negligible.

(v) Operation and maintenance of these schemes are easy and simple.

(vi) The equipments used for these plants can be standardized easily.

(vii) A low-head small hydro plant suffers much less from silt erosion and cavitation of turbine blades.

(viii) These are suitable for isolated loads.

(ix) The transmission losses are less.

(x) These are free from hazards of pollution.

(xi) These have no environmental problems like the big hydro power plant's development. No submergence of land, no loss of agricultural lands, no dislocation of habitation, no inter-state problems take place during the development of these plants.

(xii) Ecology of the region does not get affected.

(xiii) It helps to generate employment for local people as the local people can install, operate and maintain the units.

Demerits:

(i) Generally, high head-low discharge configurations require skilled manpower for operation and maintenance of the plant.

(ii) Remoteness of sites, particularly in the hilly regions and its adverse geological conditions.

(iii) General lack of awareness regarding its benefits.

8.8 PRESENT THOUGHT ON SMALL/MICRO HYDRO POWER

Ministry of Non-conventional Energy Sources of India has recently been encouraging NGO's for their active participation in the development of small hydro power projects. The first project of the country of 800 KW capacity has been installed in Himachal Pradesh and commissioned on 31st December, 2001. It has been set up by an NGO, namely Sai Engineering Foundation. It has been set in a remote area of Pooh sub-division of District Kinnaur. The project has been conceived as grid connected project and the power is being sold to HPSEB. But it is also providing power to 5 remote villages of the area.

Presently it is thought that many villages in developing countries participating in remote and hilly regions will remain off grid because of cost of centralized rural electrification. The main alternative, over the next 20 years, lies in renewable energy. A range of decentralized energy options with mature technologies are now available. But the choice of the best option, offering a sustainable and affordable supply of mechanical and electrical power, is complex and involves a range of technical, social and economic parameters. However, the available small or micro hydel system is a cost effective option.

EXERCISE

1. State the hydro power potential scenario for India and the globe. What do you mean by a small hydro power system? Classify it.
2. With the help of a neat schematic diagram, discuss in detail the utilities of the basic components used in a typical small hydro electric power plant.
3. With the schematic sketches, discuss in detail the functioning of a bulb turbine and a tube turbine generally used for small hydro power systems.
4. Mention the factors which affect the technical feasibility criterion in formation of small hydro power systems.
5. State the merits and limitations of mini, micro and small hydro power systems.

SUMMARY

This chapter has highlighted a brief discussion on the hydro energy conversion technology and its applications as higher unit capacities and larger sites oriented hydro power plants, and smaller size and smaller sites oriented hydro electric power plants. The reasons for the recent acceptance in favouring the small hydro power system have been pointed out. The classification of small hydro power system in accordance with the Central Electricity Authority and Bureau of Indian Standards have been outlined in details. The description of each of the basic components of a small hydro electric power system along with the respective design criterion have been discussed in details. A detailed discussion has been presented on the important devices e.g. turbines and generators, practically used for small hydro electric power systems. A brief discussion has been carried out on the technical feasibility criterion for small hydro power system. This chapter has also covered a brief discussion on mini and micro hydel power systems which are being presently used by the Indian Government in context to the high hydro potential with low head for this country. Merits and limitations of mini, micro and small hydro power systems have been pointed out. This chapter concludes with a brief highlight of the present thought of Govt. of India on small/micro hydro power development in the country.

REFERENCES

1. L.V. Kumar, Small Hydro: A Catalyst for Social Change, IREDA NEWS, vol. 12, no. 1, pp. 49-57, 2001.
2. L.V. Kumar, Managing Risks in Small Hydros, IREDA NEWS, vol. 11, no. 2, pp. 39-47, 2000.
3. V. Bakthavatsalam, Small Hydro Power: An Indian Experience, IREDA NEWS, vol. 12, no. 2, pp. 15-24, 2001.
4. O. Paish, et al., The Development of Traditional Hymalyan Water Mills for Sustainable Village Scale Micro-Hydro Power, 1st Int. conference on Renewable Energy-Small Hydro, Hyderabad, Feb '1997.
5. H. Werszko, Identification of Hydro Electric Power Potential in Bangladesh, Feasibility study of Mahamaya Chara Multi purpose Project Final Report, Bangladesh Water Development Board, Sept '1999.
6. R.K. Khitoliya, Small Hydro: A Potential and Alternate Source of Energy, Proc. of 1st Int. conference on Energy Management beyond 2000, Lucknow, pp. 321-326, Oct '2000.
7. R.K. Khitoliya and S.V. Rao, Small Hydro-A Boon for Energy Sustainance in Hilly Regions, Proc. Int. Semianr on Decentralised Energy: Options and Management, Kolkata, India, pp. 230-236, Feb 22-24, 2002.
8. J. Guthrie Brown, Hydro Electric Engineering Practice, vol. I & II, Second Edition, London, Blackie and Son Ltd.-Glasgow, 1964.
9. Emil-Mosonyi, Water Power Development, vol. I & II, Publishing House of the Hungarian Academy of Sciences, Budapest, 1957.
10. B.L. Theraja and A.K. Theraja, A text book of Electrical Technology, Nirja Construction and Development Co. (P) Ltd., 1994.
11. Adam Harvey, Micro-hydro Design Manual: Guide to Small Scale Water Power Schemes, Intermediate Technology Publication, 1993.
12. S.C. Arora and S. Domkundwar, A Course in Power Plant Engineering, Dhanpat Rai & Co. (P) Ltd., 1998.
13. Ministry of Non-conventional Energy Sources, 2001-2002, Annual Report.
14. J.N. Chakraborty, Sources of Energy and Indian Scene, National Seminar on Energy, Madras, July '1978,.
15. J.L. Gorden, Hydroturbine Cavitation Erosion, Jl. of Energy Engineering (ASCE), vol. 118, no. 3, p. 194, 1992.
16. P.J. Clark, Microhydel Power Projects, Indian and Eastern Engineer, pp. 67-70, Sept '1983.
17. K.M. Mittal, Non-conventional Energy Systems, Principles, Progress and Prospects, Wheeler Publishing, 1997.
18. R.J. Griesbach, The Idikki Project, Water Power, pp. 12-14, Jan '1975.
19. R.J. Hothersak, Micro-hydro Turbine Selection, WPDC, pp. 26-28, Feb '1984.
20. F.J. Parker, Generator Design for Small Hydro Applications, WPDC, pp. 56, Feb '1984.
21. MNES Annual Report, 1995-96, New Delhi.
22. C.S. Sinha, Micro Hydel: A Promising Source of Rural Energy, Urja Bharti, 3(3), pp. 47-49, 1993.
23. J.P.C.F. Dauchel, Small Hydro Schemes for Low Heads and Isolated Networks, Urja, pp. 51-52, March 27(3), 1990.
24. N.C. Bose, Siltration Problem Solved in Jaldhaka, Irrigation and Power, pp. 165-172, July '1987.
25. Patra, et al., Optimum Energy Generation Planning for Run-off River Plant, IJPRVD, pp. 1-4, Jan-Feb '1990.
26. V.B.H. Harish, Mini-Hydel Power: A Technology Insight, Urja, 26(6), pp. 48-50, Dec '1989.
27. Devdas and Lulla, Bulb type Hydro Generating Units, Irrigation and Power, pp. 8-17, Oct '1987.
28. O.D. Thapar, Microhydel – Potential and Technology, Proc. of NSEC-98, Univ. of Roorkee, pp. 1-12, Dec '1998.

9

Other Renewable Energy Sources

INTRODUCTION

This chapter deals with ocean thermal energy conversion, wave energy conversion, tidal energy conversion techniques and their respective systems. In the ocean thermal energy conversion, the basic idea has been elaborated. Different types of systems have been discussed. Its site selection procedure and prospects have also been covered.

In wave energy conversion, energy sources and its conversion devices have been explained. The tidal energy conversion area consists of basic principles involved, its operation, assessment of energy and power, site requirements, its storage and prospects.

9.1 OCEAN THERMAL ENERGY CONVERSION (OTEC)

The absorption of solar radiation by the seas and oceans causes the moderate temperature difference between the upper and lower levels of water, especially in tropical oceans which collect and store very large amount of solar energy. This stored energy as heat with its associated temperature difference can be converted into work with the help of established thermodynamic cycle. This is called ocean thermal energy conversion. Ocean waters are heated by the sun and they contain virtually an inexhaustible source of energy. As the temperature difference even in the tropics is less, the efficiency of this energy conversion is low. It has very high capital cost.

9.1.1 Working Principle

In the ocean thermal energy concept, the upper layer of ocean is considered as an infinite heat storage reservoir because the upper surface collects the solar energy as heat like a solar collector. Another fact is that the temperature difference between water at the surface and lower levels of the ocean in the tropics is around 20ºC and occurs over a few hundred metres. A typical temperature profile is shown in Fig. 9.1.1.1. The temperature of the water at the surface becomes fairly constant for the first few meters because of mixing. Subsequently it decreases and achieves the value at the lower level asymptotically. The warm surface water is used to heat some low boiling organic fluid in the boiler. The vapour of the working fluid runs a heat engine. The vapour coming out from the engine is condensed by the cold water brought up from the deeper regions. An illustration diagram is shown in Fig. 9.1.1.2.

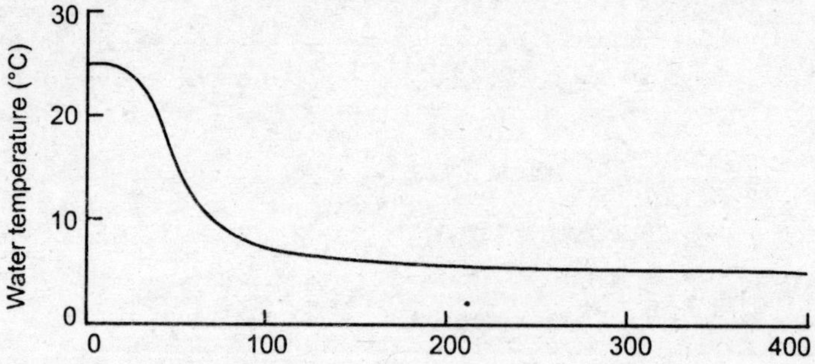

Fig. 9.1.1.1: Typical Variation of Temperature with Ocean Depth

In the tropics, the ocean surface acts as an infinite source of heat reservoir at around 27°C. Some 1 km directly below, the region acts as an infinite heat sink at about 4°C. Both the reservoirs are maintained annually by solar incidence. The ocean thermal energy conversion system uses this temperature difference in a heat engine in producing power. The surface temperatures and the temperature differences vary with latitude and season. Both becomes maximum in tropical sub-tropical, and equatorial waters, i.e. between the two tropics. These waters are considered to be most suitable for OTEC systems.

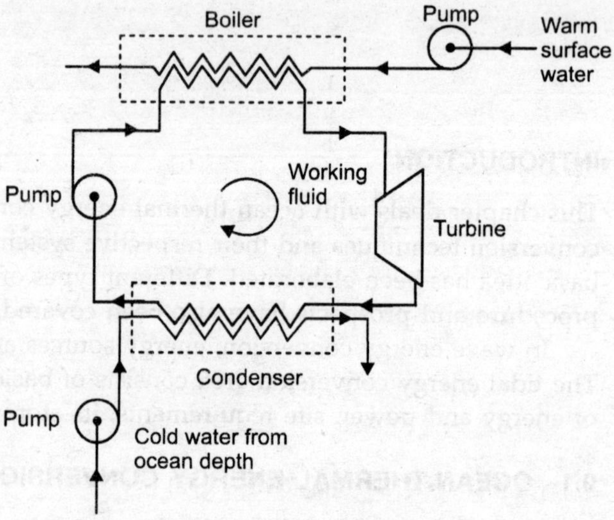

Fig. 9.1.1.2: Schematic Diagram of Rankine Cycle OTEC Plant

Since the temperature difference between the hot and cold streams is low, the efficiency of the Rankine Cycle used for OTEC system will be low. For very small temperature drops of around 4° to 5°C across the boiler and the condenser, the Rankine Cycle efficiency for most of the working fluids will range between 2 and 3 percent. This is comparatively very low with respect to the net efficiency of a conventional power plant. In spite of this drawback, OTEC systems are becoming more economically attractive because for both the collection and storage of energy, no cost will be required as this is done by nature.

The absorption of solar energy by the water follows Lambert's law of absorption. It states that each layer of equal thickness absorbs the same fraction of light that passes through it.

Thus, $-dI/dx = kI$, or, $I = I_o e^{-kx}$

Where I is the intensity of radiation at a distance x below the surface and I_o is the intensity of radiation at the surface, i.e. at $x = 0$, k is the absorption coefficient which is having unit of L^{-1}. The values of k depend on nature of the water. Followings are the typical values of k for some types of water.

$k = 0.05\ m^{-1}$ for very clear fresh water

$k = 0.27 \text{ m}^{-1}$ for turbid fresh water

$k = 0.50 \text{ m}^{-1}$ for very salt water

The ocean thermal energy is harnessed by two methods. One is open cycle system, known as the Claude cycle, and the other is closed cycle system, also known as Anderson cycle.

9.1.2 Open Cycle System (Claude Cycle)

In the open cycle, the sea water is used as the working fluid. A schematic flow and the corresponding T-S diagrams of the open cycle system are shown in Fig. 9.1.2.1 and 9.1.2.2 respectively. The warm surface sea water, say at 27°C having saturation pressure of 0.0356 bar, is sent to an evaporator where the pressure is maintained at a value slightly below the saturation pressure corresponding to the water temperature. After entering the evaporator the water gets partially flashed to an equilibrium two phase condition steam at the new pressure and temperature, say at 0.0317 bar having saturation temperature at 25°C. The low pressure in the evaporator is maintained by the vacuum pump. It also facilitates in removing dissolved non-condensable gases from the evaporator. The steam is dried and saturated steam is produced. The remaining warm water is discharged back to the ocean. The steam usually with very low pressure and high specific volume in the order of 0.0317 bar and 43.4 m^3/kg is passed through a specially designed turbine which can handle the said conditions. A direct contact condenser is used in the open cycle system. In the condenser, the exhaust steam gets mixed with the cold water taken with the help of a pump from the deep layer. After condensation, the final saturated water is allowed to be discharged to the ocean. The condenser pressure and temperature are 0.017 bar and 15°C respectively. The cooling water for the condenser at 13°C is taken from deep water at 11°C. This rise in temperature is due to heat transfer between the progressively warmer outside water and the cooling water inside the pipe as it ascends the cold-water pipe. In the open cycle system, large mass and volume flow rates of water are used. The turbine handles very low pressure and high specific volume steam which leads to the larger size of turbine. Furthermore, degasifiers or deaerators must be used to remove dissolved gases in the ocean water. But in this system, the bio-fouling control gets minimized as no heat transfer problems persist in the evaporator.

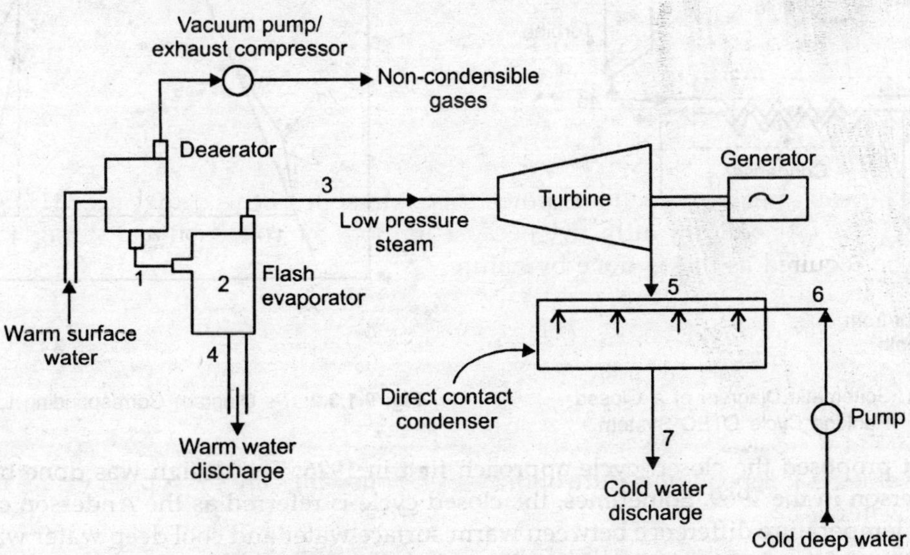

Fig. 9.1.2.1: Schematic Diagram of the OTEC Open Cycle

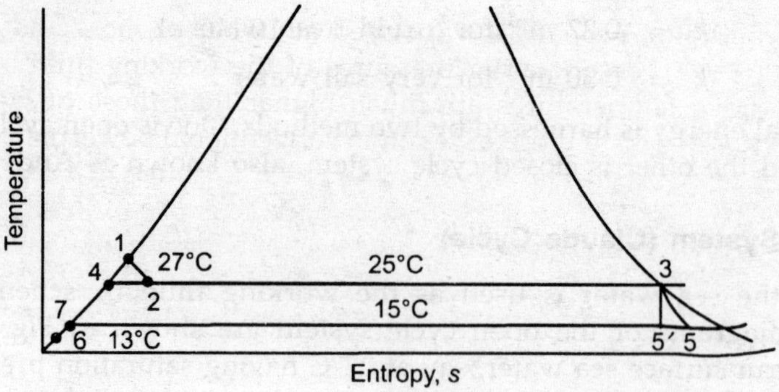

Fig. 9.1.2.2: *T-s* Diagram Corresponding to Fig. 9.1.2.1

9.1.3 Closed Cycle System (Anderson Cycle)

In the closed system, ammonia, propane, R22, etc. having low boiling point temperatures are used as working fluids. Warm surface sea water is passed through an evaporator with the help of a pump. In the evaporator, the working fluid in the form of a high pressure liquid absorbs heat and gets converted into vapour. The vapour drives the turbine connected to generator in producing electricity. After expansion in the turbine, the vapour is condensed with the help of cooling water brought up from the ocean depths through a cold water pipe. The condensed liquid is pumped back to the evaporator, thereby closing the cycle. A schematic diagram and thermodynamic T-s diagram of a closed Rankine cycle OTEC system are shown in Figs. 9.1.3.1 and 9.1.3.2 respectively.

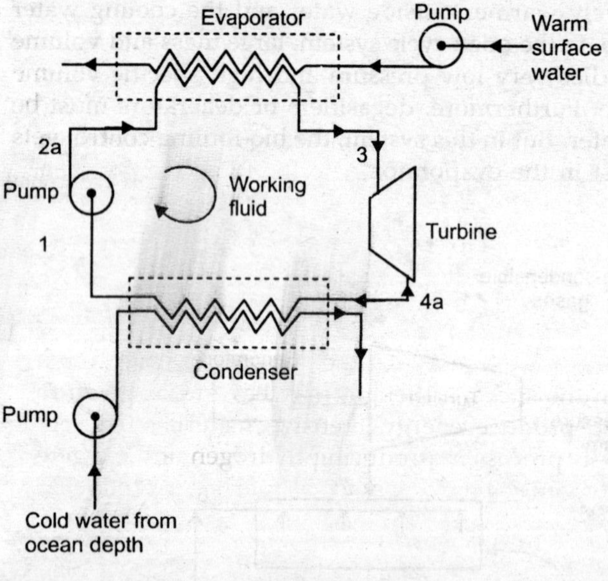

Fig. 9.1.3.1: Schematic Diagram of a Closed Rankine Cycle OTEC System

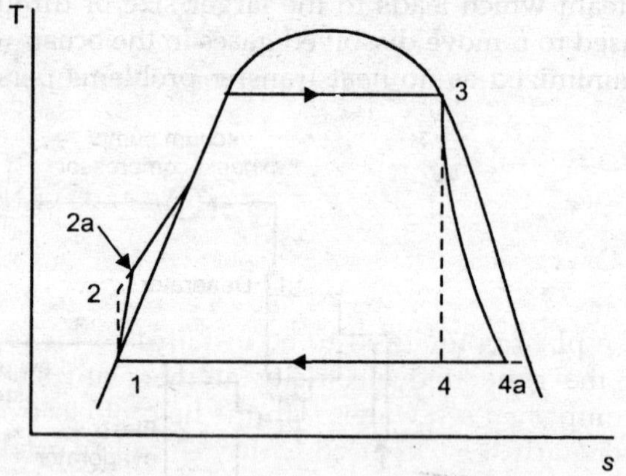

Fig. 9.1.3.2: *T-s* Diagram Corresponding to Fig. 9.1.3.1

Bajrot proposed the closed cycle approach first in 1926. The design was done by Anderson and Anderson in the 1969. Sometimes, the closed cycle is referred as the Anderson cycle. In this cycle, the temperature difference between warm surface water and cool deep water was 20°C. The cold water used for the condenser was taken from about 600m depth. Propane was used as the working fluid. In the evaporator, propane is vaporized at 10 bar or more and in the condenser it is condensed at about 5 bar. The operating pressures of the working fluid at evaporator and condenser at their respective temperatures are much higher than those of water. The specific volumes are much lower compared to those of steam in conventional powerplants. Such pressures and specific volumes result in smaller and hence less costly turbines. This OTEC cycle power plant requires very large evaporator and condenser. This is because of the low efficiency of around 2 percent. In the Anderson OTEC system, light and low cost thin plate type heat exchanger was used instead of the usual heavier and more costly shell and tube heat exchanger.

9.1.4 Bio-Fouling

In the ocean environment, the deposition and growth of micro-organisms, called biological fouling or bio-fouling is likely to occur on cooling water side of the heat exchanger. The gradual occurrence of micro-fouling will be the reason for the attachment of macro-organisms. Moreover, mineral deposits and a film of corrosion can also accumulate on the water side. In totality, the bio-fouling, corrosion, and so on, is referred to as "fouling" or "scaling". It inhibits heat transfer in the heat exchanging process in the heat exchanger. The thermal resistance of the film i.e., r_f is called the fouling factor. This thermal resistance is the reciprocal of the corresponding heat transfer coefficient of the film i.e. 'h_f'. The bio-fouling effects are more significant for the evaporator because the warmer water is conductive to the growth of marine organisms.

For copper or copper-alloy heat exchangers, bio-fouling is less. In this case, traces of dissolved copper act as a biocide. In practice, some provisions must be made to combat the formation of fouling layers and to remove the film, if formed. These are accomplished by chemical process i.e., chlorination or by mechanical cleaning with the help of brushes or rubber balls periodically or by both means. By increasing the flow rate of the water, deposition and growing of micro-organisms can also be reduced. But the increase in mass flow rate can cause erosion.

9.1.5 Criterion for Site Selection

During selection of a site for an OTEC system, a significant temperature difference between surface water and deep sea water is considered as an important criterion. The temperature difference should be at least about 20°C for 700-900m depth or more. If the temperature difference increases, the cost of generation of electricity reduces. The favourable sites are in the tropical belt between 20°N and 20°S latitude. Bio-fouling factor is also considered during selection of a site.

The OTEC plant should be located offshore for having the deep colder water. The distance between the offshore plant and the shore should be minimum (should be less than about 30 km) to facilitate transmitting the electricity to the land by submarine cable with less cost. However, if the plant is required to be installed so far from the shore, a thought may have to be given to use the generated electricity at the plant site to produce energy-intensive materials, e.g. for decomposing sea water with the help of electrolysis process in producing hydrogen and oxygen. Particularly the produced hydrogen can be used for various purposes, e.g. as fuel or for producing ammonia for the use as fertilizer.

9.1.6 Ocean Thermal Energy Conversion in India

A preliminary feasibility study for establishing a 1 MW OTEC plant in Lakshadeep Island at Minicoy has been completed by IIT, Madras. The hydrographic survey of the proposed site was conducted by National Hydrographic Office, Dehradun. Preliminary oceanographic studies on eastern site of Lakshadeep Island offered the positive observations for the establishment of shore based OTEC plant at the Island. For the proposed OTEC plant the water would be brought up from 1000 m depth. The Island has large lagoons on the western side. The lagoons are very shallow with hardly any nutrient in the sea water. In the plant, a part of the cooling water, after coming out from the condenser, is proposed to be diverted to lagôons for the purpose of aquaculture.

9.2 WAVE ENERGY CONVERSION

Ocean and sea waves are caused indirectly by solar energy. The wind, caused by the uneven solar heating and subsequent cooling of the earth's crust, generates the waves. The wave energy conversion devices produce much higher power densities than solar energy conversion devices due to higher concentration of the wave energy than the incident solar energy at the latter's peak. The wave energy is a free and renewable energy source. Wave energy conversion devices do not use up large land masses. These are relatively pollution free and they leave the water in a relatively placid state in their wakes. In the zones where there is a large amount of wave activity, a string of devices in the staggered array concept are installed not only to produce electricity but also to protect coast line from the destructive actions of such waves, minimizing erosion, and even help in creating artificial harbours. But the waves lack dependability. There is also relative scarcity of accessible sites of large wave activity. The devices are relatively complicated and lack the necessary mechanical strength to withstand the enormous power of stormy seas. Economic factors such as the capital investment, costs of maintenance, repair and replacement, as well as problems of biological growth of marine organisms, are all relatively unknown and seem to be on the large side.

9.2.1 Wave Motion, Energy and Power

The parameters that characterize a two-dimensional wave which has a free surface and is acted upon by gravity (shown in Fig. 9.2.1.1) are as:

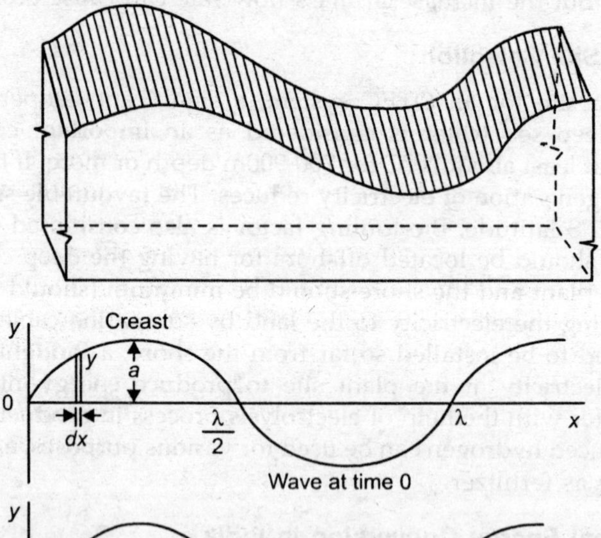

Fig. 9.2.1.1: A Typical Progressive Wave, a = Amplitude, λ = Wavelength, showing Two-dimensional Wave and Amplitudes at Time 0 and θ

λ = wave length = $c\tau$, *m*

a = amplitude, *m*

2a = height (from crest to trough), *m*

τ = period, *s*

f = frequency = $1/\tau$, s^{-1}

c = wave propagation velocity λ/τ, m/s

n = phase rate = $2\pi/\tau$, s^{-1}

The period τ and wave velocity c depend upon the wavelength and the depth of water that are depicted with the help of Table 9.2.1 and Fig. 9.2.1.2.

The range below the dashed line in the figure and table represents the most common waves. The relationship between wavelength and period can therefore be well approximated by:

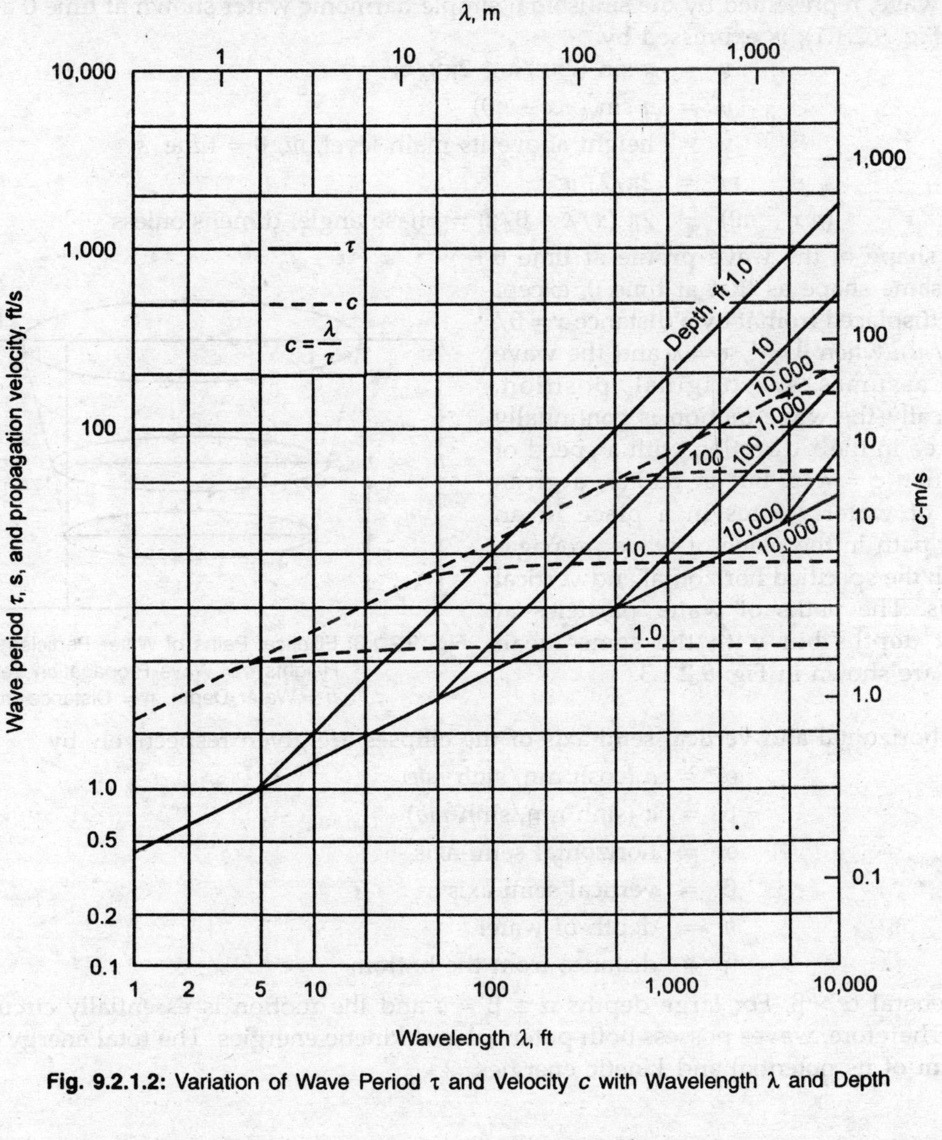

Fig. 9.2.1.2: Variation of Wave Period τ and Velocity c with Wavelength λ and Depth

Table 9.2.1: Wave Periods

Water depth h, ft	*Wavelength. ft*				
	1	*10*	*100*	*1000*	*10,000*
1	0.442	1.873	17.64	176.33	1763.30
10	0.442	1.398	5.923	55.80	557.62
100	0.442	1.398	4.420	18.73	176.45
1000	0.442	1.398	4.420	13.98	59.23
10,000	0.442	1.398	4.420	13.98	44.20

$$\lambda = 1.56\,\tau^2 \ (\lambda \text{ in } m,\ \tau \text{ in } s) \qquad ...(9.2.1.1)$$

The wave, represented by the sinusoidal simple harmonic water shown at time 0 and at time θ (refer Fig. 9.2.1.1), is expressed by

$$y = a \sin (2\pi x/\lambda - 2\pi\theta/\tau) \qquad ...(9.2.1.2a)$$

or

$$y = a \sin (mx - n\theta) \qquad ...(9.2.1.2b)$$

where y = height above its main level, m, θ = time, s

$m = 2\pi/\lambda,\ m^{-1}$

$(mx - n\theta) = 2\pi\ (x/\lambda - \theta/\tau)$ = phase angle, dimensionless

The shape of the wave profile at time θ has the same shape as that at time 0, except that it is displaced from it by a distance $x = \theta/\tau = \theta\ (n/m)$. when $\theta = \tau$, $x = \lambda$ and the wave profile assumes its original position. Theoretically the wave motion is continually lateral, i.e., in the x direction with a speed of propagation $c = \lambda/\tau$. But in reality, a given particle of water rotates in a place in an elliptical path in the plane of wave propagation, with the specified horizontal and vertical semi-axis. The paths of water particles at different depths but with the same mean position are shown in Fig. 9.2.1.3.

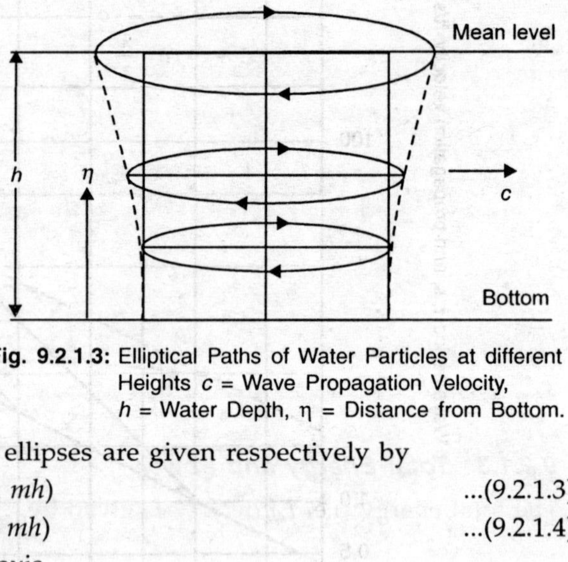

Fig. 9.2.1.3: Elliptical Paths of Water Particles at different Heights c = Wave Propagation Velocity, h = Water Depth, η = Distance from Bottom.

The horizontal and vertical semi-axis of the ellipses are given respectively by

$$\alpha = a\ (\cosh m\eta/\sinh mh) \qquad ...(9.2.1.3)$$

$$\beta = a\ (\sinh m\eta/\sinh mh) \qquad ...(9.2.1.4)$$

where α = horizontal semi-axis

β = vertical semi-axis

h = depth of water

η = distance from the bottom

In general $\alpha > \beta$, For large depths $\alpha \cong \beta \approx a$ and the motion is essentially circular at the surface. Therefore, waves possess both potential and kinetic energies. The total energy of a wave is the sum of its potential and kinetic energies.

9.2.1.1 Potential Energy

It arises from the elevation of water above the mean level (i.e. $y = 0$). For a differential volume $y\,dx$ (shown in Fig. 9.2.1.1) with a mean height $y/2$, the potential energy can be expressed by:

$$dPE = m[(y/2).(g/g_c) = (\rho\ ydx\ L).(yg/2g_c) = (\rho L.y^2\ dx.g)/2g_c \quad ...(9.2.1.5)$$

where
- m = mass of liquid in ydx, kg
- g = gravitational acceleration = 9.81 m/s^2
- g_c = conversion factor i.e. 1.0 kg.m/(N.s^2)
- ρ = density of water, kg/m^3
- L = arbitrary width of the two-dimensional wave, perpendicular to the direction or Wave propagation x

Considering Eqns. (9.2.1.5) and (9.2.1.2), the potential energy i.e. PE in J is expressed by:

$$PE = (\rho La^2g/2g_c)\int_0^\lambda \sin^2(mx - n\theta)\,dx = (\rho La^2g/2mg_c)\,(1/2\ mx - ¼ \sin 2mx)_0^\lambda$$

$$= \frac{1}{4}\rho a^2\lambda\ Lg/g_c \quad ...(9.2.1.6)$$

The potential energy density per unit area is given by PE/A in J/m^2 where $A = \lambda L$.

$$\therefore\ PE/A = \frac{1}{4}(\rho a^2 g)/g_c \quad ...(9.2.1.7)$$

9.2.1.2 Kinetic Energy

The Kinetic Energy i.e. KE of the wave is assessed from the energy of the liquid between two vertical planes perpendicular to the direction of wave propagation x and placed on wave length apart. The derivation of KE is rather complex and beyond the scope of this book. However, from the hydrodynamic theory, KE is given by:

$$KE = \frac{1}{4}(i\rho Lg)/g_c \int w\,\overline{dw} \quad ...(9.2.1.8)$$

where w is a complex potential given by $w = \{ac/\sinh(mh)\}\{\cos(mz - n\theta)\}$ and z is distance measured from an arbitrary reference point, after integrating eqn. (9.2.1.8) over the cross-sectional area bounded between two vertical planes. The result is given by:

$$KE = \frac{1}{4}\rho a^2\lambda Lg/g_c \quad ...(9.2.1.9)$$

The kinetic energy density is $KE/A = \frac{1}{4}(\rho a^2 g/g_c)$ ...(9.2.1.10)

9.2.1.3 Total Energy and Power

The total energy (i.e. E) density is given by

$$E/A = \frac{1}{2}(\rho a^2 g)/g_c \quad ...(9.2.1.11)$$

The power i.e. P (energy per unit time) is given for a wave by energy times frequency. Thus the power density in W/m^2, is given by

$$P/A = \frac{1}{2}(\rho a^2 fg/g_c) \quad ...(9.2.1.12)$$

Example: A wave with a height of 2m has a period of 6s and occurs at the surface of water 100m deep. Find the wave length, the wave velocity, the horizontal and vertical semi-axis for water motion at the surface and the energy and power densities of the wave. Assume water density as 1025 kg/m³.

Solution:

	Wave length	=	$1.56 \times 6^2 = 56.16$m
	Wave velocity c	=	$\lambda/\tau = 9.36$m /s
	Wave height 2a	=	2m
	Amplitude a	=	1m
	m	=	$2\pi/\lambda = 2\pi/56.16 = 0.1119\ m^{-1}$
At the surface	η	=	h = 100m
Horizontal semi-axis	α	=	$1 \times \cosh 11.19/\sinh 11.19 = 1$m
Vertical semi-axis	β	=	$1 \times \sinh 11.19/\sinh 11.19 = 1$m
Wave frequency	f	=	$1/\tau = 1/6,\ s^{-1}$
Energy density	E/A	=	$\frac{1}{2} \times 1025 \times 1^2 \times 9.81/1 = 5027.6\ J/m^2$
Power density	P/A	=	$(E/A)\ f = 5027.6 \times (1/6) = 837.9\ W/m^2$

9.2.2 Wave Energy Conversion Systems

The concepts on some devices that convert wave energy into mechanical or electrical energy are described briefly as follows:

9.2.2.1 *Wave-Energy Conversion by Floats*

Wave motion is primarily horizontal, but the motion of water is primarily vertical. The vertical motion of water is exploited in obtaining mechanical power with the help of floats. The concept envisages a large float that is driven up and down by the water within relatively stationary guides. This reciprocating motion is converted to mechanical and then electric power. The system is shown in Fig. 9.2.2.1. A square float moves up and down with the water. It is guided by four vertical manifolds that are part of a platform. Four large under water floatation tanks stabilizes the platform. The platform is supported by buoyancy forces. No significant vertical of horizontal displacement of the platform due to wave action occurs. Thus, even in heavy seas, the platform is expected to be relatively stationary in space. A piston, attached to the float, moves up and down inside a cylinder which is attached to platform and is therefore relatively stationary. The piston cylinder arrangement acts as a reciprocating air compressor in which the down-ward motion of the piston helps to draw the air into the cylinder through an inlet check valve and the upward motion of the piston helps to compress and send it through an outlet check valve to the four under water floatation tanks via the four manifolds. The four floatation tanks thus serve the dual purpose of buoyancy and air storage and the four vertical manifolds serve the dual purpose of manifolds and float guides. The

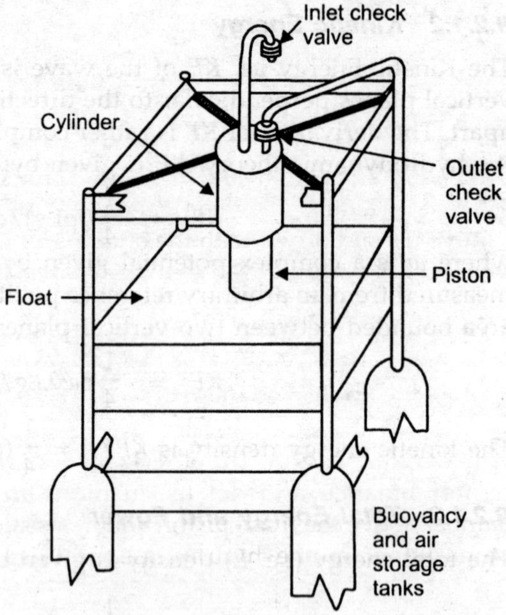

Fig. 9.2.2.1: Schematic Diagram of a Float Wave-power Machine

compressed air is utilized to run an air turbine which in turn drives an electrical generator. The generated electricity is transmitted to the shore via an under water cable.

A linear array of such modules perpendicular to wave motion is recommended as the modules attenuate the wave amplitude in the direction of wave motion and would affect other modules in that direction. To produce 100MW or more, miles of linear arrays are required. This scheme is having some problems some of which are shared by other schemes. The problems are:

(1) Waves not perfectly sinusoidal;
(2) Aspiration of water into intake and even submersion by large waves;
(3) Water entering turbine;
(4) Materials problems, such as cost and corrosion (which may be partly overcome by the use of molded plastic);
(5) Design to withstand storms;
(6) Marine growth;
(7) Power transmission to shore.

9.2.2.2 High-pressure Accumulator Wave Machines

A typical hydraulic accumulator wave machine is shown in Fig. 9.2.2.2. In this machine, the water itself is pressurized and stored in a high pressure accumulator or pumped to a high-level reservoir. Then it flows through a water turbine to produce work output. A composite piston is used to transform large volumes of low-pressure water at wave crest into small volumes of high-pressure water. The composite piston is composed of a large diameter main piston and a small diameter piston at its center. The main piston is allowed to move up and down with the entering of the wave water through the opening. Above the small piston, there exists a closed water loop. During upstroke, the pressure on the main piston is magnified on the small piston. Then high pressure water passes through a one-way valve to a hydraulic accumulator at the top of the generator. Two air (or other gas) volumes counter balance and act as cushions in a chamber above the main piston and in a sealed compartment in the hydraulic accumulator. The later also maintains the high water pressure. Part of pressurised water is allowed to flow through a Pelton wheel or Francis hydraulic turbine that drives an electrical generator and is then discharged to a storage chamber below the turbine.

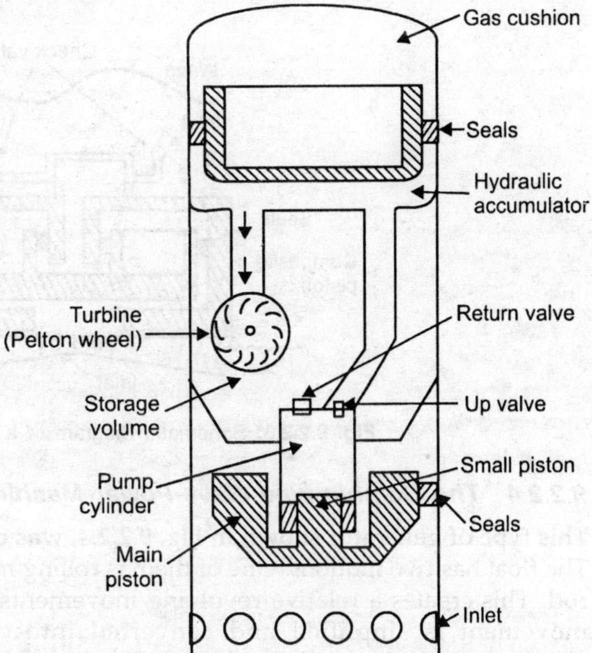

Fig. 9.2.2.2: Schematic Diagram of a Hydraulic-Accumulator Wave Machine

A 500W prototype hydraulic accumulator generator was constructed in Germany by Harold Kayser in 1975 for powering a navigation buoy. A liquid, other than water, may be chosen for the closed loop to eliminate corrosion and freezing. Alcohol is an example for such liquid. Attention must be paid to the seals between pistons and cylinders that must be gas tight and have low friction. The piston and cylinders are to be constructed of glass fiber-reinforced plastic, ferro-concrete, aluminium or steel.

9.2.2.3 High Level Reservoir Wave machine

In this concept, the pressurized water is elevated to a natural reservoir above the generator, which would have to be near a shoreline, or to an artificial water reservoir. The schematic of this wave machine is shown in Fig. 9.2.2.3.

The hydraulic accumulator is a free-floating device but the high level reservoir machine is stationary. The water in the reservoir passes through a turbine back to sea level. Typically a 20m diameter generator of this type can produce 1MW power.

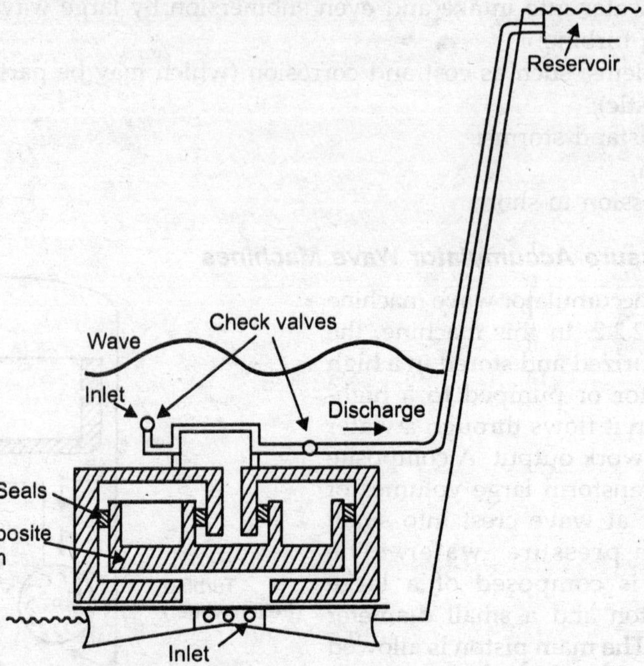

Fig. 9.2.2.3: Schematic Diagram of a High-level Reservoir Wave Machine

9.2.2.4 The Dolphin-type Wave-Power Machine

This type of generator shown in Fig. 9.2.2.4, was designed by Tsu Research Laboratories in Japan. The float has two motions. One of them is rolling motion about its own fulcrum with the connecting rod. This creates a relative revolving movements between the float and the connecting rod. The movement is amplified and converted into continuous rotary motions which drives the corresponding electrical generator. The other motion of the float is nearly vertical or having motion about the connecting rod fulcrum. It produces relative revolving movements between the connecting rod and the stationary dolphin.

A typical experiment on a (3m × 1m × 0.5m) float showed the linear power output to be given by:

$$P/L = 1.74\, a^2 \tau \qquad \text{...(9.2.2.1)}$$

Where P/L = Power per unit length perpendicular to the wave, KW/m

a = Wave amplitude (half the height), m

τ = Wave period, *s*

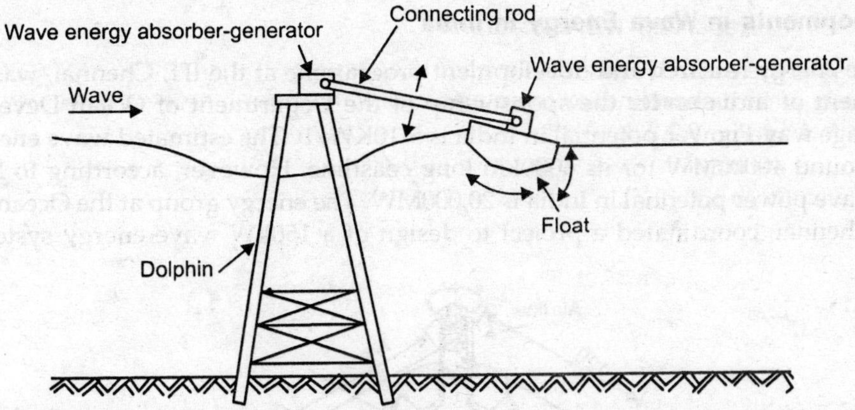

Fig. 9.2.2.4: Schematic Diagram of the Dolphin-type Water Generator (Tsu Research Laboratories)

For a wave height of 2m and the period of 6s, the linear power output is about 10 KW/m. Therefore, a linear array of units totalling 1km would generate about 10MW.

The system can be used for electric power generation, pumping for desalination equipment, or uranium extraction from the sea etc. Since it completely eliminates waves, it also can provide sites for fish farming, port facilities, etc.

9.2.2.5 The Dam-Atoll Concept

The dam-atoll wave power conversion device, shown in Fig. 9.2.2.5, was designed by Wirt and Morrow of the Lockheed Corporation, Burbank, California. The device is a massive and robust one. It overcomes complexity and fragility in heavy sea. It is also strong enough to counter any ocean storm. It incorporates some of the characteristics of both dams and atolls. On the basis of the observed action of waves as they approach atolls (small volcanic islands) in an ocean the principle of its operation is decided. Around the atolls from all sides, the waves wrap themselves. It creates a spiral motion in the center and drives a turbine before discharging laterally outward. A module, 80 m in maximum diameter and 20m high, is capable of generating 1 to 1.5 MW in 7 to 10s period waves.

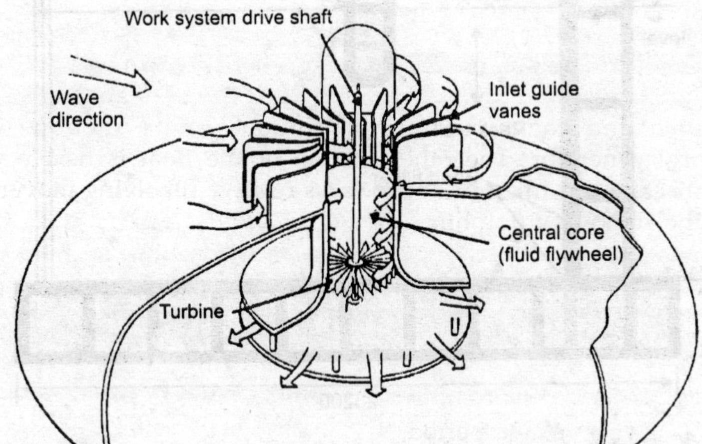

Fig. 9.2.2.5: Cross Section of a Dam-atoll Wave Machine showing Water Flow (Lockheed.)

9.2.3 Developments in Wave Energy in India

In 1982, wave energy research and development programme at the IIT, Chennai, was initiated by the Government of India under the sponsorship of the Department of Ocean Development.

The average wave-power potential in India is 5-10KW/b. The estimated wave energy potential in India is around 40000MW for its 6000km long coastline. However, according to MNES broad estimation, wave power potential in India is 20,000MW. The energy group at the Ocean Engineering Centre, IIT,Chennai, coordinated a project to design of a 150kW wave energy system shown in Fig. 9.2.3.1.

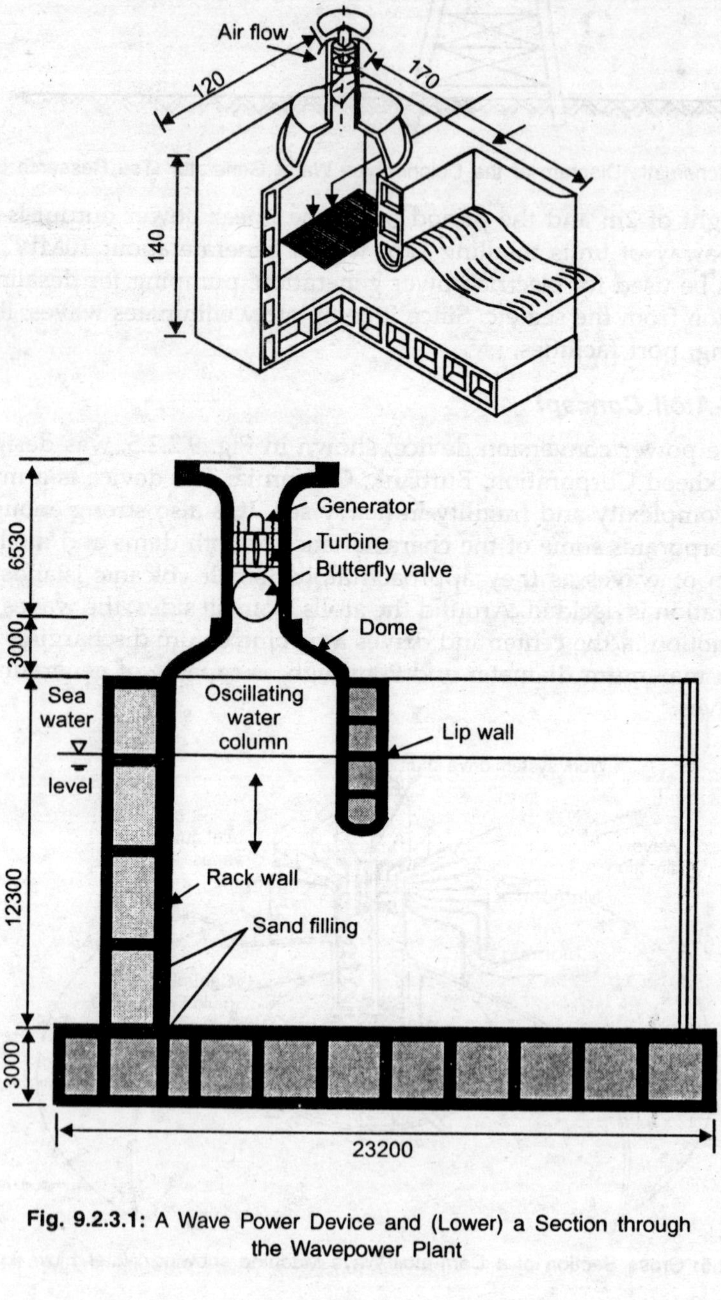

Fig. 9.2.3.1: A Wave Power Device and (Lower) a Section through the Wavepower Plant

Vizhinjam, a place 15 km from Thiruvananthapuram, was chosen as the site for installing India's first 150 kW capacity wave energy caisson (large water tight chamber used in laying foundations under water). The simplified view of the system is shown in Fig. 9.2.3.2. The system can deliver average power output of 75kW during the eight month period from April to November and 25 kW during December to March. During June to September, the system can generate 120 kW with a peak power of 150 kW. The cost of the power was estimated to be in the range between Rs. 0.73/ kWh and Rs. 2.2/ kWh depending on whether the power generation is linked with harbour construction or not.

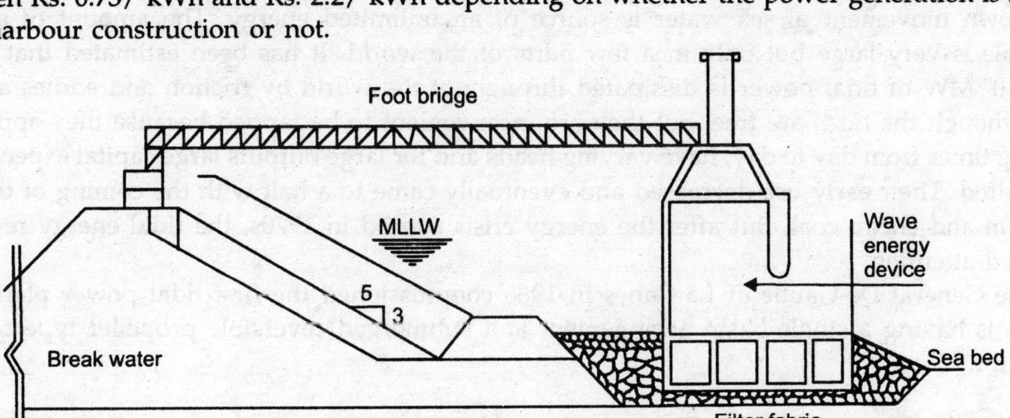

Fig. 9.2.3.2: Schematic View of Oscillating Water Column developed by Ocean Engineering Centre (IIT, Chennai)

In Andaman and Nichobar islands, a 1 MW power plant is proposed to be set up on 'build, own and operate' basis by a Swedish company under the overall supervision of MNES.

According to the study conducted by National Institute of Oceanography, Goa, the wave power potential along the Indian Coastline is shown in Fig. 9.2.3.3. The Indian Coast line has been divided in six zones such as A,B,C,D,E,F. It is noted that the annual wave energy potential along the Indian Coast line varies from 175-200 MWh/m.

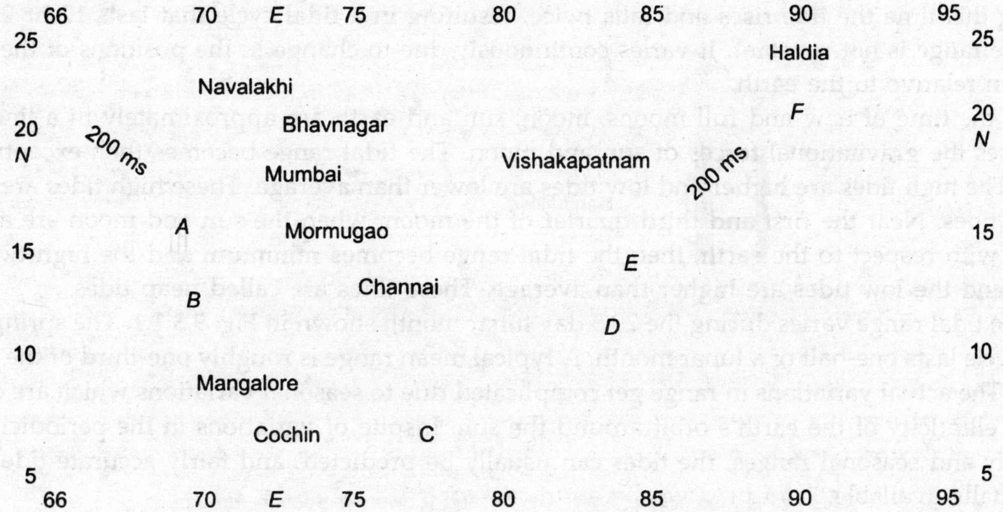

Fig. 9.2.3.3: Map showing different Wave Power Potential Zones

9.3 TIDAL ENERGY CONVERSION

Among the various renewable energies tidal energy can be effectively harnessed for generating electricity. Tide is a periodic rise and fall of the water level of sea which are carried by the action of the Sun and Moon on the water of earth. The tidal energy can be tapped from coastal waters by building dams which entrap the water at high tide and release it at low tide back to the sea. Power can then be obtained by turbines from both in-out flows of the water. The large scale up and down movement of sea water is source of an unlimited energy. The amount of energy available is very large but only in a few parts of the world. It has been estimated that about 2.4×10^6 MW of tidal power is dissipated throughout the world by friction and eddies alone.

Although the tides are free, but they are inconvenient to be tapped because they appear at varying times from day to day, have varying heads and for large outputs large capital expenditure is required. Their early use decreased and eventually came to a halt with the coming of the age of steam and cheap coal. But after the energy crisis started in 1970s, the tidal energy received renewed attention.

The General De-Gaulle at La Ranes in 1966 commissioned the first tidal power plant. The system is having a single basin arrangement and submerged reversible propeller type turbine generator.

9.3.1 Basic Principle of Tidal Power

Tides are generated mainly by the gravitational attraction force of the moon and the sun on the water of solid earth and the oceans. In the tide formation, the moon is considered to be the major factor because moon produces about 70 percent of tide producing force and sun produces 30 percent of the force.

The tides are rhythmic but not constant, nor do they occur on a regular daily schedule, as the earth rotates, the position of a given area relative to the moon changes. There are a periodic succession of high and low tides.

The tidal schedules vary from day to day because the orbit of the moon does not occur on a regular, 24 hrs daily schedule. Instead, the moon rotates around the earth every 24 hrs, 50 min. During this time the tide rises and falls twice, resulting in a tidal cycle that lasts 12 hr 25 min.

The range is not constant. It varies continuously due to change in the positions of the moon and sun relative to the earth.

At the time of new and full moons, moon, sun and earth are approximately in a line. This enhances the gravitational forces of sun and moon. The tidal range becomes then exceptionally large. The high tides are higher and low tides are lower than average. These high tides are called spring tides. Near the first and third quarter of the moon, when the sun and moon are at right angles with respect to the earth, then the tidal range becomes minimum and the high tides are lower and the low tides are higher than average. These tides are called neap tides.

The tidal range varies during the 29.5 day lunar month shown in Fig. 9.3.1.1. The spring-neap tidal cycle lasts one-half of a lunar month. A typical mean range is roughly one-third of the spring range. The actual variations in range get complicated due to seasonal variations which are caused by the ellipticity of the earth's orbit around the sun. Inspite of variations in the periodicity and monthly and seasonal ranges, the tides can usually be predicted, and fairly accurate tide tables are usually available.

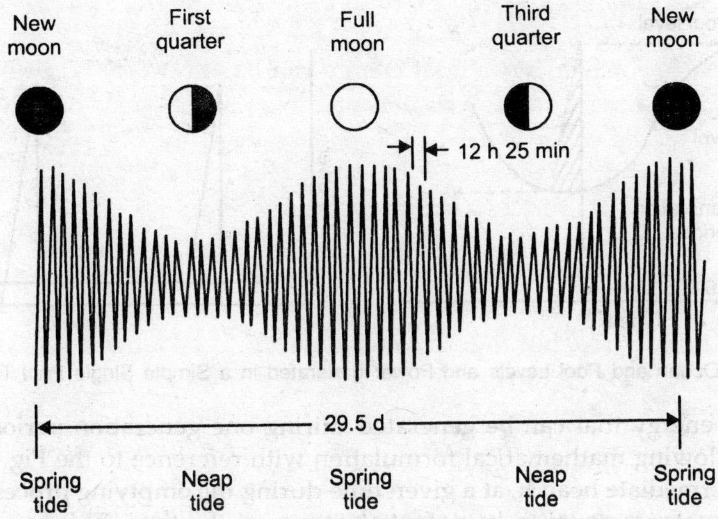

Fig. 9.3.1.1: Relative High and Low Tides showing Variation in Range during Lunar Month

From one earth location to another, the tidal ranges vary. The profile of the local shoreline and water depth influence the tidal ranges. When these are favourable, a resonance like effect causes very large tidal ranges.

9.3.2 Tidal Energy Conversion Systems

There are three schemes for the tidal energy conversion systems for the production of electric energy. These are:

(1) Simple Single – Pool or Basin
(2) Modulated Single – Pool or Basin
(3) Two-pool or Basin

In a single basin arrangement, only on basin interacts with the sea. Single pool or basin schemes can generate power only intermittently, but a double-pool scheme can produce power continuously, or on demand, which is a tremendous advantage. The draw-back is that the civil works become more extensive. In a double basin scheme there must be a dam between each basin and the sea, and also a dam between the basins, containing the power house. One basin is always maintained at a lower level than the other. The lower reservoir gets emptied at low tide, and the upper reservoir is replenished at high tide. For higher capacity generation, the size of the reservoirs will be large and dams will be long.

9.3.2.1 Simple Single-Tidal System

The tidal system is having one pool or basin behind a dam. The basin is filled from the ocean at high tide and emptied to it at low tide. The filling and emptying processes occur during short periods of time. The filling process takes place when the ocean is at high tide while the water in the pool is at low-tide level. The emptying process occurs when the ocean is at low tide and the pool is at high-tide level. This is depicted with the help of Fig. 9.3.2.1. The flow of water in both the directions is utilized through a number of reversible water turbines to produce power. Each turbine drives an electrical generator. The electrical power would be generated during two short periods during each tidal period of 12 hr, 25 min, or once every 6 hr, 12.5 min.

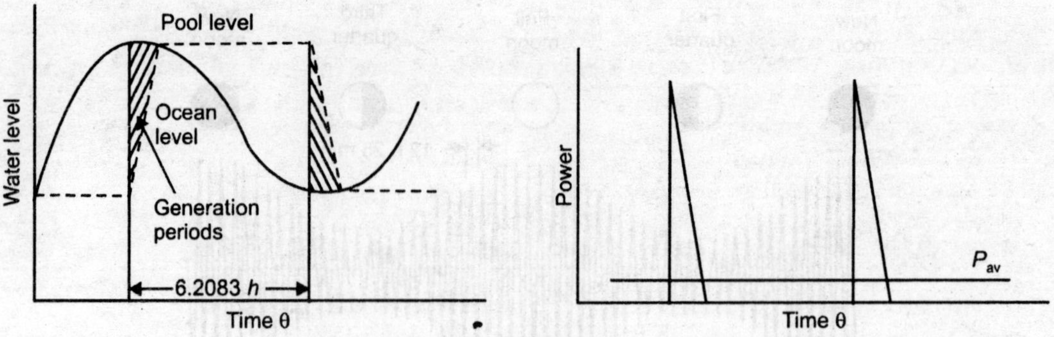

Fig. 9.3.2.1: Ocean and Pool Levels and Power generated in a Simple Single Pool Tidal System

The maximum energy that can be generated during one generation period can be assessed with the help of following mathematical formulation with reference to the Fig. 9.3.2.2, for a tidal range R, and an intermediate head h, at a given time during the emptying process. The differential work done by the water is equal to its potential energy at the time. This can be expressed by:

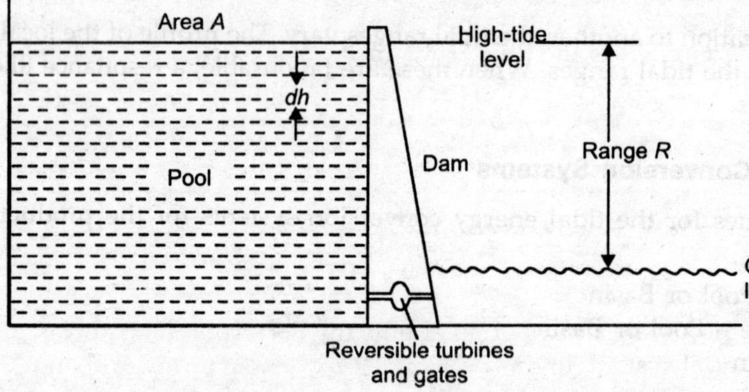

Fig. 9.3.2.2: Level changes during Power Production in a Single-pool Tidal System

$$dw = (g/g_c)\, dmh \qquad \text{...(9.3.2.1)}$$

But,
$$dm = -\rho A\, dh \qquad \text{...(9.3.2.2)}$$

∴
$$dw = -(g/g_c)\, \rho\, Ah\, dh \qquad \text{...(9.3.2.3)}$$

where,

w = Work done by the water, J

g = gravitational acceleration, 9.81 m/s^2

g_c = conversion factor, 1.0 kg-m/(N.S^2)

m = mass of water flowing through the turbine, kg

h = head, m

ρ = water density, kg/m^3

A = surface area of pool, considered constant, m^2

During a full emptying or filling, period, the total work done can be equated as

$$w = \int_R^0 dw = -(w/g_c)\rho A \int_R^0 h\, dh \quad \text{or } w = \frac{1}{2}\,(g/g_c)\,\rho\, AR^2 \qquad \text{...(9.3.2.4)}$$

The power generated during each of the periods can be assessed by dividing W by the time duration of that period. Zero power is generated during the rest of the time (shown in Fig. 9.3.2.1). Therefore, the average theoretical power produced by the water is W divided by the total time it takes each period to repeat itself i.e. 6 hr, 12.5 min or 22,350s. Thus the average power is given by:

$$P_{av} = (1/44{,}700).\ (g/g_c)\ \rho\ AR^2 \qquad ...(9.3.2.5)$$

Where P_{av} = average theoretical power, W. If the average sea water density is 1025 kg/m^3, then the average theoretical power per unit pool area can be expressed by:

$$P_{av}/A = 0.225\ R^2,\ W/m^2 \text{ or } MW/km^2 \qquad ...(9.3.2.6)$$

But in practice, during the assessment of the actual power generation, the average theoretical power is to be multiplied by the efficiency of the system which is only in the range of 25 to 30 percent.

9.3.2.2 Modulated Single-Pool Tidal System

With the help of a simple single-pool system, two high peaks, short-duration power outputs can be obtained in each tidal period. For such high peaks, large turbine-generators are required. But they remain out of service much of the time. The power peaks also occur at different times every day (50 min. later each successive day). Therefore, all times high and low tides will not always correspond to times of peak power demand. These poses a burden on the electric-power grid where they are connected to. In the modulated single-pool system, these deficiencies are partially corrected by producing power more uniformly at a lower average head. For some periods, there will be no power generation. Since the average head h is lower and work and power are proportional to h^2, the turbine generators become much smaller and they operate over much longer periods. But total work gets reduced.

Fig. 9.3.2.3 shows the ocean and pool level and the corresponding power diagrams of the modulated single-pool tidal system. In this system, the reversible turbines operate during periods of pool filling and pool emptying instead of at high and low levels only. They cease to operate when the head is too low for efficient operation. Period C_1 begins with both pool and ocean at low-tide level (1), the ocean at the beginning of tide rise, and all gates closed. When the head is sufficient (2), gates to the turbines are opened and water from the ocean is allowed to pass through the turbines. So during this period G_1 power is generated as both ocean and pool levels increase. The ocean level reaches its peak and begins to decrease but the pool level will still increase until, at 3, the head is too low for efficient generation. At this state, the gates to the turbines are closed and bypass gates are opened to fill up the pool during period F to 4. At 4, all gates are once again closed and the pool level remains constant while the ocean level decreases during period C_2. At 5, the head is once again sufficient to flow the water through the turbine in the opposite direction and a second generation period G_2 begins. At 6, generation stops but the pool is allowed to be empty during period E and the system goes back to point 1, repeating the cycle.

The total work can be evaluated by assessing the characteristics of the system, i.e. the variation of water mass flow and head with time θ, and integrating their product over the time span when generation is taking place. During the consideration of the first generation period (refer Fig. 9.3.2.4), the general relationships are given by

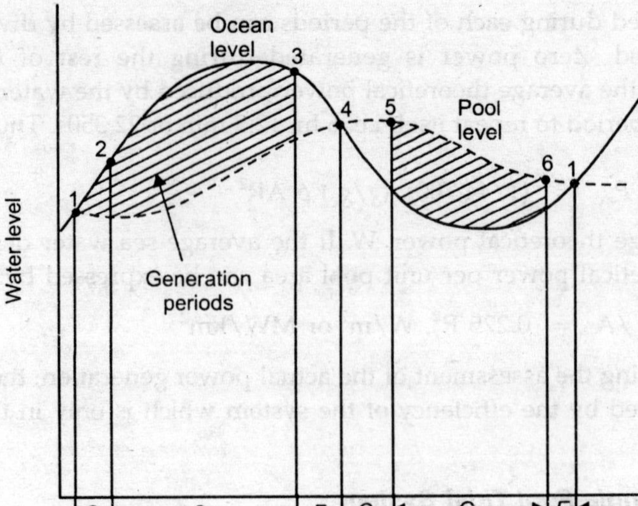

Fig. 9.3.2.3: Ocean and Pool Levels in a Modulated Single-pool Tidal System. C = Gates Closed, G = Generation, F = Oool Filling, E = Pool Emptying

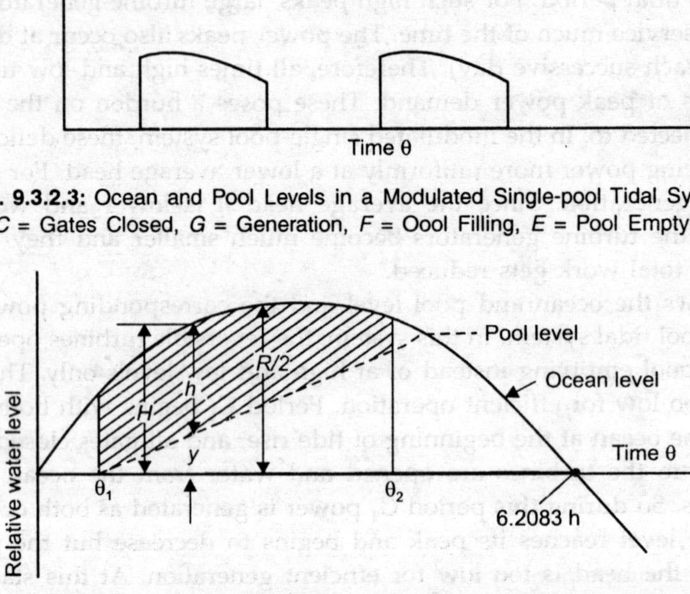

Fig. 9.3.2.4: Relative Ocean and Pool Levels in the First-generation Period of a Modulated Single-pool Tidal System

$$H = f_1(\theta) \qquad ...(9.3.2.7)$$

$$y = f_2(\theta) \qquad ...(9.3.2.8)$$

$$dw = (g/g_c)\ \mathrm{dm} \cdot \mathrm{h} = (g/g_c)\ dm\ (H - y) \qquad ...(9.3.2.9)$$

but

$$dm = \rho\ A\ dy \qquad ...(9.3.2.10)$$

Therefore,

$$dw = (g/g_c)\ \rho A\ [f_1(\theta) - f_2(\theta)]\ d\ f_2(\theta) \qquad ...(9.3.2.11)$$

and

$$w = (g/g_c)\ \rho \cdot A \int_{\theta_1}^{\theta_2} [f_1(\theta) - f_2(\theta)]\ df_2(\theta) \qquad ...(9.3.2.12)$$

where H = ocean level above mean or other appropriate datum.

y = pool level above mean or datum

θ = time

H is closely approximated by a sinusoidal function of θ such as

$$H = f_1(\theta) = (R/2)\sin(\pi\theta/6.2083) \qquad ...(9.3.2.13)$$

where θ is in hours and 6.2083 in hours is one-half of a tidal period.

y may be approximated by a linear function of θ, starting at 0 at θ_1 for a constant mass flow rate such as

$$y = f_2(\theta) = a\,R\,(\theta - \theta_1) \qquad ...(9.3.2.14)$$

where a is a constant having the dimension (time)$^{-1}$, e.g. h^{-1}, or y could be a function of $h = (H\text{-}y)$ for a constant flow resistance or some other function determined from operational data. Using the relationships for H and y of eqns. (9.3.2.13) and (9.3.2.14) the work during period G_1 (or G_2) would be assessed from

$$dw = (g/g_c)\,\rho A\,[R/2\sin(\pi\theta/6.2083) - a\,R(\theta - \theta_1)]\,d\,[a\,R(\theta - \theta_1)] \qquad ...(9.3.2.15)$$

and

$$w = (g/g_c)\,\rho A)\int_{\theta_1}^{\theta_2}[R/2\sin(\pi\theta/6.2083) - a\,R(\theta - \theta_1)]\,a\,R\,d\theta$$

$$= (g/g_c)\,\rho AR^2\,[0.988a\,\{-\cos(\pi\theta_2/6.2083) + \cos(\pi\theta_1/6.2083)\} - a^2/2\,(\theta_2^2 - \theta_1^2)] \qquad ...(9.3.2.16)$$

This work is also a function of R^2 like in the simple single pool system.

Example: Estimate the total energy and average power for (i) a simple single-pool system, and (ii) a modulated single-pool system taking R = 8 m, a = 0.0625 h^{-1}, θ_1 = 1 hr, θ_2 = 4 hr, A = 20,000 km^2 and P = 1025 kg/m^3. If the efficiency of the system is 27%, determine the actual power for both the cases.

Solution: (i) For a simple single-pool system:

The total work done $w = \frac{1}{2}(g/g_c\,\rho AR^2)$.

$\therefore \quad w = \frac{1}{2} \times (9.81/1.0) \times 1025 \times 2.0 \times 10^{10} \times 8^2 = 6.43 \times 10^{15}$ J

and P_{av} = 0.225 × 8^2 × 20,000 MW = 288,000 MW.

Actual average power = 288,000 × 0.27 = 77,760 MW

(ii) For a modulated single-pool system

$$w = (9.81/1) \times 1025 \times 2.0 \times 10^{10} \times 8^2\,[0.988 \times 0.0625\,(0.43795 + 0.87468) - (0.0624^2/2)\,(16 - 1)]$$

$$= 6.66 \times 10^{14}\ \text{J}$$

The average power during the generation period of 4 hr is

$$P_{av}\text{,gen} = [(6.66 \times 10^{14})/(4 \times 3600)]\ \text{J}$$

$$= 4.625 \times 10^4\ \text{MW}$$

The average power during the total period of 6.2083 hr is

$$P_{av} = (6.66 \times 10^{14})/(6.2083 \times 3600) = 29{,}790\ \text{MW}$$

Actual average power = $P_{av} \times n_{system}$ = 29790 × 0.27 = 80433 MW

9.3.2.3 The Two-Pool Tidal System

Fig. 9.3.2.5 shows the schematic of a two-pool tidal system. In this system an inland basin is enclosed by a dam i.e. A (ref. Fig. 9.3.2.5). The basin is divided into a high pool and a low pool again with the help of a dam B. If the proper arrangement for gating in dam A is provided, the high pool can be filled periodically at high tide from the ocean and low pool can be emptied periodically at low tide. The water is allowed to flow from the high to the low pool through the turbines which are to be situated in dam B. Intentionally, the capacities of these two pools are kept large enough in relation to the water flow between them so that a continuous and much more uniform power can be generated. Moreover, the fluctuations in the head also get minimized.

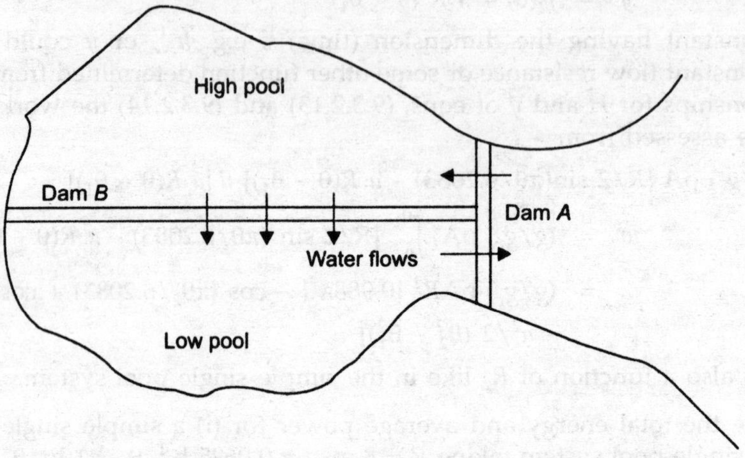

Fig. 9.3.2.5: Schematic diagram of a two-pool tidal system

9.3.3 Tidal Energy In Indian Context

India is having a vast 6000 km long coast line. There may be some promising sites where the tidal power plants can be set up. The possible sites for tidal power generation should be those where high tidal ranges are available. For example the possible sites are Gulf of Cambay (Bhavnagar Son rai), Gulf of Kutch (Kandla, Navalakhi), Sagar and Diamond Harbour etc.

Central Electricity Authority (CEA) and different electricity board of different states particularly those of West Bengal, Gujrat and Orissa have taken effective steps in creating tidal power facilities in their respective states.

The most promising locations are Gulf of Cambay and Gulf of Kutch on the West Coast. The delta of the Ganges in Sundarbans is also considered as promising location for small-scale tidal power development. Out of 8000-9000 MW power in India, about 7000 MW is in the Gulf of Cambay, 800-1000 MW is in the Gulf of Kutch and less than 100 MW is in Sundarbans. The first site of India's tidal power plant will be a creek on the Gujrat Coast. The power plant when commissioned will have a capacity of 900 MW with 36 bulb turbines, each of which will have capacity of 25 MW. The total scheme is proposed to be completed before the year 2000.

A feasibility study have already been carried out for harnessing tidal energy of 2 to 3 MW in Durgaurai creek at Sundarbans.

In 1986, Industrial Development Bank of India (IDBI) sanctioned a grant of Rs. 1.35 lakh for a research project on tidal power to be undertaken by M/s. Kaninga Engineers Ltd., Bhubaneswar (Orissa). This project was for design and fabrication of a 10 KW pilot plant at the Aharabnaki creek.

EXERCISE

(1) Discuss in brief the working principles for Ocean Thermal Energy Conversion techniques in generating electricity. Classify the OTEC systems used in practice. Mention some critical criterions considered during the design of an OTEC system.

(2) Deduce necessary expressions for assessing total energy and power of a wave machine. Present a critical comparision of the technical features of various wave machines.

(3) Discuss briefly the merits and demerits among:

(a) Claude Cycle and Anderson Cycle of OTEC systems

(b) Wave Energy Conversion by floats and high pressure accumulator wave machines.

(c) Modulated Single-pool or basin and two-pool or basin Tidal Plants.

(4) An OTEC power plant of the claude type has an ideal turbine operating between 25 and 15°C. In the case of turbine output of 1 MW, estimate (a) the mass flow rate of steam in Kg/s, (b) the mass flow rate of warm water in Kg/s. The evaporator temperature drop is 2°C, (c) the mass flow rate of cold water in Kg/s when the condenser temperature drop is 2°C and (d) the cold water pipe diameter, in meters, if the water velocity in it is not to exceed 1 m/s.

(5) A 2.5m ocean wave has a 60m wave length in 10m deep water. Calculate (a) the probable period in seconds, (b) the major and minor axes, in meters, of the water motion at the surface and at mid-depth and (c) the wave power in kilowatts per meter width of the wave front. Assume water density as 1025 Kg/m^3.

(6) An ocean wave power system of the float type uses a 25cm thick float in 2m waves that have a 60m wave length. The cylinder has diameter of 1.5m. The compressor pressure ratio is 9.5. The air is at 1 atm and 24°C. Calculate the power produced by the turbine if the compressor-turbine system has an efficiency of 75 percent of ideal.

(7) In an estuary, which is considered for tidal power generation, during the tide cycle, the observed difference between the high and low tide was 10.2m. The estuary is having an area of 8.5 sq.km. which can generate power for 3 hours in each cycle. If the average available head is 10m and the overall efficiency of the generation system is 75%, calculate (a) the power at any instant in MW, and (b) the total energy generated in the year, in KW hr/year. Sea water density is 1025 Kg/m^3.

(8) Calculate the total energy and average power of a modulated single-pool tidal plant operating with a range of 12m. The tidal pattern is assumed to be sinusoidal. The water level in the pool may be approximated by the relation $0.625R(\theta_2 - \theta_1)$, where θ is the time, in hours. Power generation occurs between $\theta_1 = 1$ and $\theta_2 = 4$ hr. The pool has a constant area of 15,000 km^2. The water density is 1025 Kg/m^3.

SUMMARY

This chapter deals with the detailed discussion on the energy sources from the sea, e.g. ocean thermal energy, wave energy and tidal energy and their applications in producing electricity. Basic principles and the different methods in harnessing ocean thermal energy have been outlined. The problems caused by the ocean environment on the equipments and their probable remedies have been mentioned. The site selection criterion for an ocean thermal energy conversion (OTEC) system has been pointed out. The present status of the ocean thermal energy conversion in India has been highlighted. The chapter includes basic concept of wave energy conversion and the assessment of respective energy and power from the wave motion. The different devices used in converting wave energy into mechanical or electrical energy have been described. The development in wave energy in India has also been focussed. Basic principles of power generation from tides have been briefly discussed. The chapter contains the detailed discussion on different tidal energy conversion systems with relevant mathematical formulations, leading to estimates of the average power generated by these systems. The potential of tidal energy sources prevailing particularly in India has been highlighted.

REFERENCES

1. APCTT, Ocean Energy, Vatis Update, Asia Pacific Centre for Technology Transfer, New Delhi, 2(4), pp. 12, 1994.
2. S.Z. Qasim, Endless Energy from the Sea, Science Today, p. 17, Sept'1980.
3. TERI, Generation of Power from Ocean Waves, TIDE, 3(2), p. 138, 1993.
4. APCTT, Inventions/New Products, VATIS, Update, 2(14), p. 6, 1995.
5. M. Ravindran and Raju Abraham, Ocean Power – Would the XXI century realize the Immense Potential?, IREDA NEWS, Millenium Special, vol. 10, no. 4, pp. 71-79, 1999.
6. M. Ravindran, 'The Indian 1MW Floating OTEC Plant – An Overview, IOA'99 conference IMARI, Japan, 1999.
7. A.V. Luis, Economics of Ocean Thermal Energy Conversion, Ocean Energy Recovery, pp. 152-181, 1992.
8. H. Uehara and Y. Ikegami, Optimization of a Closed Cycle OTEC System, ASME Journal of Solar Energy Engineering, vol. 112, no. 4, pp. 247-256, 1990.
9. M.M. EI, Wakil, Power Plant Technology, Mc Graw-Hill Book Company, New York, 1985.
10. S. Walters, Power in the year 2001, Part-2 – Thermal Sea Power, Mech. Engg., vol. 93, no. 10, pp. 21-25, 1971.
11. S.P. Sukhatme, Solar Energy: Principles of Thermal Collections and Storage, 2nd Edition, Tata McGraw Hill Publishing Co. Ltd., New Delhi, 1996.
12. R.C. Rao, and D.E. Othmer, Controlled Flash Evaporation, Mech.Engg., vol. 93, no. 5, pp. 27-31, 1971.
13. P.M. Koola, et. al., Model Study of OWC Wave Energy Device, Journal of Energy Engg., 121(1), pp. 14-27, 1995.
14. U. Natesan, Energy from Ocean Waves, Energy Mgt., Jan-March, pp. 31-36, 1994.
15. M.D. Martin, Power from Ocean Waves, The American Society of Mechanical Engineers paper no. 74-WA/Pwr-5, presented at the ASME Winter annual meeting, New York, 1974.
16. K.D.C. Thompson, One Dimensional Modelling of Tidal Power Scheme, 2nd Int Symposium on Wave and Tidal Energy, BHRA, England, 1981.
17. S.S. Rao, Tidal and Wave Energy: A Review, TIDE, 4(3), pp. 257-269, 1994.
18. V.K. Das, et. al., Wave Power Potential Along the Indian Coasts, Urja, 27(2), p. 60, 1990.
19. M. Baba and J. Mathew, Ocean Wave Energy Along Kerala Coast, Urja, 25(2), pp. 60-64, 1988.
20. T.J. Gray, and O.K. Gashus, Tidal Power, Proc. of the Tidal Power conference, Nova Scotia, 1970, Plenum Press, New York, 1972.
21. H. Jeffreys, Tidal Friction, Nature, 246, p. 346, 1973.
22. N.S. Lingayat and S.A.Khaparde, Potential Assessment of a Small Scale Site for Tidal Power Generation, Proc. of NREC-2000, pp. 415-418, 2000.
23. H.M. Chaudhury, Open Channel Flow, Prentice Hall, 1993.
24. G. Chandra, Design and Analysis of Tidal Power Plant, M.Tech Thesis, Energy Systems Engineering, IIT, Mumbai, 1999.
25. D. Prandde, Simple Theory for Designing Tidal Power Scheme, Advanced Water Resources, vol. 27, pp. 21-27, 1984.
26. APCTT, Ocean Energy, VATIS Update, New Delhi, 2(4), Feb '1994.
27. R. Marshall and D.Merrick, Energy: Present and Future Options, John Wiley and Sons Ltd., N.Y., 1980.
28. S.K. Roy Bardhan, Duragaduani, Mini Tidal Power Project – A case study, Proc. of Int. Seminar on Decentralized Energy: Options and Management, Kolkata, India, pp. 143-158, Feb '22-24, 2002.

10

Hybrid Energy Systems

INTRODUCTION

Renewable energy systems described in the preceeding chapters are essentially *distributed systems* which are eminently suitable for the 800,000 villages of India, some of which are located at large distances from the utility grid line. The supply of electricity at these places is made further difficult owing to the high cost incurred in the process of conventional fuel delivery. It has so long been projected in the earlier chapters that non-conventional gadgets like PV arrays, PV pumps, Wind Energy Conversion Systems (WECS), Solar Cookers, SWHS along with judicious cultivation of Bio-mass resources when planned and developed with minimum storage facilities can certainly be applied to cure some of the long standing ailments pertaining to the aforesaid remote locations. However, one can feel slightly enthusiastic when his or her attention is focussed on various Indian sites powered by the hybrid energy sources. It is not possible for a single renewable energy source to act as year long dependable energy and economic source at most of the selected sites. The reason has to be understood clearly and has been exemplified in the following sections.

10.1 NEED FOR HYBRID SYSTEMS

It has been observed that the PV arrays sizing (discussed in section 5.12 of Chapter 5) has a seasonal dependence, which in fact leads to monthly values of array sizes. Naturally, going by such exercises, larger array-sizes are normally estimated for the months of June to September, during which (i) there are maximum number of no sun days and (ii) charge recovery period of the battery is also high. As a result, a designer is compelled to *select very large size PV arrays and battery banks to maintain a safe limit* for a Stand – alone and also Grid Interactive Systems. Needless to mention, this imposes a HIGH CAPITAL COST with regard to the number of modules and the battery bank. Even then, the aforesaid design of Stand – alone PV systems is not a reliable one as specific periods of *No Sun days* may deep discharge the battery bank to a substantial extent during the months of August and September. Recovery of the State of Charge (S.O.C) of the battery bank is only possible by running an additional back-up source regularly for a very small duration during these critical months. A similar situation can also arise for a Stand-alone WECS during the months of November to February when the free wind velocity is below the optimum value for starting a rotor. Such a back-up source may be a (i) Conventional ones like diesel, coal-fired, *propane generators* (using conventional fuels), (ii) Non-conventional ones like Wind, Bio-gas, Micro-hydel generator etc. Thus, the Hybrid Energy System, always signifies co-generation of power by a Non-conventional source of energy with a conventional or non-conventional one, each of which runs on a mutually decided time sharing basis. This basis of time sharing has been clearly exemplified in section 10.3, which introduces the various types of Hybrid Energy sources with emphasis on the Indian scenario.

It is now obvious from the aforesaid discussions that the *hybrid energy system has a definite edge over the Stand-alone and Grid Interactive Systems* with regard to *LOWER CAPITAL COST* and *HIGHER RELIABILITY* at all seasons. Here, the exact proportion of the back-up source, as an example, will depend upon the *sufficiency or insufficiency of the constituent energy sources at optimum energy demand.* An attempt has been made in the next section to determine the cost effective sizing ratio of the principal to the back-up power source.

10.2 COST EFFECTIVE SIZING RATIO

An optimum hybrid system will be one in which the total cost (TC) can be expressed as:

$$TC = C_B \,.\, B + C_{MB}\,(1 - t).B + C_p \,.\, P + t\, C_{MP}.\, P \qquad ...(10.2.1)$$

Where C_B, C_p denote the fixed cost/KWh of the energy generated by the back-up power source and the principal power source respectively. C_{MB} and C_{MP} denote respectively the corresponding marginal or variable cost/KWh, where the variable cost includes all maintenance costs incurred during the 1st year of operation only. Here, 't' is a sizing ratio used to define the energy fraction of the principal power source (total capacity P_o KWh) and back-up power source (total capacity B_o KWh) according to the conditions:

$$P = t, P_o \qquad ...(10.2.2)$$

and

$$B = (1 - t)\, B_o \qquad ...(10.2.3)$$

Thus, (10.2.1) becomes

$$TC = B_o\,(1 - t)\, C_B + C_{MB}\, B_o\,(1 - t)^2 + P_oC_pt + P_oC_{MP}t^2 \qquad ...(10.2.4)$$

Now, the condition for obtaining minimum total cost is

$$\frac{dTC}{dt} = 0 \text{ whence } t = \frac{B_oC_B + 2B_oC_{MB} - P_oC_p}{2(C_{MB}B_o + P_oC_{MP})} \qquad ...(10.2.5)$$

A hybrid system is economically feasible if $t \leq 1$.

Thus,

$$B_oC_B \leq 2\, P_oC_{MP} + P_oC_P \qquad ...(10.2.6)$$

It is evident that eqn. (10.2.6) suggests that at any site, a hybrid system is feasible when the fixed cost of the stand-alone back-up source is less than the fixed cost of stand-alone principal source plus twice its maintenance cost.

Using the estimated sizes of the principal and the back-up power source that are jointly able to meet the energy demand in all seasons on an hourly basis, the balance of energy keeps the battery bank within the specified S.O.C. limits, only when the sizing ratio (t) is an acceptable one. Otherwise, iterations with different fractional values of t have to be carried on. Alternatively, a designer may also check with various estimates of 't' to select one which yields acceptable Loss Of Load Probability (LOLP) values.

10.3 TYPES OF HYBRID SYSTEMS

It is now easy to understand that a hybrid energy system may be opted with some typical combination of the following energy sources as shown:

(A) SPV + WECS
(B) SPV + WECS + Diesel Generator set
(C) SPV + Bio-gas Generator
(D) SPV + Diesel Generator set
(E) Solar thermal + Bio-mass gasifier.

Several other combinations have been opted in different sites throughout the globe. But the combinations (A) to (E) are very common in view of the prevailing Indian site specific conditions. In all the aforesaid options, a battery bank of suitable size continues to serve as the storage tank during the NO SUN and evening periods. The choice of any one of the options will depend on the known factors like available solar insolation at optimum array-tilt, the free wind velocity at a specified height or the sufficiency of the cattle population in a community or village. As a first example, the deficit of solar insolation in certain specified months at a site may be compensated by adequate prevailing wind velocity during the months. In such a case, the two non-conventional energy sources complement each other and option (A) or (B) may be the answer depending on the variable cost involved (C_{MP} of section 10.2) in fuel delivery at the site. According to option (A), a hybrid system design was planned by the Orissa Renewable Energy Development Agency (OREDA) for two villages in Orissa situated by the marine drive connecting Puri and Konarak. Here, the monthly average value of the energy output from the various non-conventional energy sources and their combination can be critically compared with the monthly load profile in Fig. 10.3.1. The need for an hybrid system is obvious *as the output energy from the solar PV array is always lower than the load demand throughout the year*. As a second example, if the system designer has to erect a hybrid energy system in a remote location where wind velocity is insufficient during the low sunshine months, the obvious choice is option (C) or (D). Some details of such a case study have been presented briefly in the next section. As another example, let us consider a remote village placed far away from the utility grid line. The pattern of energy consumption here is of two types principally:

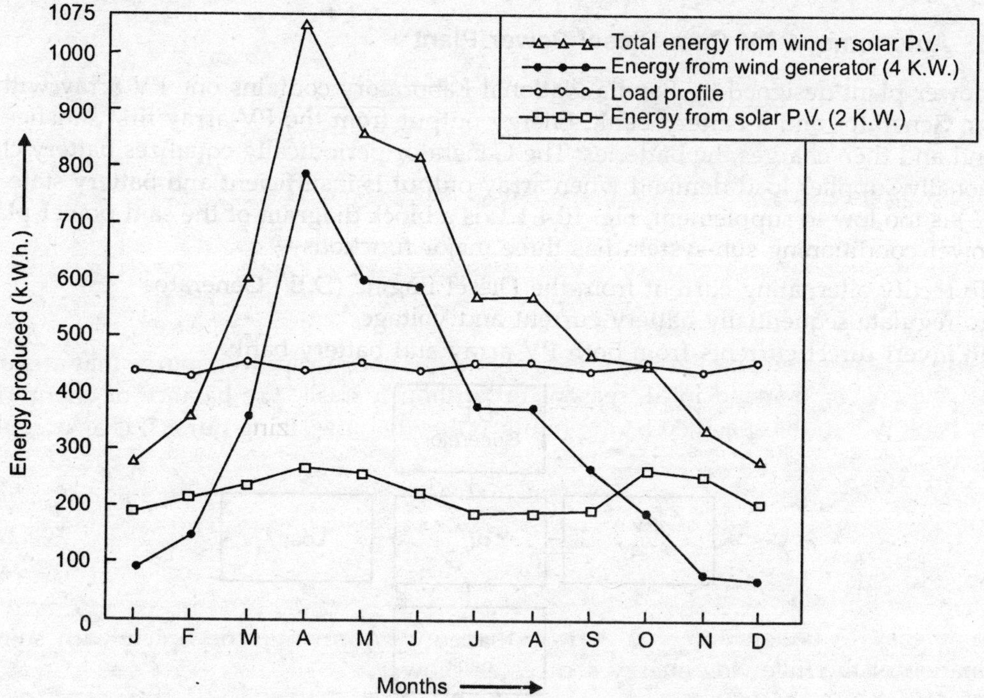

Fig. 10.3.1: Monthly Values of Wind Generator and Solar PV Energy Output against Load Profile

(a) A constant lighting load during the evening hours.
(b) A variable agricultural load principally centered around operating pumps for irrigation and small motors for rice-husking machine/spraying of insecticides.

Here, the energy requirement (a) Can be reasonably met-up by rightly sizing of a bio-gas generator while a PV pump and a PV operated battery system can cater to the requirement (b) In this case, the obvious choice of the system designer should lie in the option (c).

10.4 SITE SPECIFIC EXAMPLES

A number of Hybrid Energy Systems options have been identified in the last section and some site-specific parameters also have been discussed with regard to the particular choice of few such options. Now, it will be interesting and worthwhile to consider the design-based technical specification of some site-specific combinations which includes:

(A) PV Diesel Generator along with storage battery bank.
(B) PV with Propane gas Generator with storage battery bank for a residential system in Northern N.Y. state.
(C) PV with Diesel Generator along with battery bank for a tourist lodge in Orissa, India.
(D) Bio-mass gasifier based thermal back-up for a solar chilly drier at Sarvodaya Masala Industries, Tarapur, Gujarat.
(E) A natural convection solar dryer and a simple bio-mass burner combined for application in small scale driers in non-electrified areas of developing countries.
(F) A milk chilling unit powered by combination of bio-gas and solar energy at the village Vasna Margia, Gujarat.
(G) Power generation from a combustion of rice husk plus diesel and rice-husk plus immature paddy with diesel.

10.4.1 Autonomous PV Cum Diesel Power Plant

This power plant designed by Sandia National Laboratory contains one PV array with a Diesel Electric Generator and a battery bank. Energy output from the PV array first satisfies any load demand and then charges the batteries. The Generator periodically equalizes battery charge and occasionally supplies load demand when array output is insufficient and battery state of charge (S.O.C.) is too low to supplement. Fig. 10.4.1.1. is a block diagram of the said power plant where the power conditioning sub-system has three major functions—

(a) To rectify alternating current from the Diesel Engine (D.E.) Generator
(b) To regulate sequentially battery current and voltage
(c) To invert direct currents from both PV array and battery bank.

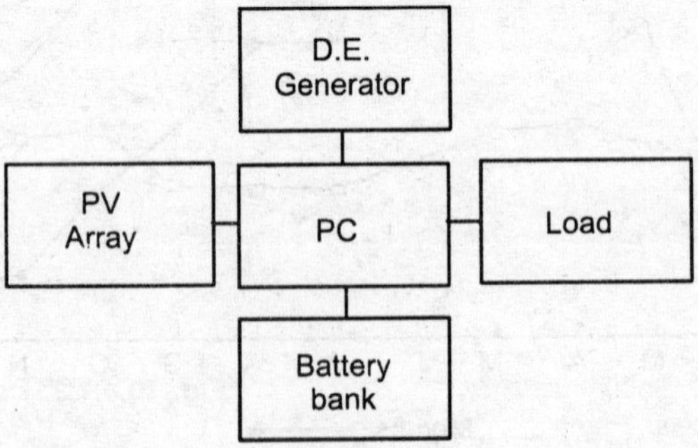

Fig. 10.4.1.1: A Block Diagram of PV-Diesel Power Point

The design procedure starts from the simulation of system performance based on the location where average daily solar insolation for each month is graphed in Fig. 10.4.1.2. The computer simulation of system performance uses data for each hour of a year and generates a list which also includes diesel fuel consumed. Sizes of the PV array area ($A_A \times 10^3$ m^2) and the battery bank of size E_o (storage capacity in MWh) are iterated until the list indicates that fractional diesel use is 10% i.e., solar PV array has been used to replace 90% of the diesel fuel that would be consumed for a diesel only system. Each maximum allowable depth of discharge (DOD) defines an *ISO-FUEL* curve (Fig. 10.4.1.3.) each of which represents combination of battery and array with 10% fractional diesel use. Life cycle cost is graphed in Fig. 10.4.1.4 where the minimum cost point corresponds to the most cost effective design where 90% of the diesel fuel required by a diesel only system has been replaced. Combination of Fig. 10.4.1.4 and 10.4.1.3 for this assumption yields an array size of 5000m^2 and a battery bank of storage capacity slightly above 4 MWh. It is evident from Fig. 10.4.1.4 that the Life Cycle Cost escalates rapidly as soon as array area below the optimum value is chosen. It is also worthwhile to check the net present benefit at this stage, which is simply the cost difference between PV – diesel system and diesel only system. This analysis included plausible values of different fuel cost escalation rates and various discount rates. Interesting features of the aforesaid design procedure are:

(A) At the optimal design point, a 10% reduction in array size calls for a 60% increase in battery capacity.

(B) A 10% increase in array size necessitates only a reduction of battery capacity by 20%.

(C) PV – diesel system is more cost effective than a stand-alone diesel or stand alone PV system when high levels of availability are required and fuel supply is difficult to obtain.

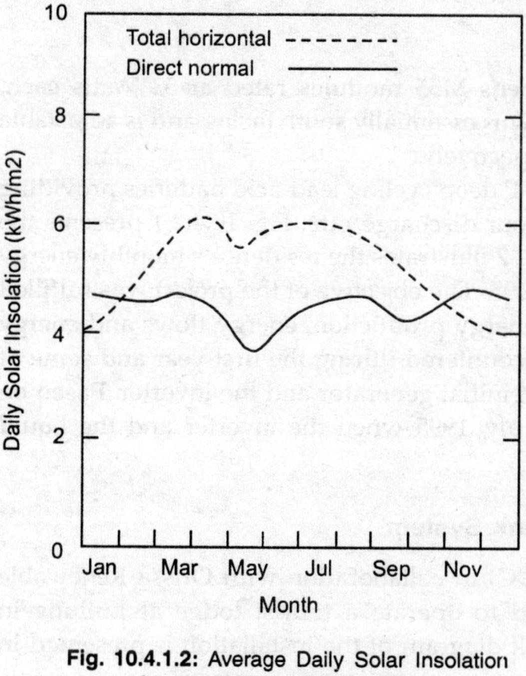

Fig. 10.4.1.2: Average Daily Solar Insolation for the PV-Diesel Power Plant

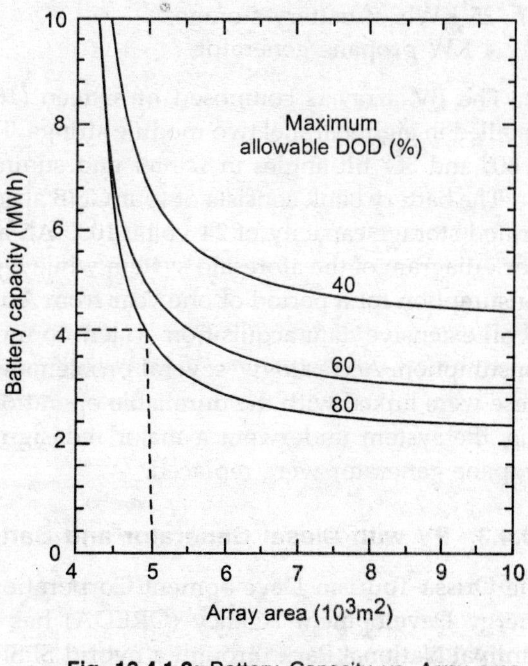

Fig. 10.4.1.3: Battery Capacity vs. Array area for 10% Diesel Consumption

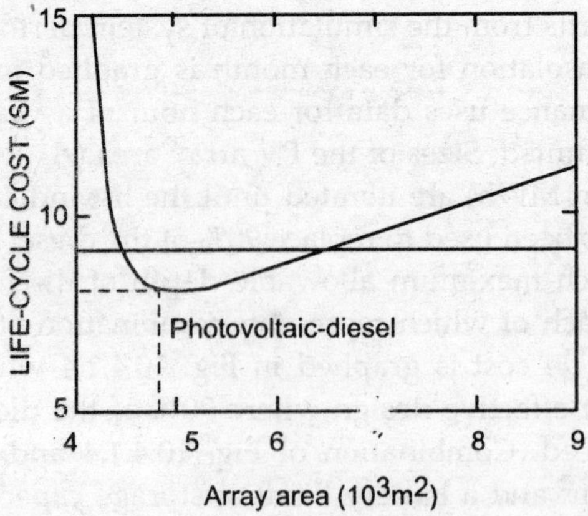

Fig. 10.4.1.4: Photovoltaic-diesel System Cost for 10% Diesel Consumption

10.4.2 PV – Propane Gas Hybrid System for Residential Applications

In August, 1994, a prototype PV-generator hybrid system rather than a PV system has been installed as a line extension alternative at a residence located in Northern New York State. This system replaced an existing diesel powered electric generator and was sized to meet the known lighting and plug loads of the residence but excluding the cooking, refrigeration and heating loads. Essential components of this hybrid configuration were:

(a) 850 Watt PV array;
(b) 25 KWh of battery storage;
(c) 4 KW propane generator;

The PV array is composed of sixteen (16) Siemens M55 modules rated at 53 Watts each, installed in eight parallel two module strings. The array is essentially south facing and is adjustable at 60° and 30° tilt angles in winter and summer respectively.

The battery bank consists of four GNB absolyte II-P deep cycling lead-acid batteries providing a rated storage capacity of 24 volts/1055 Ah at a 7 hour discharge rate. Fig. 10.4.2.1 presents the block diagram of the aforesaid system while Fig. 10.4.2.2 illustrates the residence's monthly energy consumption for a period of one year from August, 1995. The objective of the project was fulfilled by an extensive data acquisition system to monitor energy production, energy flows and energy consumption. Accordingly, several problems were encountered during the first year and some of these were linked with the unreliable operation of the initial generator and the inverter. Based on this, the system underwent a major redesign after July, 1995 when the inverter and the liquid propane generator were replaced.

10.4.3 PV with Diesel Generator and Battery-Bank System

The Orissa Tourism Development Corporation (OTDC) in collaboration with Orissa Renewable Energy Development Agency (OREDA) has decided to operate a tourist lodge at Lullung in Simlipal National Park through a hybrid SPS. A block diagram of the installation is presented in Fig. 10.4.3.1. Essential components the system were:

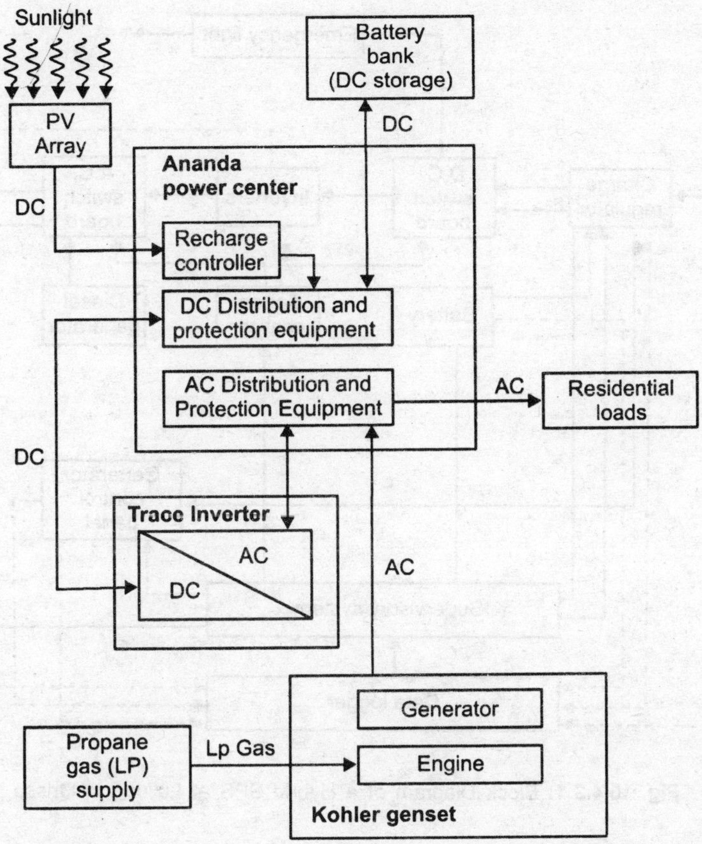

Fig. 10.4.2.1: Hybrid System Combining a PV Array and Propane Generator

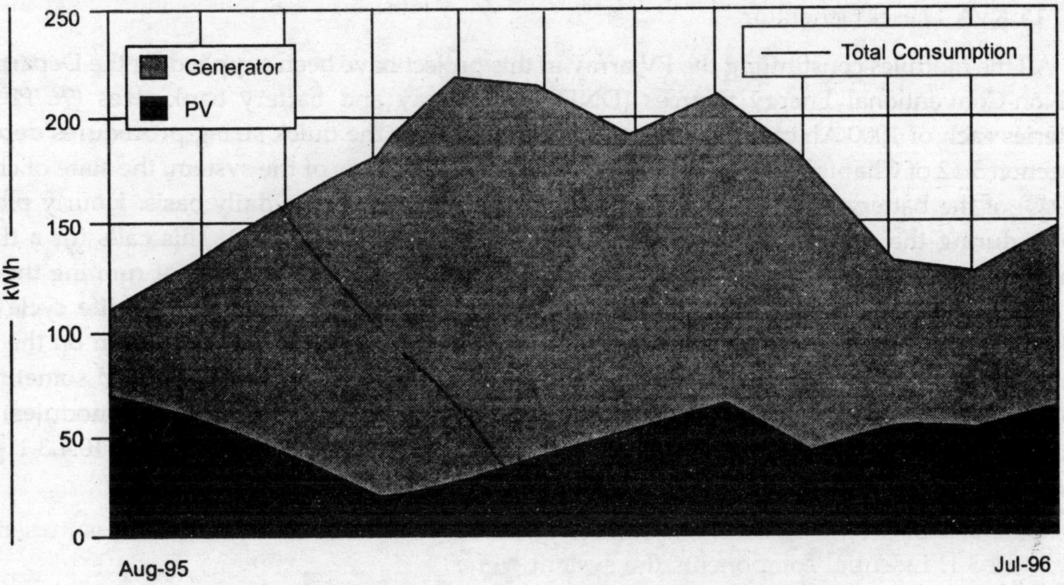

Fig. 10.4.2.2: Residence's Energy Consumption and Respective Contribution of PV Array and Propane Generator

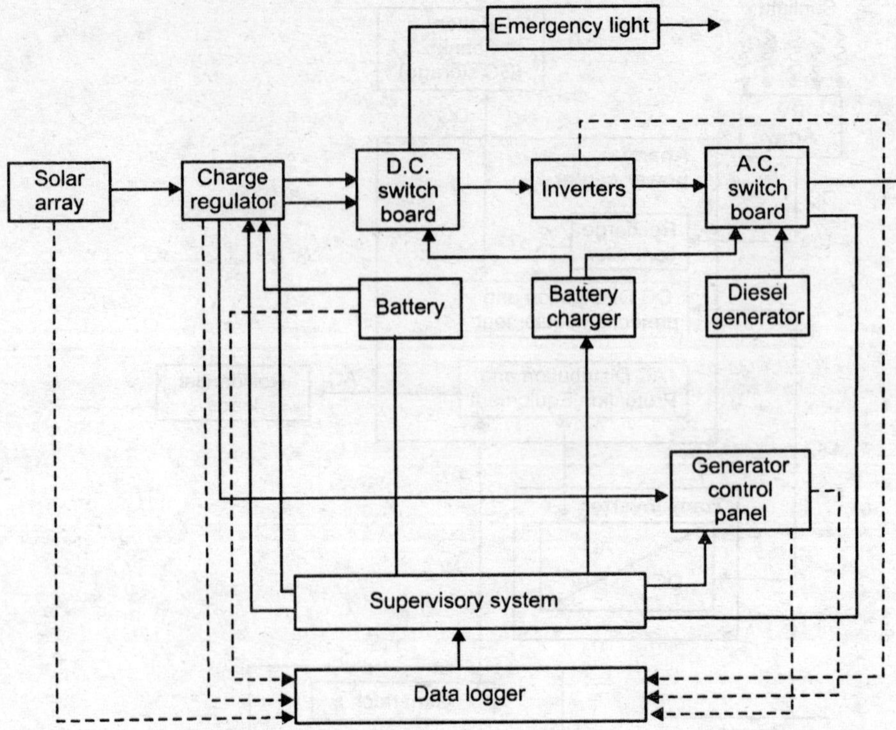

Fig. 10.4.3.1: Block Diagram of a Hybrid SPS at Lullung in Orissa

(a) 24.5 KWp PV array
(b) 192 volts battery bank
(c) 15 KVA Diesel Generator

All the modules constituting the PV array in this project have been supplied by the Department of Non-Conventional Energy Sources (DNES). The array and battery bank sizes (96 Pb acid batteries each of 1000 Ah capacity) have been computed by the quick sizing procedures depicted in section 5.12 of Chapter V. As regards the performance analysis of the system, the state of charge (S.O.C) of the battery bank was monitored on a monthly as well as daily basis. Hourly plots of S.O.C. during the critical month of September is shown in Fig. 10.4.3.2. This calls for a diesel-generator whenever the S.O.C. sinks below the 20% limit and the frequency of running the back up source has been considered critically from such performance curves. Finally, life cycle costs have been carried out for three different array-sizes and the optimum value chosen on the basis of minimum value of TLC (refer 5.13 of chapter V). This project was commissioned sometime in 1988 with an array size of 24.5 KWp (comprising of 768 standard 30 pW CEL modules). The optimization of the number of modules have been possible based on the Table 10.4.3.1 given below:

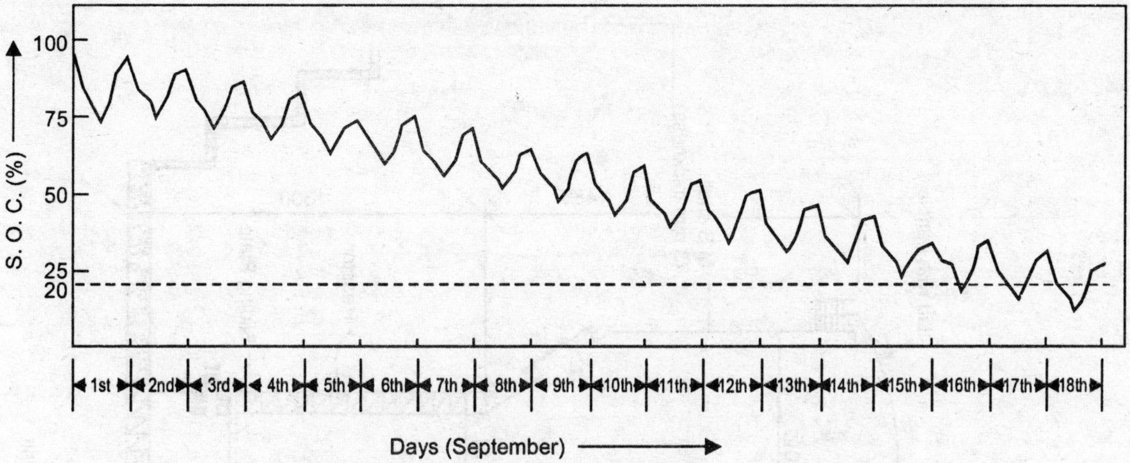

Fig. 10.4.3.2: Hourly and Daily Computation of Battery Cycle

Table 10.4.3.1: Life Cycle Cost Estimate for Different Array Sizes
(All Costs in Thousand Rupees)

Number of Modules	*768*	*672*	*576*
First Capital Cost	6136.86	5685.60	5241.24
Annual Maintenance Cost	60.42	63.71	71.58
Annual Fuel Cost	7.24	14.00	23.58
Present worth of battery	Nil	517.90	793.91
Present worth of Diesel	Nil	Nil	77.69
Total Life Cycle Cost	6204.52	6281.21	6208

10.4.4 Bio-mass gasifier based thermal back-up for a Solar Chilly drier

A natural draft based gasifier-burner system suitable for drying applications has been designed, fabricated and retrofitted with the Sardar Patel Renewable Energy Research Institute's Solar Chilly Drier at Sarvodaya Masala Industries, Tarapur. The system consists of a natural draft type gasifier of maximum Bio-mass consumption capacity of 8 Kg. per hour, an aspirated type producer gas burner and a shell and tube air-gas heat exchanger as the major components. The heat exchanger has been coupled with the drier as seen from schematic of the unit presented in Fig. 10.4.4.1. For sizing of the gasifier, R & D efforts in the laboratory stage have yielded:

(a) Specific gasification rate of 60 Kg/h sq.m. and
(b) Reactor height to diameter ratio of 2:3

The hopper capacity has been fixed up taking into consideration the batch operation hours and the average bulk density of the feedstock, a proximate analysis of which is shown in Table 10.4.4.1. Air turbulators have been provided with the producer gas burner to achieve swirling action for better air gas mixing.

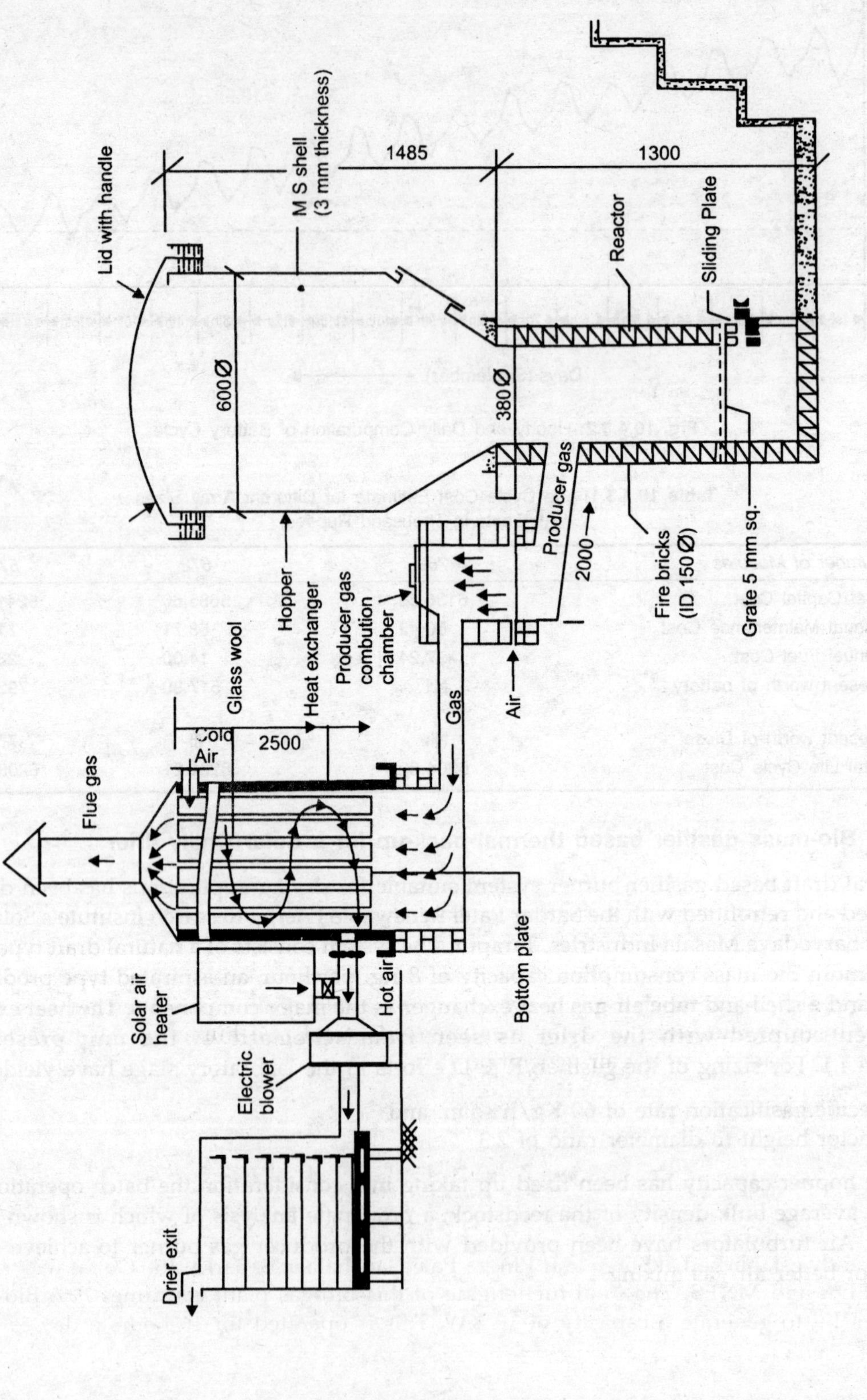

Fig. 10.4.4.1: Schematic of the Biomass Gasifier-based Thermal Back-up Unit for Solar-based Chilly Dryer at Tarapur.

Table 10.4.4.1: The Proximate Analysis of Various Feed Stocks used

Fuel Type	*Volatile Matter (VM)%*	*Fixed Carbon (FC)%*	*Ash %*
Arhar Stalks	87.68	10.69	1.63
Ipomoea	86.77	10.59	2.64
Saw mill woody waste	87.37	11.53	1.10
Babul wood	82.68	15.85	1.47

10.4.4.1 Performance Evaluation of the System

Several parameters were monitored to evaluate the system performance. These include type and size of the feedstocks, the fuel consumption rate, the flame temperature, flue gas temperature at the exit of the heat exchanger, hot air temperature inside the drier and relative humidity of inlet air and exhaust air of the drier. Details of a typical system performance test conducted have been summarized in table 10.4.4.1.1. The gasifier took only 15 minutes to generate producer gas of adequate quality. It consumes 5 to 6 Kg. of woody Bio-mass to generate around 750m^3 of hot air at an average temperature of 80°C. The average temperature of the flame and the exhaust of the heat exchanger were found to be 833°C and 110°C respectively.

Table 10.4.4.1.1: Summary of System Performance Tests

Agro products dried	Green chilly
Load per batch	50 Kg
Initial moisture content	88% (wet basis)
Final moisture content	8% (wet basis)
Drying duration	8 hours
Optimum drying temperature	60°C
Ambient air temperature	20 to 25°C
Relative humidity	60 to 80%
Hot air flow rate	750m^3/hour
Bio-mass consumed	52 Kg.
Electricity consumption	6 KWh
Manpower required	2 persons
* Thermal efficiency of the gasifier heat exchanger unit	55%
* Thermal efficiency of the complete system	33%

10.4.4.2 On-line Instrumentation

Aluminium chromel and copper constantan thermocouples were used for recording the temperature across the gasifier and drier respectively with the aid of a multi channel temperature recorder. Portable hygrometers were used for measurement of relative humidity. Anemometers were employed to quantify the hot air generation rate.

10.4.4.3 Dual-fuel Bio-gas Engine

More and more hybrid energy systems will continue to dominate the scenario as these are the right options considering the reliability and capital cost of a system vis-a-vis site specific parameters. In fact, the dual fuel engine of a Bio-gas plant is the simplest hybrid energy source which has catered to maximum number of applications on the Indian scenario. A 60m^3 Bio-gas plant has been recently established at Shree Idar Pingra Pale Gaushala in Sabarkantta, Gujrat with subsidies from GEDA and MNES. The dual fuel engine of this Bio-gas plant consumes 70% Bio-gas and 30% diesel – to generate a capacity of 10 KW. This is operated for 16 hours a day to produce power to the tune of 130 KWh/day. The generated power is utilized for chaff cutter, submersible pumps and internal lighting of the campus.

10.4.5 Natural Convection Solar Driers in Combination with Bio-mass Back-up Heater

The brace design of solar driers is appealing particularly to small scale producers because it is easy and inexpensive to construct, simple to use and can produce a good quality dried product under favourable climatic conditions. One significant disadvantage of the drier is that it is normally not used with any form of back-up heating. Obviously, for commercial production, this factor limits their ability to process a crop when there is inadequate solar radiation. For commercial producers, the ability to process continuously with quality and reliability is important to satisfy market requirements. There have been some attempts to overcome this limitation in simple natural convection solar dryers with the exception of a dryer reported by Bassey et al. (1987) which used a saw dust burner to provide heat during poor weather and at night. This saw-dust burner was used as a separate component rather than being integrated with the drying cabinet and provided 400 W/m^2 of energy to the drying cabinet using steam as the heat transfer medium. Bio-mass, particularly fuelwood is the most common source of energy in rural areas of developing countries and is often utilized very inefficiently. Thus, there existed a need for simple, affordable combustion device complementing the Brace dryer.

A simple Bio-mass burner, which can be fabricated with easily available materials, tools and skills has been demonstrated in combination with a Brace direct type solar dryer very recently by Bena and Fuller (Sol-Energy, vol-72, No-1, 2002). From a series of evaluation trials of the system, the capacity of the drier was found to be 20-22 kg. of fresh pine apple arranged in a single layer of 0.01m thick slices. Other trials estimated the efficiency of the burner in producing useful heat for drying to be 27%, with the drying efficiency of the Brace type solar drier alone to be 22%. Certain key features of this hybrid design which contributed to aforesaid acceptable thermal efficiency values are

(A) The thermal mass on the top surface of the drum;
(B) An internal baffle plate to lengthen the exhaust gas exit path and;
(C) A variable air inlet valve.

Since the burner can be operated by the two renewable energy sources independently or in combination, there exists a scope for a comparison between the two energy sources as projected in Table 10.4.5.1.

Table 10.4.5.1: Comparison of Key Operating Parameters using Solar and Bio-mass Energy

	Solar thermal	*Bio-mass*
Energy use (MJ)	112	463
Moisture removal (%)	47	53
No. of hours of operation	44	39
Drying efficiency (%)	22	06

It is interesting to note from Table 10.4.5.1. that although approximately equal amounts of moisture was removed during the day (44 hrs.) and night time (39 hrs.) periods, Bio-mass energy utilized was four times more than solar energy. This indicates that direct absorption of solar energy compared to the transfer of heat from the drum surface by convection is a much more efficient process – a factor which is also reflected in the big difference in the drying efficiencies.

Fig. 10.4.5.1. shows that the hybrid design of drier consists of a drying cabmet mounted on top of a brick chamber that encloses a simple Bio-mass burner. The outside base dimensions are 1.2m by 1.2m. There are three drying trays, each with a wire-mesh base providing an effective drying area of nearly 3m^2. Interestingly, the top surface is inclined at an angle of 28° (to maximize the capture of solar radiation at the Melbourne latitude of 38° during the trial period. The cabinet was covered with a single layer of 0.15mm thick UV stabilized polyethylene film.

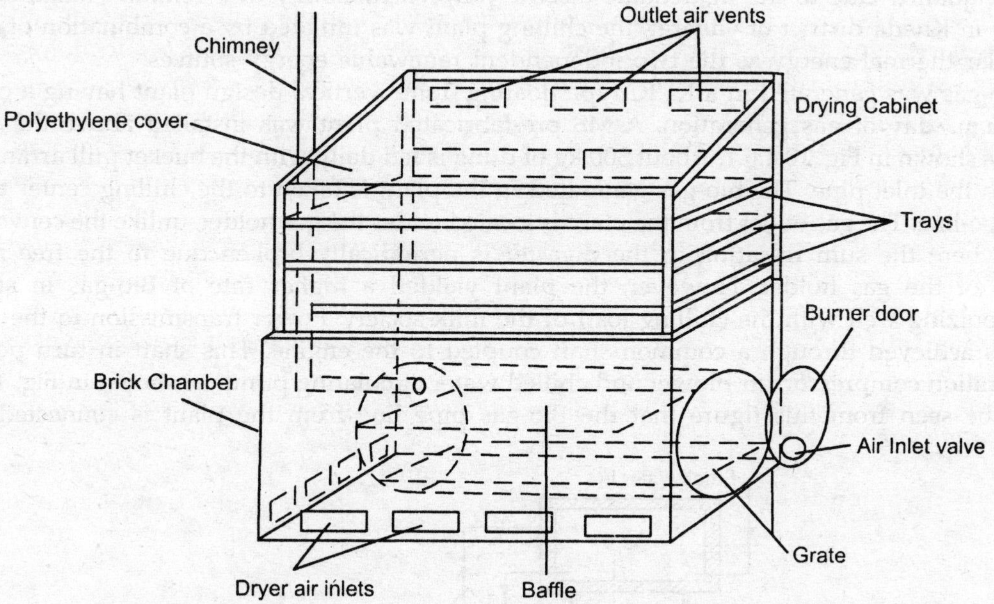

Fig. 10.4.5.1: Natural Convection Solar Dryer with Biomass Back-up Heater

The bio-mass burner based on fuelwood was fabricated from a 0.2m^3 drum laid on its side. The burner-grate was constructed from a perforated tray supported on rails riveted to the inside walls of the drum. A 0.2m diameter hole was cut in the center of the bottom of the door below the grate and fitted with a simple spinning valve to regulate the flow of air into the burner. The exhaust air exits via a 0.09m diameter and 1.8m long flue pipe located on the other end of the drum. In order to lengthen the flow path of the combustion gases and maximize the transfer of heat to the drum walls, a metal baffle was inserted above the grate and below the exhaust gas exit.

10.4.5.1 Dryer Efficiency

Most of the performance trials have been carried out and measurements reported specific to a particular dryer in a single climate drying only one crop. This makes it difficult to compare the performance of driers. Brenderfer et al. (1985) suggested that the two most significant parameters for evaluating the performance of driers are overall thermal efficiency and pick-up efficiency. However, it is difficult to measure the later efficiency in a multi-tray direct type solar dryer as there is an energy gain at each tray level. Therefore the thermal efficiency (η_{th}), over an entire drying trial, is the ratio of the energy used to evaporate the moisture from the crop to the energy supplied to the drier. As both solar radiation and Bio-mass supplied the input energy,

$$\eta_{th} = WL/(HA + CM) \qquad ...(10.4.5.1)$$

Where W is the mass of water evaporated (kg), L is the latent heat of vaporization (MJ/kg). H is the insolation on the dryer (MJ/m^2). A is the effective area exposed to solar radiation (m^2), C is the calorific value of wood (MJ/kg) and M is the mass of wood (kg).

10.4.6 Village Level Milk Chilling Center Powered by Bio-gas and Solar Energy

Undoubtedly, chilling is the most trusted method practiced in dairy industry today for preservation of milk quality. Due to the inadequate electric power availability at a remote village of Vasna Margia in Kheda district of Gujarat, the chilling plant was initiated by a combination of bio-gas and solar thermal energy as the two independent renewable energy sources.

Bio-gas was generated in a KVIC type floating dome vertical design plant having a capacity of 12cu.m./day of gas generation. A MS pre-fabricated plant was installed above the ground level, as shown in Fig. 10.4.6.1. About 300 kg of dung is fed daily with the bucket pull arrangement through the inlet pipe. The bio-gas generated in the plant is taken to the chilling center through a GI pipe-line. The gas outlet from the plant is located under the gas holder, unlike the conventional plant where the sum resulting in the digester is periodically broken due to the free rotation facility of the gas holder. However, the plant yielded a higher rate of Bio-gas in summer, synchronizing well with the chilling load of the milk society. Power transmission to the chilling plant is achieved through a common shaft coupled to the engine. This shaft in turn powers a refrigeration compressor, air blower and chilled water circulating pump as shown in Fig. 10.4.6.1. It can be seen from this figure that the Bio-gas pipe line from the plant is connected to the

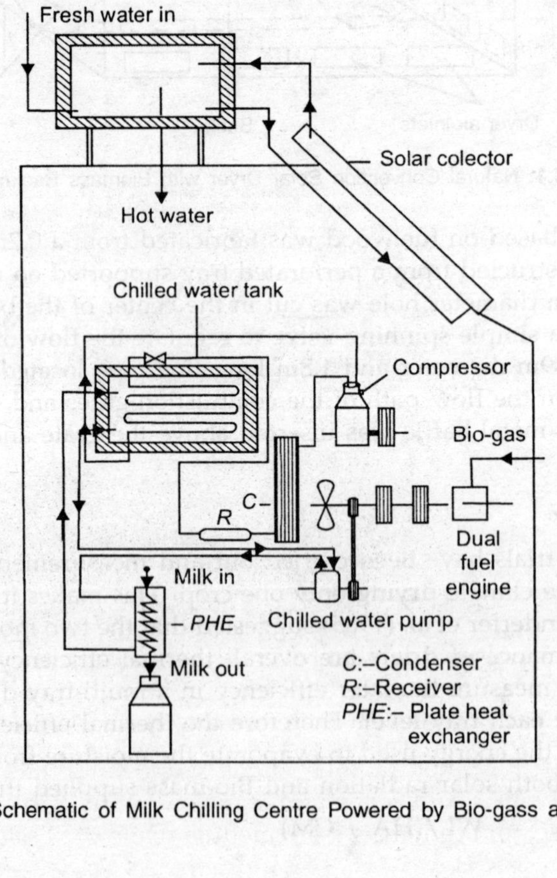

Fig. 10.4.6.1: Schematic of Milk Chilling Centre Powered by Bio-gass and Solar Energy

air-inlet chamber of a dual fuel engine through a valve. To begin operation, this Kiroloskar make 5 hp dual-fuel engine is started on 100% diesel as the power source. The engine switches naturally to the dual fuel mode after the gas valve is opened. The fuel ratio maintained henceforth is 80% Bio-gas and 20% diesel. Another salient feature that can be observed from Fig. 10.4.6.1 is that the roof of the building housing the chilling plant is partially substituted by flat plate solar collectors. An insulated water storage tank placed on the society building is connected to the collector. The collector generates about 100 litres of hot water (60°C) per day by thermosyphon mechanism. This hot water is utilized for cleaning of cans and the milk chilling equipment/components.

Prior to the setting up of the plant, the milk co-operative society used to incur a loss of Rs. 12,000/- to Rs. 14,000/- per year on account of sourage. With this plant commissioned only for the summer/monsoon months from March to October, the net average savings generated varies from Rs. 3500/- to 4000/- per year (average sourage loss of Rs. 13,000/- per year). In fact, if a chain of such centers can be planned on a milk truck route, it is possible to despatch the milk to the processing dairy once a day only.

10.4.7 Power Generation from a Combustion of Rice Husk Plus Diesel and Rice-husk Plus Immature Paddy with Diesel

In several regions of rural India, rice-husk and immature paddy is treated as a waste material. Recently, the School of Energy Studies, Jadavpur University has carried out extensive number of field survey works to assess the potential of both rice-husk and immature paddy in producing electrical energy. Table 10.4.7.1. gives some details about the production of immature paddy and rice-husk in some of the surveyed districts of West Bengal in course of 2001-02.

Table 10.4.7.1: Districtwise Production of Immature Paddy and Rice-husk

Name of district	*Total rice-husk (tonnes)*	*Total immature paddy (tonnes)*
Dakshin Dinazpur	148569.346	60138.339
Uttar Dinazpur	214910.59	80948.255
Birbhum	389795.612	54788.321
Bankura	363396.033	61314.99
Murshidabad	342700.05	55772.639
Burdwan	619274.45	102567.50
24 Parganas (s)	353201.38	70360.05

Interestingly, it has also been estimated by Ghosh and Saha (M.Tech thesis in Energy Science and Technology, J.U.'2002) that

(i) 1.2 kg of rice husk + 0.1 litre of diesel or

(ii) 0.6 kg of immature paddy + 0.6 kg of rice-husk combined with 0.09 litre of diesel can generate 1 KWh power. The results of aforesaid studies also point to the future potential that if all agricultural residues like the rice-husk and immature paddy are utilized for HYBRID POWER GENERATION, in the state of West Bengal, more than 630 power plants, each of 1000 KWh capacity can be planned for installation.

10.5 INTEGRATED ENERGY SYSTEM

It is well understood that utilization of solar energy is possible through two well established technologies – solar photovoltaic conversion and the solar thermal conversion. The basic unit that caters to solar photovoltaic applications is a PV module while the basic unit employed for solar thermal route is a flat plate collector intended for solar water/space heating applications. An integrated energy system arising out of the aforesaid technologies is a photovoltaic/Thermal (PV/T) collector which evolves as their end product both electrical and thermal energy. The concept of using the same module for both photovoltaic and thermal conversion arises from the well conceived difficulty that efficiency of solar photovoltaic cells used in PV modules range typically from 10 to 15% only. (This is the status of efficiency of silicon PV cells manufactured by CEL, BHEL, Webel (SL) Energy Systems Ltd., Tata BP Ltd. on the Indian Scenario). As a result, the single crystal silicon PV modules enjoy a conversion efficiency of around 10% - which means that nearly 90% of the solar radiation incident on a module is lost as heat energy to surroundings. Thus, there is a definite need for reducing such huge energy losses through intelligent design of a combination of a PV module and a flat plate collector on the same gadget. This integrated gadget can then be placed at the roof top a residential building to provide—

(a) Electrical energy for running tullu pumps for raising water.
(b) Charging small battery banks for night time requirement of few lighting loads.
(c) Hot water to the greyser of one or two toilets.
(d) Hot water to the kitchen to reduce the demand on conventional energies like electricity, kerosene and cooking gas.

It can easily be identified that out of the aforesaid applications in a residential building items (a) and (b) are SPV applications while (c) and (d) are SWHS applications. Since typical thermal efficiencies of flat plate collectors are in the range of 30 to 50%, overall PV/T efficiencies are in the range of 50 to 60% for integrated energy systems involving a combination of SPV and SWHS. So, besides a substantial amount of conventional energy-savings, such integrated systems can also reduce the huge energy losses to the surroundings occurring from a simple PV module. Possible classifications and related merits/demerits of such integrated systems are pointed out briefly in the next section.

10.5.1 Various Forms of PV/T Systems

In the very first choice, these systems can occur in flat plate and concentrated forms. It is convenient to use a flat plate system at different sites as the latter type adds to system complexity and is also responsible for additional capital cost which increases the BOS cost remarkably over the PV module cost.

Now a flat plate PV/T system inevitably uses a medium like liquid or air to extract the thermal energy. This, can be served in two ways in a flat plate PV/T system:

(A) Solar air heater combined with a PV module.
(B) Forced water circulation SWHS combined with a PV module.

A brief description of both these systems is presented below:

i) Air PV/T systems, consisting of two air channels, a top 'static' channel for insulation and a bottom flow channel for heat transfer. A sheet of glass with PV cells attached to its underside forms the common boundary between the two channels. The other side boundaries of the top and bottom channels are constituted by a top cover glazing and a bottom secondary absorber. The arrangement is illustrated in Fig. 10.5.1.1 through a cross-sectional view only.

Liquid PV/T systems have been reported to have higher values of thermal and overall efficiencies. Here, the solar cells are pasted directly on the metallic absorber plate by using a material which must be thermally conductive but electrically an insulator. PV efficiency of such a system principally depends on factors like:

(i) Packing fraction of the module.
(ii) Thermal energy transmitted to the absorber plate.
(iii) Emissivity of the absorber.
(iv) Mean absorber temp.

Since the efficiency of the solar cells increases as the temperature decreases, a convenient design strategy will be to identify the factors responsible for reduction of absorber temperature. Out of these, the most crucial aspect is perhaps the amount of water/liquid in the storage tank. A common practice followed during the operation of these systems is to lower the absorber plate temperature by removing some hot water at a suitable high temperature and replace it with equal amount of cold water.

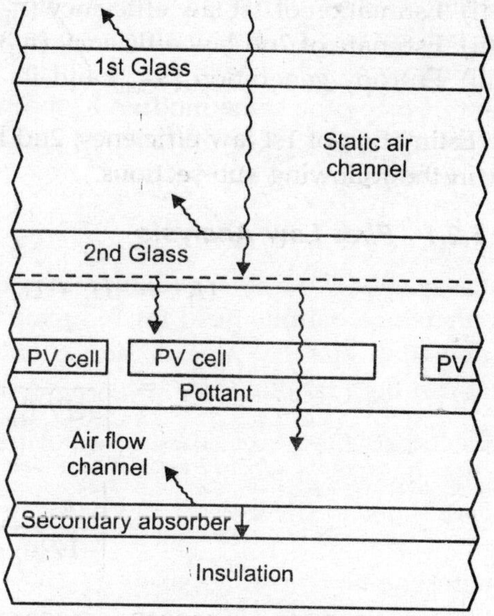

Fig. 10.5.1.1: Cross Section of a View of Air PV/T Systems

10.5.2 Liquid PV/T Systems

One typical experimental arrangement of this type of integrated energy system is shown in Fig. 10.5.2.1. Obviously, the intention of incorporating a SWHS here is (i) to harness the unused solar energy towards photovoltaic generation in the form of thermal energy and (ii) to maintain the solar cell temperature in the PV module as minimum as possible to obtain higher PV efficiency from the module. Hence, it is essential to identity the optimum operating parameters of the solar thermal conversion part of the integrated energy system to achieve this goal. Since thermodynamic optimization is the only answer, the following methodology can be commonly envisaged:

Fig. 10.5.2.1: An Experimental Set-up of Liquid PV/T Integrated Energy System at B.E. College., Howrah

(i) Estimation of 1st law efficiency (η_I) and its dependance on the collector efficiency function.
(ii) Estimate of 2nd law efficiency (η_{II}) and its dependance on the collector efficiency function.
(iii) Entropy generation (S_{Gen}) and its dependance on collector efficiency function.

Estimation of 1st law efficiency, 2nd law efficiency and entropy generation have been carried out in the following sub-sections.

10.5.2.1 First Law Analysis

$$U_L = U_t + U_b + U_s \qquad ...(10.5.2.1)$$

$$U_t = \left| \frac{N}{(C/T_{cm})\{(T_{cm} - T_a)/(N+f)\}^{0.33}} + \frac{1}{h_w} \right|^{-1}$$

$$\left| \frac{\sigma(T_{cm}{}^2 + T_a{}^2)(T_{cm} + T_a)}{1/\{\epsilon_p + 0.05N(1-\epsilon_p)\} + (2N+f-1)/\epsilon_g - N} \right| \qquad ...(10.5.2.2)$$

Where, $C = 365.9(1 - 0.00883\beta + 0.0001298\beta^2)$ and $f = (1 - 0.04h_w + 0.0005\, h_w{}^2)\,(1 + 0.091N)$

$$Q_u = F_R\,.A_C\,[\{(\rho\,\gamma)\,(\tau\,\alpha)_e\,\}.\, I_T - U_L\,(T_i - T_a)] \qquad ...(10.5.2.3)$$

$$T_m = T_i + \frac{Q_u(1 - F_R/F^1)}{U_L F_R A_C} \qquad ...(10.5.2.4) \text{ after } [10.5.2.3]$$

$$T_{cm} = T_i + \frac{Q_u(1 - F_R)}{U_L F_R A_C} \qquad ...(10.5.2.5) \text{ after } [10.5.2.3]$$

First law efficiency is given by,

$$\eta_I = Q_u/(A_C\,.\,I_T) = F_R\,(\rho\,\gamma)\,(\tau\,\alpha)_e - U_L F_R\,[(T_i - T_a)/I_T)] \qquad ...(10.5.2.6)$$

Optical efficiency is given by, $\eta_{op} = (\rho\,\gamma)\,(\tau\,\alpha)_e$ ...(10.5.2.7)

The symbols used in this section from Eqs. (10.5.2.1) to (10.5.2.7) have been introduced in chapter 3.

10.5.2.2 Second Law Analysis

The energy function of solar thermal radiation is given by,

$$\eta_e = [1 - 4/3\,(T_a/T_s) + 1/3\,(T_a/T_{s)}{}^4] \qquad ...(10.5.2.8) \text{ after } [10.5.2.4]$$

Energy of primary radiation on unit area of collector aperture is,

$$e_1 = I_T\,.\,\eta_e \qquad ...(10.5.2.9)$$

Energy loss due to optical imperfection,

$$e_{L,\,op} = [1 - (\rho\,\gamma)\,(\tau\,\alpha)_e]\,.\,I_T.\eta_e \qquad ...(10.5.2.10)$$

Energy of absorbed radiation,

$$e_a = (\rho\,\gamma)\,(\tau\,\alpha)_e \,.\, I_T\,[1 - T_a/T_{cm}] \quad ...(10.5.2.11)$$

Energy in utility heat,

$$e_u = F_R\,[\{(\rho\,\gamma)\,(\tau\,\alpha)_e \,.I_T - U_L\,(T_i - T_a)]\,.\,(1 - T_a/T_m) \quad ...(10.5.2.12)$$

Energy loss due to collector-ambient heat loss,

$$e_{L,\,a} = (U_L/T_{cm})\,.\,(T_{cm} - T_{a)}{}^2 \quad ...(10.5.2.13)$$

Destruction of energy due to sun collector temperature difference,

$$e_{D,SC} = e_1 - (e_a + e_{L/op}) = (\rho\,\gamma)\,(\tau\,\alpha)_e \,.\, I_T\,[T_{a/}T_{cm} - T_a/T_s] \quad ...(10.5.2.14)$$

Destruction of energy due to collector- fluid temperature difference,

$$e_{D,\,Cf} = e_a - (e_u + e_{L/a)} = F_R\,[(\rho\,\gamma)\,(\tau\,\alpha)_e \,.\, I_T - U_L\,(T_i - T_a)]\,.\,[T_a/T_m - T_a/T_{cm}] \quad ...(10.5.2.15)$$

Second law efficiency:

$$\eta_{II} = \frac{\text{Energy out in product}}{\text{Energy in}}$$

From equations (10.5.2.4) and (10.5.2.12), the second law efficiency can be written as

$$\eta_{II} = e_u/e_1 = (\eta_1/\eta_{e)}\,(1 - T_a/T_{m)}) \quad ...(10.5.2.16)$$

10.5.2.3 Entropy Generation

The overall heat loss Q_0 is given by

$$Q_0 = U_L\,.A_C.\,F_R\,(T_i - T_a) \quad ...(10.5.2.17)$$

The total rate of entropy generation in the collector is

$$S_{gen} = [(1 - h_{op})\,I_T.A_C]/T_a + Q_u/T_m + Q_o/T_a - (I_T.A_C)/T_s \quad ...(10.5.2.18)$$

The following assumptions are made during the optimization of the system:

(i) Collector thermal inertia is ignored;

(ii) Collector thermal performance is described by Hottel-Whillier Bliss (HWB) equation over the whole operating range;

Irreversibility due to friction is negligible compared to that of heat transfer.

A recent experimental study carried out by Chakrabarti and Mukherjee at B.E.College (D.U.) premises on the basis of the aforesaid theoretical results have revealed that:

(a) The optimum operating point of the integrated energy system for all the cases occurs corresponding to an optimum value of collector efficiency function;

(b) The optimum value of the collector efficiency function can be used by any system designer to evaluate the optimum fluid inlet temperature corresponding to any fixed value of insolation.

10.5.3 A Photovoltaic/Thermal (PV/T) Collector with a Polymer Absorber Plate

Here, a combined PV/thermal collector was fabricated by pasting single crystal silicon solar cells on to a black plastic solar heat absorber. The PV cells are extremely brittle and have considerably lower thermal expansion coefficient than the polymer material. A special silicon adhesive, that was sufficiently elastic to absorb the difference in thermal expansion between the cells and the plastic absorber, was employed and a thin adhesive layer (0.5 mm) ensured acceptable thermal contact.

Thirty pv cells (1.5Wp each) were arranged in six rows of five cells as shown in Fig. 10.5.3.1., resulting in an effective module area of 0.32m^2. The six rows were further separated into two equal units of 15 series connected cells, allowing the two units to be connected in series. (30 cells in series or alternatively used in parallel.) Three top rows can be disconnected from the three bottom rows to form two submodules. Also, measurement points for the PV cell temperatures T_1, T_2 and absorber plate temperature T_3 are shown in the Fig. 10.5.3.1. The cells characterized by highest measured short circuit current (I_{sc}) values were placed at the top of the module; followed by cells exhibiting decreasing values of I_{sc} towards the bottom. This design was intended to minimize the mismatch losses occurring due to ΔI_{sc} values, thus improving the overall efficiency (due to lower values of Relative Power Losses), when the cells are connected in series.

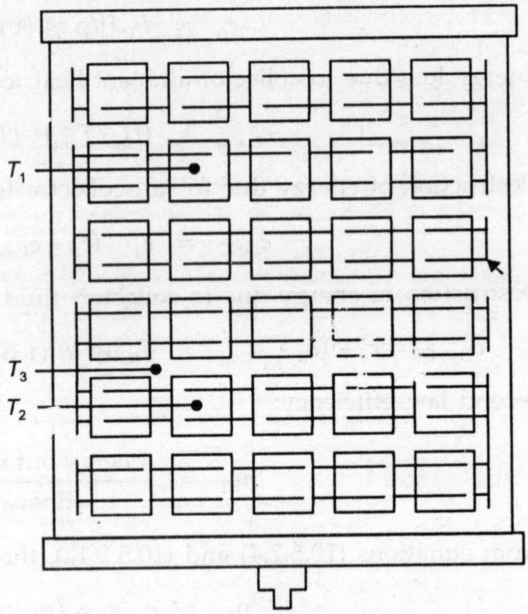

Fig. 10.5.3.1: Interconnected PV Cells Pasted on the Absorber Plate in Six Rows of Five Cells

EXERCISE

1. What do you mean by Hybrid Energy Systems ? Write down an expression for the total cost of such a system and deduce a simple condition for the feasibility of the system.
2. What are the various types of Hybrid Energy Systems ? Draw typical curves to justify the choice and role of renewable energy sources constituting a hybrid system.
3. Explain clearly the concept of integrated energy system. How are these systems different from hybrid energy systems ? Clarify by at least two specific examples.
4. What do you mean by collector efficiency function of an integrated energy system? Deduce mathematical expressions for 1st law efficiency and 2nd law efficiency of the collector of such a system. Show clearly a list of symbols and mention any assumptions used.
5. Write short notes on:
   (A) Potential of Power gneneration from Rice husk/Immature Paddy
   (B) Air PV/T systems.
   (C) Milk chilling centers powered by two different renewable energy sources.
   (D) Advantages of a PV/T collector with a polymer absorber plate.

SUMMARY

Since a single renewable energy source in isolation can no longer be considered dependable and economic at most of the selected sites, the renewable energy scenario in most of the developing countries has welcomed the design and installation of a large number of Hybrid Energy Systems. Such systems principally incorporates two or more energy sources (at least one of them is Renewable) in tandem on a time sharing basis. The need for such systems along with a cost-effective sizing ratio is first outlined. Interestingly, the design, material science, fabrication and performance analysis of various site specific examples catering not only to Indian sites but also to typical recent sites installed in a few developed countries have been reviewed systematically through a number of sub-sections. Next, the concept of integrating two or more renewable energy modules into the same gadget has been introduced and classified on the basis of current literature. Detailed thermodynamic analysis for optimizing the solar thermal part of one such gadget has been carried out. Careful consideration need to be focussed on the line of distinction between the concepts of hybrid energy systems and integrated energy systems. While integrated energy systems/gadgets incorporate more than one renewable sources in parallel operation on a decentralized basis, hybrid systems are designed and tested on the basis of multiplexing (time sharing) of more than one renewable/non-renewable energy sources. Case studies of few centralized power stations of hybrid nature have illustrated the above distinct features.

REFERENCES

1. H. Saha, D. Mukherjee, R.K. Sathpathy and M.K. Mukherjee, 'Photovoltaic System Design – A critical appraisal through a case study in India' – 19th IEEE Photovoltaic Specialist Conference, p. 102, 1987.
2. H. Saha, DPR on Power Generation by Solar photovoltaic System for Simlipal Forest Bungalow, Implemented by OREDA 1987.
3. H. Saha, D. Mukherjee, R.K. Sathpathy, Design of a 25kwp hybrid Solar photovoltaic System Power Supply for a forest lodge in India, Proc. of 1st International Conference on Power Supply and Sources, London (U.K.), April '1987.
4. Highlights of Sandia's Photovoltaic Program, Quarterly Report, vol. 1, SAND 98-0048, April '1998.
5. Andrew Rosenthal et al., Economic Analysis of PV hybrid Power System: Pinnacles National Monument, 26th IEEE PVSC, Sept. 1997.
6. S. Durand et al. Photovoltaic Hybrid System performance comparisions: Prediction versus Field Results, 25th IEEE PVSC, May 1996.
7. S. Fuller and S. Peterson, Life-cycle costing manual for Federal Energy Management Program, NIST Handbook, p. 135, February '1996.
8. D. Mukherjee, Ph.D. (Engg) Thesis, Jadavpur University, 1988.
9. B. Sandnes and J. Reekstad, A Photovoltaic/Thermal (PV/T) collector with a polymer absorber plate – Experimental Study and Analytical Model, Sol. Energy, vol. 72, no. 1, pp. 63-73, 2002.
10. T. Bergene and O.M. Lovvik, Model Calculations on a flat plate Solar Heat Collector with integrated Solar Cells, Sol. Energy, 55, pp. 453-462, 1995.
11. J. Rekstad et al., Effective Solar Utilisation – More dependent on system design than solar collector efficiency, Proc. of ISES, 1999, Solar World Congress, Jerusalem, Israel, 1999.
12. B. Bena and R.J. Fuller, Natural Convection Solar Drier with Bio-mass back-up heater, Solar Energy, vol. 72, no. 1, pp. 75-83, 2002.
13. M.S. Sodha, et al., Solar Crop Drying, vol. II, CRC Press, Florida, 1987.
14. ASAE standards, American Society of Agricultural Engineers, U.S.A., 1994.
15. Solar Nor. As Integrated Solar Systems, Product Description, 1996.
16. A. Ghosh, On line Instrumentation and Monitoring Relating to PV/Thermal Integrated Energy Systems, P.G. Thesis submitted to the Engg. Faculty of B.E.College (D.U.), 2000.

17. H.P. Garg and R.S. Adhikari, State of the Art on the development of Hybrid Photovoltaic/Thermal (PV/T) Systems, 25th NREC, Osmaina University, Hyderabad, pp. 49-53, 2001.
18. S. Chakraborty, et al., Thermodynamic Optimization of an integrated Energy System, Proceedings of 25th National Renewable Energy Conference, Osmania University, Hyderabad, India, pp. 76-79, Dec. 27-29 (2001).
19. A. Ghosh, et al., Studies and Performance Analysis of an Integrated Energy System at B.E.College (D.U.), Howrah, West Bengal, pp. 215-219, Proc. of the 24th NREC, IIT, Mumbai, Dec '2000.
20. T.S. Bhatti and D.P. Kothari, Dynamics and Control of stand-alone Wind-Diesel-Photovoltaic Hybrid Power Systems, Proc. of the 24th NREC, IIT, Mumbai, pp. 431-436, Dec '2000.
21. Tiwari and Suneja, Solar Thermal Engineering Systems, Narosa Publishing, 1997.
22. S.V. Sudhakar and M. Sharon, Fabrication and Perfromance evaluation of a Photovoltaic/Thermal Hybrid System, SESI Journal, pp. 1-7, 1994.
23. C.H. Cox and P. Raghuraman, Design considerations for flate plate Photovoltaic/Thermal Collectors, Solar Energy, vol. 35, no. 3, pp. 227-241, 1985.
24. Q.A. Ahsan and A. Hoque, Wind and Solar generating units – Potential candidates for generation of mix of an isolated area, Sol.Energy, 55(5), pp. 395-404, 1995.
25. M. Ashari and C.V. Nayar, 'An optimum despatch strategy using set points for a PV-Diesel-Battery hybrid power system, Sol. Energy 66(1), pp. 1-9, 1999.
26. M. Salameh et al., 'Methodology for optimally sizing the combination of a battery-bank and PV array in a Wind-PV hybrid system, IEEE Trans on Energy Conversion, 11(2), pp. 367-375, 1996.
27. V. Dutta, et al., A comparative life cycle energy cost analysis of Photovoltaic and Fuel Generator for load shedding application, Solar Energy Materials and Solar Cells, 60, pp. 309-322, 2000.
28. W.Z. Gandhare, et al., Power quality and harmonics generated by grid connected Windmills, National Renewable Energy Convention-99, Indore, India, pp. 334-337, 1999.
29. T. Markvart, Sizing of hybrid PV-Wind Energy Systems, Sol. Energy, 59(4), pp. 277-281, 1996.
30. G.R. Bhagwatikar, et al., Wind Power Quality Monitoring, International Conference on Renewable Energy for Rural development, Dhaka, Bangladesh, p.p.133-137, Jan'2002.
31. S. Bogdon, et al., Methodology for optimally sizing the combination of a battery-bank and PV array in a Wind-PV Hybrid System, IEEE Trans on Energy Conv., 11(2), pp. 367-375, 1996.
32. C.V. Nayar, Stand-alone Wind-Diesel-Battery Hybrid Energy System, Wind Engineering, 21(1), pp. 13-19, 1997.

11

Role of Renewable Energy with Respect to Environment

INTRODUCTION

While concept of energy use was considered in isolation more than a decade back, it has recently been recognized as one of the significant imputs for national development. Indeed, the economic development can be profiled through per capita energy use. According to a plot of extent of development vs per capita energy use, several developing countries are poised at the lowest left hand corner while developed countries like US, Germany, Canada are situated diagonically across at right hand top corner. Currently economics has established a correlation between energy consumption and our standard of living. In fact, the World Energy Conference of Montreal (1989) turned over a new leaf and outlined clearly the nexus between energy and environment – "The economical and rational use of our energy resources is essential for protecting the environment". Modern standards of living can be measured in terms of our access to the components of modern infrastructure like transportation (roads, cars, planes), communication (phones, FAX, e-mails, IT and internet browsing), entertainment (cable TV, satellite TV etc), education (institutions, open Universities and Web based learning), health care (treatment in clinics/hospitals, medical transcriptions and Robotic surgeries). All these well known sectors are concerned with flow of energy and power. Recent UNDP reports define a Physical Quality of Life Index (PQLI), which is a composite measure of living standards encompassing health, communication, employment, mortality, longevity etc. In fact, every small use of energy has an environmental consequence and every energy source has a double sided price-tag: a benefit for the user and an environmental price to be paid for. This is why today's goal of sustainable development is to maximize the former while simultaneously minimise the latter. In terms of our conventional energy reserves of coal, oil and natural gas, a higher efficiency of energy use has the twin benefits of slower resource depletion and a reduced level of pollution. Energy planning and environmental plannings are integrated together into any sustainable development model. The concept of sustainable development refers to development which is in harmony with nature and which permits future generations to enjoy a standard of living not inferior to the existing standards. This is possible through responsible planning and steward ship of our earth's available resources. In fact, some of the major changes in perception about the energy environment linkages have been highlighted. This is why global attention has been seriously focused on the adverse environmental impacts of conventional energy sources. In contrast, non-conventional energy sources, particularly the renewable ones have enjoyed a clean image vis-à-vis environmental impacts. The sections to follow will be devoted to the understanding of the (i) magnitude of adverse environmental effect by the non renewables and (ii) *whether the renewable sources are all environmentally benign as commonly believed*? Specific case studies will also be attempted relating to the aforesaid two issues.

11.1 IMPACT OF CONVENTIONAL ENERGY SOURCES

It has been mentioned in Chapter I that coal, oil, various natural gases, uranium and hydro are the commonly known commercial or conventional sources of energy. More than 90% of the world's present day commercial energy needs are met by these fossil fuels or traditional fuels. Inevitably, through the burning of these fuels, large quantities of CO_2 are released into the atmosphere. Along with CO_2, NO_x and SO_2 in comparatively small quantities are also emitted. A realistic estimate of such pollutants are presented in Table 11.1 corresponding to some specific categories of conventional coal and gas based energy conversion technologies. Notably, the highest greenhouse emission leading to global warming corresponds to pulverized coal fired steam plants. However, if standards of living are to improve along with a healthy industrial growth, any country needs a sustainable energy supply. In India, the industrial sector consumes about 50% of the total commercial energy involving large quantities of pollutants as suspended particulate matter, SO_2, NO_x, CO_2 and CO. Although all of them can be identified as health hazardous elements, it is the increasing concentration of CO_2 that is contributing to climatic changes in the long run. Thus, if a country has to make satisfactory progress in terms of HDI, it is important to shift towards an energy mix, rich in renewable energy which is not endangering the environment. The aforesaid situation can also be expressed by a convenient mass and energy flow diagram (Fig. 11.1.1) for a typical fossil fuel consuming 1000 MW electrical generating plant. Expected daily emissions from such a plant is listed in Table 11.2. From this table, an approximate extent of pollution caused by the by products can easily be estimated.

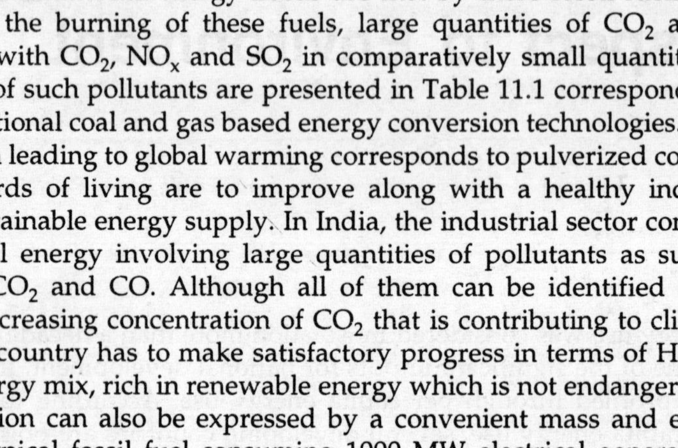

Fig. 11.1.1: Mass and Energy Flow Diagram for Fossil Fuel Combustion

It has been mentioned in chapter 8 that hydroelectric power projects are the most extensively studied power generation options vis-à-vis environmental impacts, alongside thermal power plants. There is a general belief currently that large hydroelectric hydel projects may be the most ecologically damaging of all power generation alternatives. In the range of 1950's when only a few such projects have been commissioned across the world, hydroelectric power was perceived as a clean and all purpose energy production option. With the aforesaid flavour, Jawaharlal Nehru, the then prime minister of India regarded these power plants in India as 'Temples of modern India'. Today after a span of forty five years it is an accepted global practice that environmental activities go up in arms whenever such a project is planned.

Table 11.1: Conventional Energy Technologies and Emissions

Technology	*Conversion* η%	No_x GMs/kWh	SO_2 GMs/kWh	CO_2 GMs/kWh
Pulverised Coal Fired Steam Plant (Without Scrubbers)	36	1.29	17.2	884
Pulverised Coal Fired Steam Plant (With Scrubbers)	36	1.29	0.86	884
Fluidized bed Coal Fired Steam Plant	37	0.42	0.84	861
Integrated Gasification combined cycle plant	42	0.11	0.30	758
Aeroderivative gas turbine	39	0.23	0.00	470
Combined cycle gas turbine	53	0.10	0.00	345

Table 11.2: Daily Emissions from a 1000 MW Power Plant[a]

Emission	*Quantity*
SO_x	1,100,000 kg[b]
N_xO_x	350,000 kg
CO_2	72,500,000 kg
CO	94,000 kg
Particulates	300,000 kg
Radioactivity	2.59×10^2 Bq[c]
Stack	1.35×10^{12} J
Heat Condenser	4.05×10^{12} J

[a] Assumes operation at full capacity at 40 percent efficiency.
[b] For a typical bituminous coal.
[c] Chiefly as 226.228 Ra. This figure is for coal; for oil it is about 50 times smaller. One bequerel (Bq) = one disintegration per second.

The 'Narmada Bachao' movement launched by Medha Patekar, S.L. Bahuguna and other noted intellectuals have a long history involving the fast-unto-death protests of environmentalists. In fact, major ecological impacts are caused by these projects in all the four havitats associated with the project – the reservoir catchment, the artificially created lake, the downstream reaches of the dammed river and the estuary into which the river flows. Natural environmental stresses are caused by altered timing of river flow, increased evapotranspiration, seepage water losses, barriers to acquatic organism life, thermal stratification, changes in sediment loading and nutrient levels and an increase in the occurance of water-borne diseases. Above all, most of the dams constructed are associated with serious problems of rehabitilation for those already habiting the reservoir area. According to some recent reports (Rosa and Schaeffer, 1993) greenhouse gas emissions (especially methane) from some hydroelectric reservoirs are nearly comparable to emissions from fossil fuel powered plants.

11.2 IMPACT OF RENEWABLES ON ENVIRONMENT

It is now imperative to examine the extent to which 'soft' or renewable energy sources are environmentally benign. To address this broad issue, we shall take up the context of solar energy based systems as Sun is undoubtedly the father figure from which other renewables source as wind, Bio-mass, ocean thermal energy conversion, etc. have been feasible.

11.2.1 Solar Energy

Direct solar energy based systems appear to be the easiest and cleanest means of tapping renewable energy. The routes followed for direct conversion of solar radiation to various utilizable terms are solar thermal systems, Solar Photovoltaic Systems and also solar architecture relating to intelligent building design. These need large tracts of land to house the collectors, balance of systems or the buildings. In fact, a few recent reports (Siddayao, 1993) indicate that material requirements for a controlled solar thermal power plants are larger than that of a fossil fuel plant per unit energy. There are other reports cautioning the increased concentrations of problematic water pollutants. Thus the aforesaid and other adverse impacts of solar energy systems are:

(i) Permanent use of large area without reclamation until plant is decommissioned.
(ii) Generation of non-recyclables during module fabrication like fiber-glass, glass, coolant and various insulations.
(iii) In case of certain thin film and amorphous-si solar cell modules comprising a PV array, additional disposal problems are likely to becaused due to the presence of cadmium and arsenic.
(iv) Aesthetic impacts are basically similar to steam based fossil fuel plants excluding air emissions.
(v) Soil erosion and compaction; Potential decrease in evaporation rate from soil.
(vi) Hazards to eyesights from reflectors (Parabolic Concentrators) and from toxicants in coolant fluids.
(vii) Roof top PV and solar thermal collectors are now an integral part of building design especially in developed countries. If these are used profusely in densely populated areas, changes in the albedo and weather conditions are likely.

11.2.2 Wind Energy

A major advantage of WECS relative to Bio-mass energy systems, geothermal power systems, fossil fuel and solar central receiver systems is that generation of electricity *does not need cooling water*. On the debit side, wind generators are responsible *for aesthetic degradation in a number of ways*. Lakes that are situated downward from the wind generators might become warmer as a *result of reduced evaporation from their surface*. Soil moisture surrounding a WECS might also increase due to the same scientific reason. Noise pollutants are undoubtedly severe as they tend to increase with increase in the value of prevailing/incoming wind velocity. The phenomena commonly called television interference or electromagnetic interference/microwave interference in a classified way has been mentioned briefly in chapter 7. Recently, one residential area in Block Island (U.S.A.), which is closely located to wind turbine blades, experienced chopping down of the TV signals *to an irritating degree*. Improvement in TV signal reception employing power antennas has already been attempted (Chester, 1994).

11.2.3 Bio-mass Energy – Dispersed Systems

In order to clarify the role of Bio-mass with regard to environment, the types of *dispersed systems* to be recollected are:

(a) House hold systems using *bio-mass directly as fuel*.
(b) Community sized electricity/heat producing systems based on pyrolysis, coal gasification or liquefaction.

The first category of dispersed systems are widely used in developing countries including India (Dwivedi, 1994; Sinha, 1994). It has been estimated that nearly 90% of India's cooking energy requirement and about 40% of all energy requirements are met from the aforesaid first category. In India, the current withdrawal of fuelwood from forests is of the order of 220 million tonnes where as the sustainable production capacity is only around 28 million tonnes. The adverse impact on the underground water and soil productivity can be easily realized considering the huge magnitude of firewood extraction. Secondly, use of fuelwood directly in the homes is a very major source of air pollution responsible for chronic health hazards to women and children who are exposed to this pollution for a significant period. It has been estimated and reported (Saksena et.al, 1995) that the *emission of air pollutants like carbon monoxide, sulphur dioxide, mitrogen oxide, organics and particulates are much larger compared to othe renewable sources.*

In fact, the pollution problems confronting Bio-mass based energy sources are similar to those of *large thermal power plants*. Disposal, in this case, entail serious drawbacks. If the dispersed plants dispose of the same total quantity of pollutants into the environment as a centralized plant, it is likely that a greater reach of the habitat would be affected. Further centralized systems are subjected to survelliance by *regulatory agencies (ISO 14000 certification process)* and are bound to practice certain streamlined pollution control measures. Such regulations are less likely to be enforced *in dispersed bio-mass utilization units.*

Finally, the following impacts on environment by large land based bio-mass projects have also been identified:

(a) Horticulture is a massive water consuming activity requiring profuse quantity of water per hectare contributing significantly to water pollution via the pesticides and fertilizers that are essential to sustain any intensive cultivation (Tandom, 1994);

(b) Significant amounts of nutrient losses will be incurred by harvesting of crop residues for bio-mass energy. With a corn yield of 7840 kg/hectare, the nutrients contained in both grain and residues are listed in Table 11.3 below:

Table 11.3: Nutrients Contained in Grain and Residue

Nitrogen	224 kg	Reference:
Phosporous	37 kg	Pimentel et al. (1983)
Potassium	140 kg	
Calcium	6 kg	

Significantly, nearly half of the nutrients are concentrated in the residue. Thus, it is essential that nitrogen and other nutrients have to be replaced for each subsequent crop.

11.2.4 Ocean Thermal Energy

Ocean Thermal Energy Conversion (OTEC) power plants have sufficient potential to cause a couple of adverse effects on *the quality of ocean water*. These are the prominent impacts on *marine ecosystem in terms* of altering *thermal structure*, altering water chemistry, eutrophication, algal booms and introduction of xenobiotics in the form of biocides.

The enrichment of the near surface water with the nutrient-rich cold water brought up from a typical depth of 1000 metres is of particular significance. This phenomenon gives rise to natural sites that are enormously rich in marine life – the best example probably observed in the Humboldt current off Peru, which enriches the surface water to be fertile enough to be responsible for almost one-fifth of the world's fish harvest.

In the nutrient-rich water unicellular algae grow to a density 27 times greater than the density in suface water and are, in turn, consumed by filter-feeding shellfish. However, abundance of nutrients in acquatic ecosystems is an indication of serious trouble as it invites all the adverse consequences associated with eutrophication. However, the designer of state-of-the-art OTEC systems hope that if an OTEC plants can discharge its water below the photic zone (the region in the surface water where photosynthesizing organisms live) the surface water will not be enriched. But, if nutrient-rich water is discharged anywhere near the surface water intake valves, it could cause biofouling inside the pipes.

Sea surface temperatures in the vicinity of an OTEC plant could be lowered by the discharge of cold water from the plant. This is likely to have impacts on organisms and microclimate. The pumping of large volumes of cold water from depths of the ocean to the surface is likely to relaease *dissolved gases such as* CO_2, O_2 *and* N_2 *to the atmosphere* – a factor influencing the *pH and DO status of water* causing stress to marine life. Ammonia is often the working fluid in *closed cycle OTEC systems*, the leakage of which can be of serious consequence to the ocean *ecosystems*.

11.2.5 Mini and Micro-hydel Systems

These systems essentially result from the dispersal of hydel power generation – either by constructing a member of low head dams or in-stream generators. It is estimated that the potential of small hydro systems in several small countries (Nepal, Madagascar, Papua/New guinea and a few latin American countries) exceeds the total installed power generation capacity in these countries from all other energy sources.

In fact, the problems appearing from a dispersed approach are no less serious, per kilowatt generated, than those from centralized hydropower. There shall be significant interruptions in water flow and water loss due to evaporation of *slower moving water stream*. Even more access roads shall be needed, which in turn, shall further *erode the ecology of the region*. The problems of *siltation* and *eutophication* common with large hydro systems are likely to assume more serious proportions. Further, the threat imposed by greenhouse gases (methane) are likely to endanger also from shallow reservoirs.

11.3 RENEWABLE ENERGY POTENTIAL FOR ENVIRONMENTAL MITIGATION

Sun, the mankind's biggest future source of energy is undoubtedly the father of all other renewable energy sources. Research efforts in course of last five decades have given rise to feasible technologies for producing mechanical and electrical power with solar energy. In course of the last three decades, the capability of sunlight for detoxification and disinfection of waste water, drinking water, industrial air emissions and indoor air is well established. However, only recently this technology has been cultivated with systematic and serious R & D efforts.

When sunlight is used to break up toxic chemicals by direct absorption through the chemical surface, the process is called *photolysis*. If this objective is achieved by the deployment of catalysts, the process is photocatalysis. Recent research has concentrated on *photocatalytic detoxification and disinfection* and the sub-sections to follow will cover the fundamental aspects of photochemistry, reactor designs and engineering systems relating to photocatalysis.

11.3.1 Photocatalysis

Photocatalysis or Photocatalytic oxidation is a process where a semiconductor upon absorption of a photon acts as a catalyst in producing reactive radicals. These are principally hydroxyl radicals which can oxidize organic compounds and mineralize them. Thus, organic molecules are decomposed to form carbon dioxide, water and mineral acids as final products. The most commonly used photocatalyst is TiO_2. Other semiconductors such as ZnO, CdS, WO_3 and SnO_2 have also been investigated as alternatives to TiO_2. Doping TiO_2 with metals such as Platinum or Tungsten has improved the effectiveness of TiO_2 (Magrini et.al, 1994).

When a photon with an energy equal to or more than the band-gap of TiO_2 is absorbed on its surface, it is responsible for the excitation of an electron from the valence band to the conduction band, forming a positive hole in the valence band. The excited electron and the positive hole either recombine or release heat or migrate to the surface where they can react with adsorbed molecules and cause either a reduction or oxidation of the adsorbate. Since recombination in the bulk or on the surface is the most common reaction, the quantam yields (molecules reacted/photons absorbed) of most of these reactions are slow.

The positive holes cause oxidation of the surface absorbed specie, while the electrons cause reduction. Both reactions must take place in order to maintain electro neutrality. Thus, if the objective is the oxidation of organics, the electrons must be consumed in a reduction reaction such as absorption by oxygen molecules to form superoxide, in order to keep the holes available for oxidation. On the other hand, if reduction and recovery of metals is desired, all other reducible species such as oxygen must be eliminated. The following set of equations describe the oxidation and reduction reactions:

$$TiO_2 \xrightarrow{h\nu} TiO_2\ (e_{cb}^- + hole_{cb}^-) \quad ...(11.3.1)$$

$$TiO_2\ (e_{cb}^- + hole_{vb}^-) \xrightarrow{recomb} TiO_2 + heat \quad ...(11.3.2)$$

$$H_2O \rightarrow OH^- + H^+ \quad ...(11.3.3)$$

$$hole_{vb}^- + OH_{ads}^- \rightarrow OH \quad ...(11.3.4)$$

$$O_2 + e_{cb}^- \rightarrow O_2^- \quad ...(11.3.5)$$

$$2O_2^- + 2H_{aq}^+ \rightarrow H_2O_2 + O_2 \quad ...(11.3.6)$$

Oxidation of Organics: $OH + Organics + O_2 \rightarrow Products\ (CO_2, H_2O, etc.)$ ...(11.3.7)

Reduction of metals: $ne^{-cb} + M^{n+} \rightarrow M^o$ ...(11.3.8)

Table 11.4 Shows the relative oxidation power of various species relative to chlorine.

Table 11.4: Oxidation Power for Various Species (Relative to Chlorine) (Elizardo, 1991)

Species	*R.O.P.*
Iodine	0.54
Bromine	0.80
Chlorine	1.00
Chlorine dioxide	1.15
Permanganate	1.24
Hydrogen peroxide	1.31
Atomic Oxygen (singlet)	1.78
Hydroxyl radical	2.06
Fluorine	2.23

Photocatalysed detoxification using TiO_2 as catalyst requires photons of ultraviolet (UV) radiation in the range of 295 to 395mm. The extra terrestrial radiation from the Sun reaching the earth contains only around 5 to 10% of such UV radiation. Typically, the maximum value of UV radiation is about $50W/m^2$ for a south facing surface tilted at an angle equal to the latitude of the location. SOLMET Typical Metrological Year (TMY) represents the insolation values for the entire spectrum. However, utilization of this data for modelling detoxification process requires the estimation of the ultraviolet component. Riordan et.al (1990) correlated measured UV to measured total values. Accordingly, global horizontal and direct normal UV radiation can be calculated from:

$$\frac{I_{UV,h}}{I_t} = 0.14315K_t^2 - 0.20445K_t + 0.135544 \qquad ...(11.3.9)$$

And
$$\frac{I_{UV,b}}{I_{t,b}} = 0.0688e^{-0.575M} \qquad ...(11.3.10)$$

Where the symbols K_t and M stand for cloudiness index and the air-mass respectively. However, the models (11.3.9) to (11.3.10) were based on limited measurements at only a few locations with an uncertainty greater than 20%.

For a flat plate reactor, UV radiance can be calculated by using the following equation suggested by saltiel et.al (1992) which is in fact, a modified form of Rabl's (1981) version of total solar radiation:

$$I_{UV} = I_{UV,b}\,K(\theta)\max\{\cos\theta, 0\} + 0.5K(\theta)\,[I_{UV,d}\,(1 + \cos\lambda) + \rho_{UV,gr}\,I_{UV,h}\,(1 - \cos\lambda)] \qquad ...(11.3.11)$$

where θ is the incident angle at the reactor aperture (estimated at the mid-point of each hour), $K(\theta)$ is the incident angle modifier, λ is the reactor tilt, $\rho_{UV,gr}$ is the UV reflectance of the ground. In the absence of actual values of $K(\theta)$ for the reactor under consideration, the following expression for a solar thermal collector (Duffie and Beckman, 1980) may be used, although reactor receiver covers and fluids are generally not similar to the materials used by solar thermal collectors. The flat plate incident angle modifier based on data from a single glazed flat plate collector (Duffie and Beckman), can be expressed as

$$K(\theta) = 1 - 0.10\,(1/\cos\theta - 1) \qquad ...(11.3.12)$$

The incident angle modifier used for the concentrating collectors, based on data aquired for the parabolic trough SERI Hexcel Collector, can be placed as:

$$K(\theta) = 1 - 2.02 \times 10^{-5}\theta^2 + 4.69 \times 10^{-7}\theta^3 - 1.80 \times 10^{-8}\theta^3... \qquad ...(11.3.13)$$

11.3.1.1 Design of Photocatalytic Reactors

These designs have initiated from the well-known designs of solar thermal collectors including concentrating and non-concentrating variety. The obvious differences are—(i) The fluid to be treated must be exposed to UV solar radiation. Thus the absorber must be transparent to UV solar radiation.

(ii) Since temperature does not affect photo reactions, no insulation is needed. Therefore the first outdoor reactor is essentially a parabolic trough solar thermal collector where the absorber/glazing tube combination is replaced with a simple pyrex glass tube through which contaminated water is flown (Fig. 11.3.1.1.1) When this reactor was used to treat water contaminated with Trichloroethylene (TCE), the catalyst, TiO_2 powder was mixed with contaminated water to form a slurry which was passed through the pyrex glass tube (reactor tube) located at the focal plane of the parabolic trough. In course of nearly the last decade, the reactors have been classified according to their configuration as (A) Concentrating and (B) Non-concentrating reactors. Another basis of classification is the deployment of catalyst which has evolved the following types of reactors: (C) Fixed catalyst, (D) Slurry, (E) Neutral density large particles.

A disadvantage with the concentrating reactors is that they cannot use diffuse solar radiation. It is known that solar photocatalytic detoxification with TiO_2 as a catalyst uses only the UV portion of the solar radiation, utilising hardly 5 to 6% of the total spectrum. As much as 50% of UV radiation is likely to be in the diffused form particularly in locations with high humidity and frequent cloudy periods. Another disadvantage of the reactors (A) is that they have inherently low quantam efficiency resulting in a reaction rate constant (K) depending on the intensity of UV radiation as

$$K \propto I^{1/2} \quad \text{...(11.3.14)}$$

Researchers at the University of Florida have developed a number of non-concentrating solar reactors including trickle-down flat plate, pressurized flat plate, tubular (rigid and inflatable), free falling film, shallow solar ponds. Wyness et.al (1994) described a trickle-down flat plate with a thin film glazing as shown in Fig. 11.3.1.1.2. This reactor consists of a black plate on which water trickles down from a spray bar at the top. A woven mesh, made of fiber glass, covers the back plate in order to damp out surface waves and even out the flow. A UV transparent glazing presents evaporation from the flow. This reactor can be operated in a fixed catalyst configuration by replacing the plain woven mesh with the catalyst fixed on it. Detailed experimental analysis on this model has been attempted by Wyness et. al (1994) under various conditions of sunlight and in a slurry as well as a fixed catalyst configuration. It was found that 1st order reaction rate constants depend on solar intensity as:

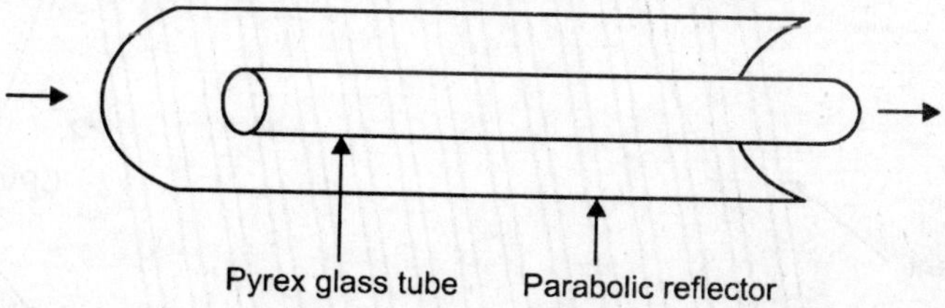

Fig. 11.3.1.1.1: Parabolic Trough Solar Photocatalytic Reactor

$$K/K_o = a\,(I/I_o) \quad \text{...(11.3.15)}$$

And

$$K/K_o = a\,(I/I_o)\,.\,(\delta_o/\delta) \quad \text{...(11.3.16)}$$

where equations (11.3.15) and (11.3.16) correspond to slurry and fixed catalyst configuration respectively. Here, a is a contant and δ is the film thickness of water. For slurry flow, value of a was found to be 0.88 for 0.1% TiO_2 concentration and 1.14 for 0.01% concentration. In the fixed catalyst configuration, the value estimated was nearly unity.

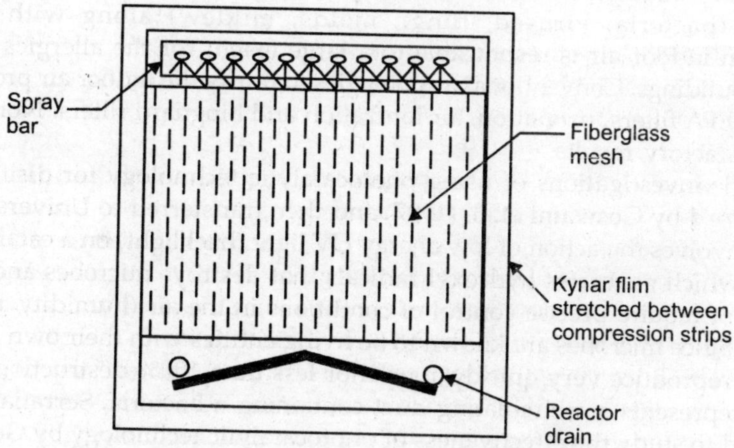

Fig. 11.3.1.1.2: Trickle-down Flat Plate Reactor

Due to the weight of water, the glazing of a pressurized flat plate reactor requires additional structural support. Goswami (1993) proposed a simple concept of such a reactor in Fig. 11.3.1.1.3. In this system, the contaminated water flows through the small channels between the lower and the upper headers. A variation of this concept was incorporated by the American Energy Technologies, substituting rigid transparent tubes in place of rectangular channels. Such tubular reactors were tested by the University of Florida to treat water contaminated with volatile organic compounds (oberg et.al, 1993). Groundwater contaminated with fuel, oil and lubricants was also passed through these reactors at the field testing center at the Tyndall Air Force Base, Fig. 11.3.1.1.3 gives the performance of this system.

Finally, solar detoxification has yielded good and reproducible results if the application involves large bodies of contaminated water, such as spills in lakes or seas. Neutral density spheres coated with TiO_2 can be floated in large bodies of water to treat the contamination in situ in the presence of the Sun.

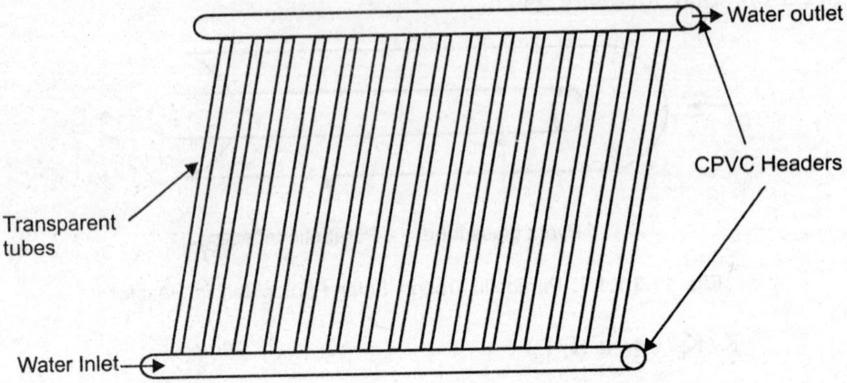

Fig. 11.3.1.1.3: Flat Plate Reactor with Rigid Transparant Tubes

11.3.2 Photocatalytic Purification of Air

A number of studies initiated on sick building syndrome have concluded that endotoxins from microorganisms (bacteria, virused, fungi, molds, mildew) along with volatile organic chemicals(VOC) in indoor air is responsible to a large extent for the allergies and sicknesses of the people in the buildings. Conventional technologies to purify the indoor air projects use activated charcoal filters, HEPA filters, ozonation, air ionization and bioguard filters. None of these options have yielded satisfactory results.

Initial scientific investigations of using photocatalytic technology for disinfecting indoor air have been performed by Goswami et.al, (1997) and then transferred to Universal Air Technology. This technology involves the action of low energy UV light (backlight) on a catalyst in the presence of water vapour which generates hydroxyl radicals that destroys microbes and VOC's in the air. All VOC's may not require precise control of conditions in the air (humidity, residence time) for their destruction. Since microbes are known to be living entities with their own defence and repair mechanisms and reproduce very quickly, partial or less than 100% destruction is not acceptable.

Fig. 11.3.2.1 represents a recirculating duct containing a bacteria, Serratia Marcescens in air and has been used to study the effectiveness of photocatalytic technology by Goswami et. al. Such an effort conducted at University of Florida achieved a 100% destruction of micro-organisms in a recirculating loop in a period of 600 minutes. This time for 100% destruction was then reduced to half by modifying the reactor design providing increased residence time (Fig. 11.3.2.2). In the latest design of reactors developed by UAT, the time has been improved by a few orders of magnitude.

Thus, it has to be born in mind that solar photocatalytic oxidation has been successfully demonstrated to treat contaminated groundwater and industrial waste water. In order to attract major commercial interest, a sizeable reduction in the cost of technology can only be triggered

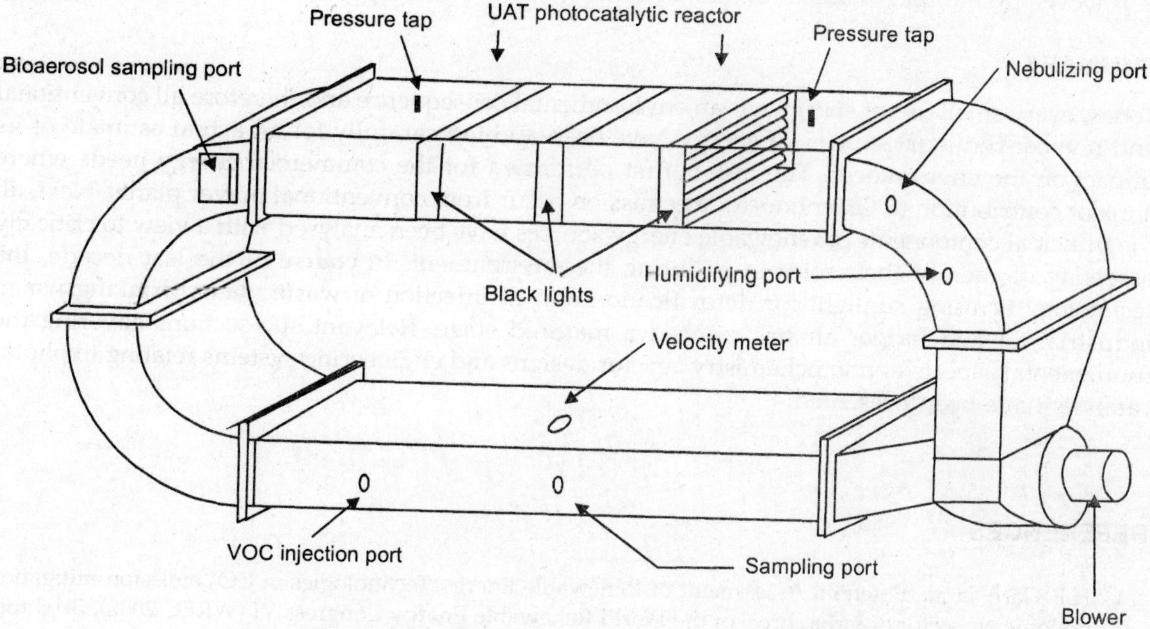

Fig. 11.3.2.1: Universal Air Technology Recirculating Duct Facility

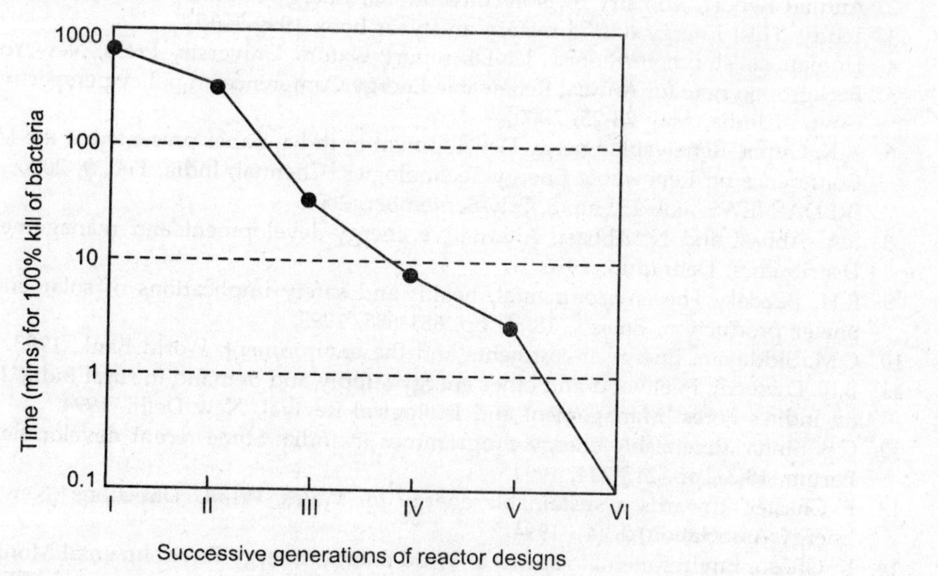

Fig. 11.3.2.2: Time for 100% Kill with Successive Generation of Reactors

through a sizeable reduction in the catalyst costs. Presently, the most common catalyst TiO_2 limits the useful range of the solar spectrum wavelengths to less than 385nm – a region in which only 4 to 6% of the solar radiation is available. Thus, todays' R & D efforts are principally concentrated on:

(A) Physical and chemical modification of TiO_2.
(B) Dye sensitization to improve the useful wavelength range of solar radiation.
(C) Development and characterization of homogeneous catalysts.

SUMMARY

Today, every small use of energy has an environmental consequence and therefore all conventional and non-conventional sources of energy have been audited carefully for an actual estimate of its impact on the environment. The task is first performed for the commercial energy needs where a major contribution to Greenhouse Gas emission occur from conventional power plants. Next, all the principal components of renewable energy sources have been analysed with a view to critically assessing of each of their roles in polluting the environment. In course of the last decade, the technology of using sunlight for detoxification and disinfection of waste water, drinking water, industrial air and indoor air has reached a matured stage. Relevant sub-sections covering the fundemental aspects of photochemistry, reactor designs and engineering systems relating to photo-catalysis have been presented.

REFERENCES

1. H.P. Garg, et al., Potential Assessment of Renewable Energy Technologies in CO_2 emission mitigation in domestic sector of India, Proc. of the World Renewable Energy Congress VI (WREC-2000), Brighton, U.K., July '2000.
2. Annual Report, Ministry of Non-conventional Energy Sources, Govt. of India, 1999.
3. Teddy, TERI Energy data directory and year book 1999-2000.
4. Human development report , UNDP report, Oxford University Press, New York, 1999.
5. Background note for Annual Renewable Energy Conference on policy perspectives – 2000-2012, MNES, Govt. of India, May 23-25, 2000.
6. A.K. Gupta, Renewable Energy Development in India, paper presented at the Millenium International Conference on Renewable Energy Technologies, Chennai, India, Feb. 9, 2000,
7. IREDA NEWS, vol. 13, no. 3, July-September, 2002.
8. S.A. Abbasi and N. Abbasi, Alternative energy-development and management, International Book Distributors, Dehradun, 1996.
9. R.H. Bezdek, The environmental, health and safety implications of solar energy in central station power production, Energy, 18(6), pp. 681-685, 1993.
10. C.M. Siddayao, Energy investments and the environment, World Bank, 1993.
11. B.N. Dwivedi, Fuelwood and other energy supply and demand in rural India, International workshop on India's Forest Management and Ecological Revival, New Delhi, 1994.
12. C.S. Sinha, Renewable energy programmes in India: Some recent developments, Natural Resources Forum, 18(3), pp. 213-224, 1994.
13. P. Chester, Towards a sustainable energy for Wales, WIND Directions (Newsletter of British Wind Energy Association) 13(4), 1994.
14. K. Ghosh, Environmental aspects of Wind energy, Energy Environmental Monitor, 11, pp. 13-19, 1995.
15. B.S.K. Naidu, Small hydro in India-environment friendly alternative energy, TERI Information Monitor on Environmental Science, 1(2), pp. 81-93, 1997.

16. D.Y. Goswami, New and Emerging Developments in Solar Energy, Proc.25th National Renewable Energy Convention-2001, Hyderabad, 2001.
17. K.A. Magrini, et al, Improving catalyst performance for the solar based Photocatalytic Oxidation of Organics, Proc. ASME International Solar Engineering Conference, pp. 163-170, 1994.
18. K.A. Magrini, et al., Photocatalyst evaluation for solar based Aqueous Organic Oxidation, Proc. ASME International Solar Engineering Conference, pp. 415-420, 1995.
19. D.M. Blake, et al., Kinetic and Mechanistic Overview of TiO_2 – Photocatalyzed Oxidation Reactions in Aqueous Solutions, Solar Energy Materials, vol. 24, pp. 584-593, 1991.
20. D. Behanemann, et al., Mechanistic Studies of Water Detoxyfication in Illuminated TiO_2 Suspensions, Solar Energy Materials, vol. 24, pp. 564-583, 1991.
21. M.R. Prairie, Solar Detoxyfication of Water Containing Chlorianted Solvents and Heavy Metals via TiO_2 Photocatalysis, Proc. ASME International Solar Energy Conference, vol. 1, pp. 1-8, 1992.
22. K. Elizardo, Fighting Pollution with Hydrozen Peroxide, Pollution Engineering, pp. 106-109, Sept'1991.
23. C.J. Riordan, et al., Influences of Atmospheric conditions and Air Mass on the ratio of Ultraviolet to Total Solar Radiation, SERI Report/TP- 215-3895, 1990.
24. C. Saltiel, et al., Performance Analysis of Solar Water Detoxyfication Systems by Detailed Simulation, Proc. ASME International Solar Energy Conference, vol. 1, pp. 21-28, 1992.
25. P. Wyness, et al., Performance of Non-concentrating Solar Photocatalytic Oxidation Reactors, Part-I: Flat Plate Configuration, Jl. of Solar Energy Engineering, ASME, vol. 116, no.1, pp. 2-7, 1994.
26. D.Y. Goswami, Recent Developments in Photocatalytic Detoxyfication and Disinfection of Water and Air, Proc. ISES, Solar World Congress, Jerusalem, Israel, 1999.
27. N.H. Anita, et al., Simple Methods for disinfecting household drinking water, Journal of Solar Energy Engineering, ASME, vol. 2, no. 2, pp. 131-137, 1992.
28. J.F. Klausner et al., Energy Consumption of Waste Water Treatment Technologies, pp. 29-36, Proc. of Solar Engineering-vol. 1, ASME, 1992.
29. R. Pandey, et al., Framework for design and evaluation of Electric Supply options for a Rural Area, Proc. of 24th NREC-2000, IIT, Mumbai, pp. 367-376, Nov '2000.t
30. J.A. Duffie and W.A. Beckman, Solar Engg. of Thermal Processes, Wiley Interscience, N.Y. 1980.

